Geographic Information Systems

Second Edition

Geographic Information Systems

Applications in Natural Resource Management

Michael G. Wing
Pete Bettinger

OXFORD
UNIVERSITY PRESS

70 Wynford Drive, Don Mills, Ontario M3C 1J9
www.oupcanada.com

Oxford University Press is a department of the University of Oxford.
It furthers the University's objective of excellence in research, scholarship,
and education by publishing worldwide in

Oxford New York
Auckland Cape Town Dar es Salaam Hong Kong Karachi
Kuala Lumpur Madrid Melbourne Mexico City Nairobi
New Delhi Shanghai Taipei Toronto

With offices in
Argentina Austria Brazil Chile Czech Republic France Greece
Guatemala Hungary Italy Japan Poland Portugal Singapore
South Korea Switzerland Thailand Turkey Ukraine Vietnam

Published in Canada
by Oxford University Press

First published 2008

Library and Archives Canada Cataloguing in Publication

Wing, Michael G
Geographic information systems : applications in forestry and natural
resources management / Michael G. Wing & Pete Bettinger.—2nd ed.

Previous eds. by Pete Bettinger and Michael G. Wing.
ISBN 978-0-19-542610-6

1. Forests and forestry—Remote sensing. 2. Natural resources—Remote
sensing. 3. Geographic information systems. I. Bettinger, Pete, 1962– II. Title.

SD387.R4W56 2008 634.9'028 C2008-902309-9

1 2 3 4 - 11 10 09 08
This book is printed on permanent (acid-free) paper ♾.
Printed in Canada

Contents

List of Tables

Preface

This second edition of *Geographic Information Systems: Applications in Natural Resource Management* is intended for introductory courses in geographic information systems or computer applications that address topics related to natural resource management. The emphasis of the book is on geographical information systems (GIS) applications in natural resource management. GIS tools are now considered core technologies for natural resource organizations and have become part of day-to-day activities in many parts of the world. In addition, many natural resource programs in higher education require that students complete at least one course that contains significant treatment of GIS. We provide detailed discussions and examples of GIS operations such as querying, buffering, clipping, and overlay analysis (and others), as well as background information on the history of GIS, database creation, editing, acquisition, and map development. The applications provided in this book can be extended to any region in the world, although the primary emphasis is on North America, as portrayed by numerous examples of natural resource management scenarios.

The contents of this book were determined largely through our experiences in natural resource management and research, as well as our extensive instructional experience over the previous decade. Many applications similar to the ones we present in this book have been performed by natural resource professionals (as well as by the authors) as part of their normal job tasks in private organizations and public agencies.

The goal of this book is to introduce students, field personnel, biologists, and other natural resource professionals to the most common GIS applications and principles associated with managing natural resources. Therefore, the book focuses mainly on GIS applications rather than on GIS theory. We would be remiss, however, if we did not provide some background on the history, technology, and theory that defines GIS. Consequently, the first part of the book provides a brief background on many of those areas as well as map development; it is not all-inclusive, however, as we wish to focus the text on GIS applications in natural resource management. For a broader treatment of GIS concepts, other resources are recommended, including more general GIS books or User Guides specific to GIS software packages.

With that in mind, who comprises the audience of this book? Students, field personnel, biologists, and other natural resource professionals (and their managers) who work in natural resource-related fields, but where GIS is perhaps not their primary job responsibility. People who already serve as GIS analysts, coordinators, or technicians will likely find many of the topics presented in this book to be familiar. We try to focus on topics and applications used commonly by field professionals, those that are essential to their management needs. There are a variety of resources that delve deeper into various subject areas of GIS that will undoubtedly be of value to GIS analysts, coordinators, and technicians. In our experiences these resources rarely consistently focus on natural resource applications typical to field professionals associated with federal, state, provincial, or private natural resource organizations.

To illustrate the applications of GIS to natural resource management, we have provided four sets of GIS databases. The first set references the hypothetical Daniel Pickett forest, one that may be familiar to those who have taken courses in forest management, as it is one of the landscapes used to illustrate management alternatives in the book *Forest Management* (Davis et al., 2001). The second set references a fictional forest called the Brown Tract. The Brown Tract represents a more realistic landscape and includes a digital orthophotograph so that users can actually see the resources being managed. The third data set represents land uses in Saskatchewan, while the fourth represents mill locations and counties of the southern US. These databases were derived from actual GIS data, but were modified significantly by the authors to make them suitable for use in this text. Each of these sets of GIS data can be accessed through a website hosted by Oregon State University (http://www.forestry.oregonstate.edu/gisbook).

Part 1 provides readers not only with the history and development of GIS, but also with a common language and perspective on GIS. Too often people using GIS have little formal training; instead, they gain knowledge and skills through trial-and-error applications, short courses, or through other means. We do not want to discourage the efforts of self-motivated GIS users; however, they usually have an abridged perspective on the history of GIS, how and why data structures are different, and in other related topics. We hope that communication among natural resource professionals as it relates to GIS processes and requests will thus be improved with a more thorough perspective, allowing work tasks to be accomplished more efficiently.

Part 2 emphasizes GIS operations and introduces readers to many of the most powerful and commonly used GIS applications in natural resource management. Each chapter in Part 2 introduces GIS techniques, and then provides applications related to the techniques. The concepts introduced in Part 2 are initially related to the management and use of vector GIS databases. The concepts build upon themselves, and culminate in a synthesis of advanced analyses presented in chapter 12. Chapters 13 and 14 provide treatments of raster GIS database uses in natural resource management.

Part 3 of the book introduces a number of topics related to the trends in the use of GIS in natural resource management, the challenges and opportunities faced by those organizations desiring to use GIS to assist in decision-making processes, and the ongoing and contentious issues related to certification and licensing of GIS users. The appendices of the book provide users with a glossary of terms, a summary of organizations and academic journals associated with the use of GIS, and references to the developers of the most common GIS software programs.

This book is dedicated to those students who have challenged us to develop coursework that relates directly to the GIS tasks that they will likely perform as natural resource managers.

References

Davis, L.S., Johnson, K.N., Bettinger, P.S., & Howard, T.E. (2001). *Forest management: To sustain ecological, economic, and social values* (4th ed.). New York: McGraw-Hill.

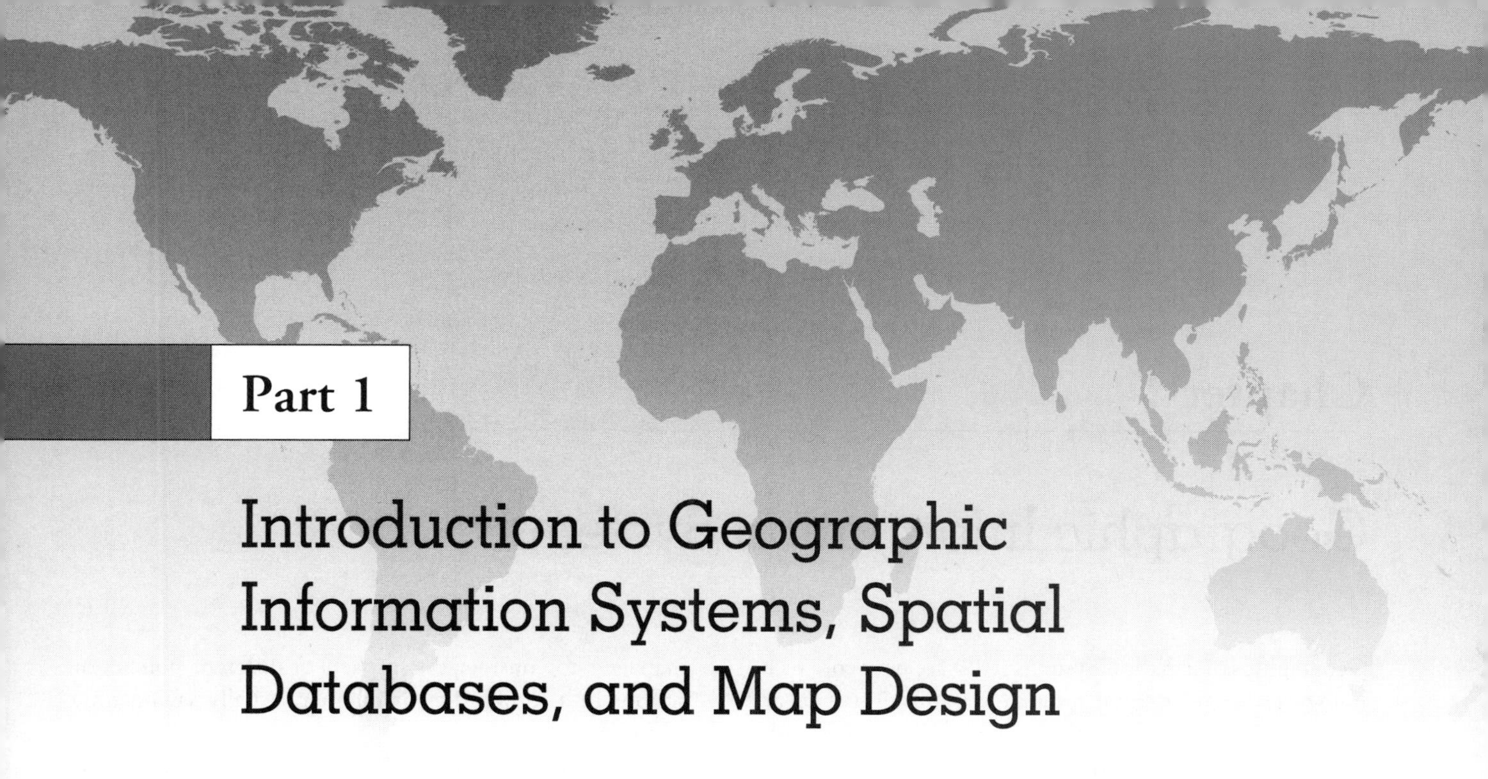

Part 1

Introduction to Geographic Information Systems, Spatial Databases, and Map Design

We hope Part 1 of *GIS Applications in Natural Resources* provides readers with a common language and perspective on geographic information systems (GIS). Frequently, people using GIS have little formal training, and they gain their knowledge and skills either through short courses or through their own initiative. While these self-motivated efforts are laudable, they usually result in an abridged perspective on the history of GIS, how and why data structures differ, and other related topics. Communication among natural resource professionals as it relates to GIS processes and requests should be improved with an informed perspective of GIS, encouraging work tasks to be accomplished more effectively.

In chapter 1, the historical development of GIS and the various tools you might use to create GIS databases are examined. The focus of chapter 2 is an essential topic for GIS: data. Chapter 2 begins by describing the ways in which we can quantify and measure the Earth's size and shape, and how results from these methods can be incorporated into GIS. Chapter 2 also includes material on how data can be structured within GIS and outlines the options that are available for these purposes. In addition, some of the more common sources of data for developing or augmenting spatial databases are presented in chapter 2. Chapter 3 builds upon the data theme introduced in chapter 2 by examining how organizations might acquire or develop databases, and discussing issues related to database editing and potential errors. Chapter 4 delves into cartography; one of many supporting disciplines from which GIS has evolved but also one of the central ways in which GIS results can be communicated to others. In addition, chapter 4 includes a thorough discussion of the concepts and components that may lead to a successful map, while at the same time identifying some common pitfalls to avoid in the map creation process.

Chapter 1

Geographic Information Systems

Geographic Information Systems (GIS) are now core technology for many natural resource organizations and are also applied in disciplines throughout society. The initial applications of GIS that demonstrated some of the power and potential of this spatial technology, however, were within natural resource applications (Wing & Bettinger, 2003). In one of the first papers on the use of GIS in natural resource management, de Steiguer and Giles (1981) describe the potential uses of GIS in natural resource management. In adapting one of their introductory remarks to the present day, you will find the relevance of GIS to natural resource management clearly stated:

> A natural resource manager is often called upon to select an area of land to designate as critical wildlife habitat, as a potential area to implement a timber harvest, or as an area to recommend a silvicultural treatment, or to evaluate a landscape under alternative management policies. The manager describes to the GIS the characteristics of the ideal area in terms of forest structural conditions, soils, or topography. Within seconds the manager receives graphic and tabular information to locate the appropriate management areas, or to compare alternative policies. (de Steiguer & Giles, 1981, p. 734)

Obviously, natural resource managers can perform the same task of identifying appropriate management areas or of analyzing policies by examining sets of paper or mylar maps, but the process becomes much more efficient and accurate when performed with GIS. Further, analyses of the impact of alternative policies are facilitated, allowing managers to consider the impacts of different policies or actions in a more efficient manner, usually saving time and money.

For many natural resource management organizations, GIS has become an irreplaceable tool to assist in the day-to-day management of land, water, and other resources. The applications of GIS vary widely among organizations and may range from using GIS primarily as a mapping tool to using GIS to model policy alternatives that may impact landscape features during the next 100 years and beyond. Regardless of how a natural resource management organization plans to use GIS, understanding the potential applications of GIS to natural resource management is essential for natural resource professionals. This text is designed to introduce readers to GIS concepts and principles and to provide examples of how to apply this knowledge in a natural resource management context. The introductory chapter begins by describing the tools and technology that comprise a GIS, and illustrating why GIS has become so important for many organizations. A brief history of the evolution of GIS and identification of significant contributors to GIS development is then provided. Toward the end of this chapter the key component of any successful GIS (spatial data) is discussed.

Objectives

This chapter represents an introduction to GIS concepts, tools, technology, history, and significant contributors. Given that the focus of this book is on the applications of GIS to natural resource management, what is provided in this chapter is a condensed version of these topics.

Nevertheless, at the conclusion of this chapter, readers should understand and be able to discuss the pertinent aspects of the following topics:

1. the reasons why GIS use is prevalent in natural resource management,
2. the evolution of the development of GIS technology and key figures,
3. the common spatial data collection techniques and input devices that are available,
4. the common GIS output processes that are typical in natural resource management, and
5. the broad types of GIS software that are available.

What is a Geographic Information System?

A geographic information system consists of the necessary tools and services to allow you to collect, organize, manipulate, interpret, and display geographic information. A GIS is more than just the hardware and software familiar to most people; it extends to the staff who operate the system, the databases, the physical facilities, and the organizational commitment necessary to make it all work. A GIS can be defined by how it is used (e.g., a land information system, a natural resource management information system), by what it contains (spatially distinct features, activities, or events defined as points, lines, polygons, or raster grid cells), by its capabilities (a powerful set of tools for collecting, storing, retrieving, transforming, and displaying spatial data), or by its role in an organization (a map production system, a spatial analysis system, a system for assisting in making decisions regarding basic geographic questions such as: Where is it? What is it? Why is it there?). The core component of a GIS however is a database that contains a geographic component. We will discuss geographic data in more detail shortly.

GIS can also be defined as **geographic information science** (GIScience). GIScience involves the identification and study of issues that are related to GIS use, affect its implementation, and that arise from its application (Goodchild, 1992). In short, GIScience both encourages users to understand the benefits of GIS technology in providing a powerful set of analysis tools and encourages users to view the technology as part of a broader discipline that promotes geographical thinking and problem solving strategies as being useful to society. The development of GIScience is an outgrowth of the fact that GIS technology is available to more users today than ever before, and that spatial categorization and analysis is applicable to many societal issues and problems.

Regardless of how a GIS is perceived or used, it is the integration of the various tools and services that leads to a successful GIS. Although other software programs perform GIS-like tasks (e.g., database management, graphics, or computer assisted drafting [CAD] software), a GIS is unique in its ability to allow users to create, maintain, and analyze geographic or spatial data. The term **spatial data** implies that a database not only describes landscape features (e.g., condition, composition, structure of forests), but also includes a geographic reference to where features can be found. A GIS allows you to manipulate and display spatial data so that questions regarding a resource and its conditions can be answered. A GIS, when used properly, is capable of analyzing a large volume of spatial data quickly and providing graphical and tabular results. A GIS stores spatial data in a digital database file; the database file may be referred to using a number of terms including themes, maps, covers, tables, layers, or GIS databases. The terminology for referring to a GIS database varies depending on the GIS software program being used and, in some cases, the version of the software being used. In most GIS software programs, similar landscape features are maintained in a single GIS database. For instance, you might have a soils GIS database that contains the soil characteristics of a landscape, a hydrographic database that shows the locations of rivers and lakes, or a wildlife GIS database that contains the nest locations of a single species or a group of species of animals. GIS allows the integration and simultaneous examination of multiple GIS databases through a process described as overlay analysis. Overlay analyses allow us to determine how features in one database relate spatially to features in another database, and provide us a powerful means of supporting landscape mapping and investigation. Overlay analysis, described in further detail later in this book, represents the essence of what many consider to be the main role of GIS in natural resource management: the ability to combine two or more GIS databases into a single database that demonstrates their spatial connectivity.

GIS is related to a number of other fields and disciplines, including computer aided drafting (CAD), computer cartography, database management, statistics, and remote sensing. In fact, GIS both contains and relies on certain aspects from each of these fields, and thus is closely related to each of them. However, the difference between GIS and any of these other allied fields is notable. For example, most CAD software programs have rudimentary

In Depth

In many ways, college and university students are examples of a living, breathing GIS. Each day you venture from your home into the world, and make decisions about where you are going, how you will get there, and what you will do when you arrive. For instance, as a typical student, you probably have a route that you usually take to campus. Chances are that you have designed this route over time and based on your experiences, so that you can arrive as quickly and easily as possible. Perhaps you have included a stop at your favorite coffee shop in your route. If you have a car, then you might elect to drive, and depending on the time of day, you might alter your usual route to avoid traffic. Road construction may force you to alter your route for a few days or weeks. You will make other adjustments to avoid unforeseen delays. Once you arrive on campus, you will have to find a parking space, and then walk another route to get to your first class. Of course, you might have decided that the troubles with parking make riding a bike to campus more attractive, but then you will still need to design a route for the bike trip. The choices you make just to get to school in the morning require you to analyze multiple layers of spatial information about your present location, your destination, and the intervening influential factors. In short, as you solve your daily transportation challenge, you are acting as a GIS. This type of example, in which location is a key component in decision making, can be applied to many activities that people engage in, ranging from how best to cross the street, to navigating a downhill skiing or snowboarding course, to arranging trips to other countries.

links to a database management system and are often limited in their ability to store and analyze descriptive information about features, whereas GIS software programs generally have strong links to a database management system. CAD spatial modelling capabilities are also limited, whereas GIS contains a wide variety of spatial modelling capabilities (these will be examined in later chapters of this book). The field of computer cartography emphasizes map production, and while the databases used may be similar to those used in GIS, computer cartography generally puts less emphasis on the non-graphic attributes of spatial landscape features than does GIS. Database management software programs have the ability to store and manage location and attribute data of landscape features, but they generally lack the power to display the locations and characteristics of features. Visual capabilities are fundamental qualities within most GIS, CAD, and cartography software programs. Statistical programs are usually designed so that users can quickly develop summaries of data, such as averages, standard deviations, or correlations that allow us to describe a large amount of data or to describe the relationship of a single variable to another. GIS software can usually not only accommodate basic statistical operations, but can show where the results of statistical operations are centered or located, and can help visually determine what other variables may be of interest in an analysis. Interestingly, several of the more powerful and commonly used statistical packages now integrate GIS-like functionality in some of their modules. Finally, remote sensing-related software programs generally focus on the manipulation and management of raster GIS data derived from satellites, scanners, or other photographic devices; they have a limited capability to handle vector GIS databases, which tend to be more commonly used within natural resource management organizations.

A Brief History of GIS

As previously mentioned, GIS is unique from other software programs in its integrative ability that enables you to process, catalog, map, and analyze spatial data. Spatial data have been collected and maintained for millennia, with records of property boundary surveys for taxation purposes in Egypt dating back to about 1400 BC. It is only within the past 40 years, however, that society has learned how to digitally capture, maintain, and analyze spatial data. Although the term 'geographic information system' was first used in the 1960s, **overlay analysis** has been demonstrated through manual techniques for over 200 years. Overlay analysis is the process of analyzing multiple layers of information simultaneously to address management issues. The layers represent different types of information but are related to each other in that the information is drawn from a common landscape area (Figure 1.1). GIS allows you to drape, or overlay, the layers on top of one another and to combine all parts into a new, integrated layer that contains all or some of

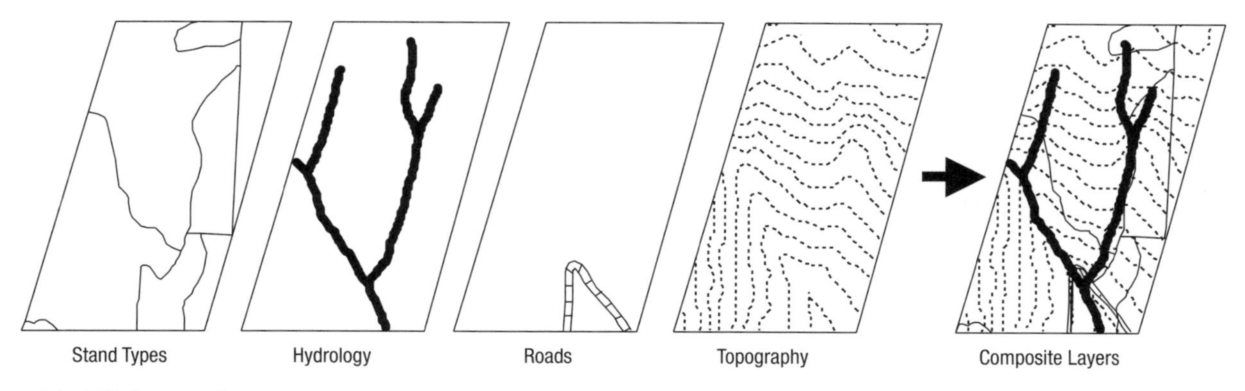

Figure 1.1 GIS theme overlay.

the parts of the original layers, depending on the type of overlay selected by the user.

The new integrated layer allows us to examine the spatial relationships of the information contained in the original layers. Although digitally-based GIS has been available for a relatively short period in history, there is a significant history of analysts using the overlay approach through manual techniques. During the American Revolution, the French cartographer Louis-Alexandre Berthier overlaid multiple maps to analyze troop movements (Wolf & Ghilani, 2002). In 1854, Dr John Snow conducted a spatial analysis by comparing the locations of cholera deaths to well locations in London. His analysis revealed that well water drawn from specific wells was a means of spreading cholera infections. The first written description of how to precisely combine multiple maps through a manual overlay process appeared in a 1954 text titled *Town and Country Planning Textbook* by Jacqueline Tyrwhitt (Steinitz et al., 1976). In 1964, Ian McHarg described how to use a series of transparent overlays to determine the suitability of areas for development in New York's Staten Island. By using a transparent overlay for each layer of interest (soils, forests, parks, etc.) and blacking-out the areas on each overlay that presented development impediments, the layers could be overlaid and the final suitable areas defined. McHarg (1969) later published examples of his overlay techniques in his seminal book, *Design with Nature*, which continues to be sold throughout the world.

In the early stages of the development of GIS technology, two facts were evident: there was little geographic or spatial data to work with, and the technology to store and manipulate the data was rudimentary (by today's standards). Some may argue that GIS technology has not evolved very much in the passing years simply because many of the computational processes used today were initially developed in the 1980s, however, advancements in computer technology and the increasing availability of GIS databases indicate otherwise. In addition, a growing number of people throughout society have heard of GIS (even though they may often confuse its purpose with that of a similar acronym, GPS [global positioning systems]).

We provide a brief history below of the development of digitally-based GIS, and note that many of its advancements were made by innovators and scientists throughout North America. During the 1960s, organizations in the United States (including the US Geological Survey and the US Department of Agriculture's Natural Resource Conservation Service) began to create GIS databases of topography and land cover (Longley et al., 2001). Students and researchers began to write computer programs and design hardware devices (such as the precursor to today's digitizing table) that would allow you to trace the outlines of landscape features on hardcopy thematic maps and transfer them into a digital format. These early programs were designed to handle specific tasks and were often limited in scope. As programmers began to bring these algorithms together to create more versatile, powerful software programs, the era of computer mapping applications began. Early examples of mapping programs include IMGRID, CAM, and SYMAP (Clarke, 2001).

In conjunction with the development of software programs, other organizations began to assemble GIS databases for mapping and analyzing features of interest to public agencies. The first example was the GIS database created by the US Central Intelligence Agency (CIA) and was called the 'World Data Bank'. Spatial layers in the GIS database included coastlines, major rivers, and political borders from around the world. The US Census Bureau designed a methodology for linking census information to locations in preparation for the 1970 US census. The 1970 US Census was the first census that was mailed, and the only piece of information that was returned to reference the location of the respondent was the address. The Census Bureau, however, was faced with the challenge of matching the response addresses to a map so that the spatial distributions of responses could be mapped and analyzed. The Census Bureau developed a system known as DIME (Dual Independent Map Encoding) in response to this challenge, which not only created digital records of all streets, but also associated addresses to street locations. The DIME system allowed the Census Bureau to understand which streets were connected to which other streets, and what landscape features were adjacent to each street. This method of associating the digital representations of landscape features to other landscape features was a critical advance because it enabled the identification of spatial relationships within a digital environment.

The description or characterization of the spatial relationships between landscape features in a GIS database is referred to as **topology**. Topology is an important concept with respect to GIS applications and will be discussed in more detail in chapter 2. Topology manages objects and requires objects to be organized and analyzed according to their location and with respect with proximity to other objects. The topological characteristics of data structures allow a determination, for example, of how water travels through a stream network, the connectivity of roads in a forest to other roads, or the identification of forest stands that share a border with other forest stands. These relationships form the basis of many resource analyses that take locational position into account in problem solving techniques.

The DIME system was the predecessor to TIGER (Topologically Integrated Geographic Encoding and Referencing System) files, which were introduced by the US Census Bureau in 1988, and are still used today to distribute spatially-referenced census and boundary data. The availability of TIGER files was instrumental in promoting GIS use in the US. The US Geological Survey (USGS) made an additional important contribution to spatial data availability when they began digitizing features from its 1:100,000 scale hard copy maps in the early 1980s. Spatial data from these maps were made available as digital line graphs (DLGs) that, like the TIGER and DIME systems, were also stored in a file format that allowed the topology of objects to be characterized. The file format was restructured in the early 1990s, and the USGS has made features from finer-resolution 1:24,000 scale maps available for small portions of the country. The USGS has since become a worldwide leader in mapping land cover resources and making maps available in both hardcopy and digital format.

To manage and analyze spatial data for their jurisdictions, Canadian and US organizations began to develop software programs in the 1960s. One of the most ambitious and noteworthy of these systems was the Canada Geographic Information System (CGIS), which, in 1964, was created under the guidance of Roger Tomlinson. A chance meeting, on a plane, between Tomlinson and Canada's Minister of Agriculture resulted in Tomlinson overseeing the creation of a national effort to inventory Canada's land resources, and developing a software program to quantify existing and potential land uses. The CGIS is recognized as being the first national-level GIS, and thus Tomlinson continues to receive recognition as a GIS pioneer for his efforts. Other early landmark efforts in the evolution of GIS include the development of the Land Use and Natural Resource Inventory System (LUNR) in New York in 1967, and the development of the Minnesota Land Management System (MLMIS) in 1969.

The success of these early systems and need for further refinements were recognized by a group of faculty and students at Harvard University's Laboratory for Computer Graphics and Spatial Analysis. The group set forth to create a versatile GIS that would map and track locations like the DIME system, while possessing the land measurement strengths of the CGIS. From this effort the Odyssey GIS (containing modules named after parts of Homer's epic work, *The Odyssey*) emerged in 1977, and pioneered the use of a data structure known as the arc/node, or vector data structure. We will discuss the vector data structure in more detail in the next chapter, however, it is important to note that the specifics of the Odyssey vector structure were first published by Peucker and Chrisman (1975) and the structure continues to influence the design of modern GIS software programs. Jack Dangermond, a Harvard Lab student, founded the Environmental Systems Research Institute (ESRI) in 1969, and earlier versions of ArcView and ArcInfo—the most widely used desktop and workstation GIS software programs—were based on the Odyssey vector data structure. ArcInfo, in fact, was introduced in 1981, marking the first major commercial venture into the development GIS technology. Both of these GIS packages have been significantly rewritten in terms of computer code support and user interface; they are now offered as different licenses within ArcGIS.

The 1980s also witnessed the proliferation of the microcomputer, today's version of the personal computer (PC). In response, software manufacturers began to produce GIS software programs that could operate on the microcomputer (see Appendix C for a list of GIS software manufacturers). In 1986, MapInfo Corporation was formed, and subsequently developed the world's first major desktop vector GIS software program for the PC. Soon afterwards, raster GIS software programs, such as IDRISI, began to appear. Some software programs, such as the raster GIS program GRASS, utilize a software architecture that was developed for workstation computer platforms.

Other significant developments in GIS included the emergence of GIS-related conferences and publications. The first AutoCarto Conference was held in 1974 and helped to establish the GIS research agenda. One of the first compilations of available mapping programs was published by the International Geographical Union in 1974. *Basic Readings in Geographic Information Systems*—a collection of papers that discussed GIS technology—was published in 1984, and in1986 the first textbook written specifically for GIS, *Principles of Geographic Information Systems for Land Resources Assessment* was published (Burrough, 1986). Finally, the first GIS-related academic journal, the *International Journal of Geographic Information Science*, was published in 1987.

More recently, Internet technology has advanced to the point where people worldwide can access and use rudimentary forms of GIS for free. Google Earth is perhaps the best example of the integration of remote sensing technology (digital orthophotographs and satellite imagery) with transportation networks and other landscape features that is available online. A limited number of geographical processing tools are available: however, Google Earth represents a significant advancement in allowing the general public to visualize the landscape. Microsoft's TerraServer is similar in this respect. MapQuest, perhaps the most widely used geographic locator online, is now similar in this respect as well.

The history of GIS continues to evolve, with GIS users providing a number of challenges. GIS users, for example, have the ability to influence the development of GIS software program features. As new and challenging natural resource management issues arise, users identify and propose processes and functions that will make the task of analyzing potential natural resource decisions more efficient and accurate. In addition, GIS users increasingly expect support and training related to specific GIS software programs, and expect that software will be mostly perfected by the time of its release to the general public. Further, as GIS databases are shared amongst organizations, the need to standardize data formats is evident, because data transformations can require an extensive commitment of resources and may lead to flawed results if not done correctly.

Society is fortunate today, on one hand, to have a variety of GIS software programs from which to choose. On the other hand, evaluating which of these programs best suits the needs of a natural resource management organization is problematic. This fact posed a significant challenge even in the creation of this book. Since each organization (natural resource management, as well as academic) may use a different GIS software program, we decided to design this book as a general reference for describing, in general, the typical types of GIS applications faced by field-level professionals associated with natural resource management organizations. Therefore, specific examples of how to address each application described in this book, those that are related to specific GIS software programs, will be made available through other means (e.g., a book-related website at www.forestry.oregonstate.edu/gisbook).

Why Use GIS in Natural Resource Management Organizations?

It is commonplace to see GIS used to assist managers make decisions in today's natural resource management environment. For example, maps are required to be submitted to state agencies in the western US in support of forest management plans. In most areas of North America, pesticide plans require a map to detail the proposed activity, as well as the nearby homes and water resources. While maps may still be hand-drawn in a handful of natural resource organizations, GIS allows map production processes to be automated and repeated, reducing a lengthy drafting exercise to only a few short minutes and likely producing far more reliable results. In addition, GIS allows some processes to be accomplished that would normally tax a person's analytical abilities. For example, a natural resource management organization in the southern US considering a fertilization project, yet operating with a limited budget, may need to locate those forested areas (stands) that would benefit most in terms of the growth of the forest from a fertilization application in order to make efficient use of their budget. If you were to assume that the stands must be dominated by pine tree species, and located on certain soil types, you can imagine the enormous task faced by a large landowner (> 500,000 acres) if paper maps (soils and stands) were the only resource available for analysis. A process such as this would have required several days to complete with paper maps, but might require only a few minutes when performed within GIS.

The application of GIS in natural resource management organizations has become standard practice during the last 10 years (Wing & Bettinger, 2003). Part of the reason for this widespread use is because of the efficiencies hinted at above, but also a result of continued technological advances in computer hardware and software. Computer prices continue to decline while processing power and storage efficiency have continued to grow. A wide variety of GIS software programs have also emerged, and the trend in GIS software program design has been to make programs more user-friendly while, perhaps, sacrificing efficiency of operations.

One of the primary reasons for the growth of GIS use in natural resource management organizations is that the collection and analysis of landscape measurements is fundamental for most natural resource analysis and management activities. GIS allows you to work with measurement information to facilitate mapping and modelling landscape features or to support the evaluation of management policies. For example, you might be interested in determining the extent of vegetation resources within a watershed, the amount of wildlife habitat within a natural area, or the potential impacts of changes in riparian management policies. GIS facilitates an efficient exploration of the information related to natural resources.

Although many natural resource management organizations employ a GIS expert (guru, manager) at a centralized GIS office, these people are often overloaded with work and unable to offer sustained assistance to field personnel. In some cases, the experts, particularly if they have a computer background as opposed to a natural resource background, may be unaware of the common types of GIS applications in natural resource management. When you additionally consider that many natural resource management organizations desire new employees to have a GIS background, the advantages for those involved in natural resource management to be familiar with the potential uses of GIS are clear. This familiarity should include the ability to communicate using basic GIS-related terminology and the ability to perform basic GIS processes, such as viewing data and making maps of features of interest.

Recent graduates of many university-level natural resource management programs complete at least one course involving GIS. Geospatial skills are currently in-demand in natural resource management, and are seen as part of an emerging industry that will experience continued growth for the near future (Wing & Sessions, 2007). One survey of natural resource management employers (Brown & Lassoie, 1998) supports these assertions, while another indicates that nearly half of industrial employers expected new employees to have obtained GIS experiences during their undergraduate education (Sample et al., 1999). What skills should students develop a proficiency prior to graduating? Merry et al. (2007) surveyed recent graduates who were employed in natural resource management-related positions and found that ESRI's ArcMap and ArcView software products were the most commonly used GIS software packages. Others types of software products were also being used (e.g., MapInfo, Google Earth, DeLorme, Landmark Systems' SoloField CE, and commercial software packages from Davey Resources), and learning how to use at least one program will make adapting to the use of others relatively easy. Basic GIS operations (heads-up digitizing, manual editing of attributes, manual editing of spatial positions, and querying of tabular attributes) were the most frequently used GIS processes. More complex processes, such as combining and erasing features, and spatial queries, were also moderately used by recent graduates. Of the types of products recent graduates created with GIS, basic locational maps, management decision-related maps (i.e., planting maps), and GIS databases (e.g., prescribed fire locations, conservation easements, invasive species distributions, soil maps) were the most common.

GIS Technology

Operating a GIS requires working with a wide range of technology and possessing, or acquiring, the skills necessary to understand and manipulate the technology. During the next part of this chapter, a description of the various technological components of GIS is presented. Although it may not be important for GIS users to be considered experts in all GIS-related technologies, familiarity with the various components may help you understand how the components are integrated.

Data collection processes and input devices

Despite the rapid advances of GIS hardware and software, one of the primary challenges for organizations using GIS relates to the development and maintenance of GIS databases. Collecting spatial data, preparing the data for GIS use, and documenting these processes continue to comprise the majority of budgets allocated for GIS processes. Spatial data quality is central to successful GIS implementation and analysis.

Data are often described in terms of their precision and accuracy, two terms that are often confused. **Precision** relates to the degree of specificity to which a measurement is described. A measurement that is described with multiple decimal places, such as an area measurement of 2.6789 hectares, is considered a very precise measurement. If this measurement were derived from a property boundary survey where distances were gathered by counting paces, and angles were measured using a handheld compass, you might question the accuracy of the measurement; however, it is inarguably presented in a highly precise manner. Precision can also be described in terms of the relative consistency among a set of measurements. For instance, if the measurements related to a property boundary were measured multiple times with a sophisticated surveying instrument and the resulting variation among measurements was small, you could then describe the measurements as being relatively precise.

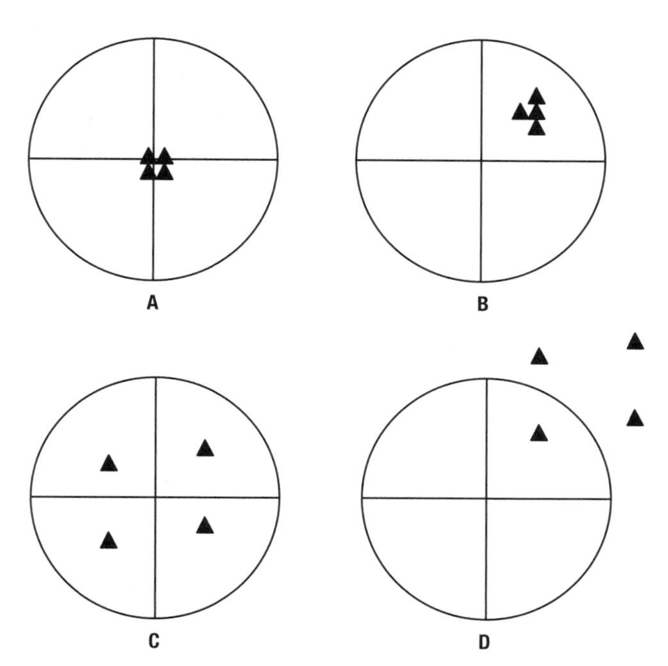

Figure 1.2 Examples of accuracy and precision. Part A shows accurate and precise locations of data around the circle center; Part B shows precise but not very accurate data; Part C shows accurate but not very precise data; and Part D shows neither precise nor accurate data around the circle center.

Accuracy refers to the ability of a measurement to describe a landscape feature's true location, size, or condition. Accuracy is typically described in terms of a range or variance that details a threshold within which we would expect to find the likely value. The assessment of accuracy attempts to answer the following question: How close are the measurements to their true value? Examples of accuracy levels include distance measurements of ± 0.5 m or angle measurements of ± 1 second. You can have measurements that are both highly precise and accurate (Figure 1.2, Part A), highly precise without being very accurate (Figure 1.2, Part B), not very precise, but accurate (Figure 1.2, Part C), or neither precise nor accurate (Figure 1.2, Part D).

Accuracy and precision may also be stated in relative terms. Suppose the length of a stream is measured twice with a 100-foot metal tape, resulting in measurements of 232.7 and 232.5 feet. The average length of the stream is 232.6 feet. If the metal tape was previously broken, say at the 10-foot mark, and spliced back together, reducing the effective length of the tape to 99.9 feet, the relative accuracy and precision of the measurements can be calculated. Since the tape was used about 2.3 times when measuring the stream, and the broken part of the tape was used each time, the accuracy of the measurements is about 2.3 × 0.1 foot = 0.23 foot off of the value you might have expected with an unbroken tape. You could express the relative precision of the measurements as (232.7 – 232.5) / 232.6 = 1:1,163 and the relative accuracy of the tape as 0.1 / 100.0 = 1:1,000. One advantage of using relative accuracy is that it provides an assessment of the expected portion of error given some measured amount. This allows for the relative comparison of these errors between different measured items and locations. Relative accuracy can also provide a means of stating or assessing required or minimal mapping accuracies. The US Federal Geodetic Control Subcommittee (FGCS) (1984) and Natural Resources Canada (1978) follow this practice whereby the relative precision can be categorized as acceptable or unacceptable, given a desired measurement accuracy.

It is important that GIS users are aware of the distinction between precision and accuracy, particularly when considering the value of using a GIS database in an analysis that leads to a management decision. These terms, despite their common usage, imply information about different qualities of a measurement or activity. Although accuracy and precision imply different characteristics, many use these terms interchangeably and, as a result, incorrectly.

There are many ways to create and collect data, and all methods require varying degrees of skill and organizational commitment. The following sections describe some of the most common methods for creating GIS databases.

Manual map digitizing

The ability to manually encode vector maps using a digitizing table (Figure 1.3) and associated software has been available since the late 1960s. Paper or mylar maps are taped down to a digitizing table, in which is embedded a fine mesh of copper wire. Known reference points on the maps are identified using the digitizing table's 'puck' (similar to a computer mouse), which sends a signal to the wire mesh within the table. Once the reference points have been identified, all other landscape features can be encoded in a Cartesian coordinate system and related to the reference points. For point features, this requires lining up the cross-hairs of the puck with the point locations and identifying the points. For line and polygon features, it involves tracing the boundaries of the lines or polygon boundaries, noting each change in a line's direction. Features can be recorded with either 'stream mode' or 'point mode' referencing processes. In stream mode, the spatial location of the digitizing puck is recorded at either regular time intervals (e.g., every second) or regular distances intervals (e.g., every 0.25 inch). In point mode, the spatial location of the digitizing puck, and hence the location of landscape features, is recorded every time a button on the puck is pushed.

Manual digitizing of maps can be a tedious process and, like many other tasks that are done by hand, subject to human error and variation. Today, digitizing is still a necessary function for many natural resource management organizations but reliance on this technique has dramatically decreased as other data collection methods have emerged or become refined.

Figure 1.3 Digitizing table.

Scanning

Scanning involves the examination of maps by a computer process that seeks to identify (and convert to digital form) changes in map color or tone, which identify landscape features. Flat-bed scanners allow a picture or map, such as an aerial photograph or a topographic map, to be converted to a digital form. The resulting images are described by the raster data structure, and include pixels or grid cells that may be encoded (or attributed) differently, depending on how the scanner interprets the color or tone of each feature. Scanners (Figure 1.4) generally move systematically across a picture or map, and record the reflectance values of the tones or colors for each grid cell. Scanned images tend to look very much like the pictures or maps that were scanned, yet there is usually some difference in quality due to the size of the grid cells assumed in the scanning process, the quality of the picture or map, or the quality of the scanner.

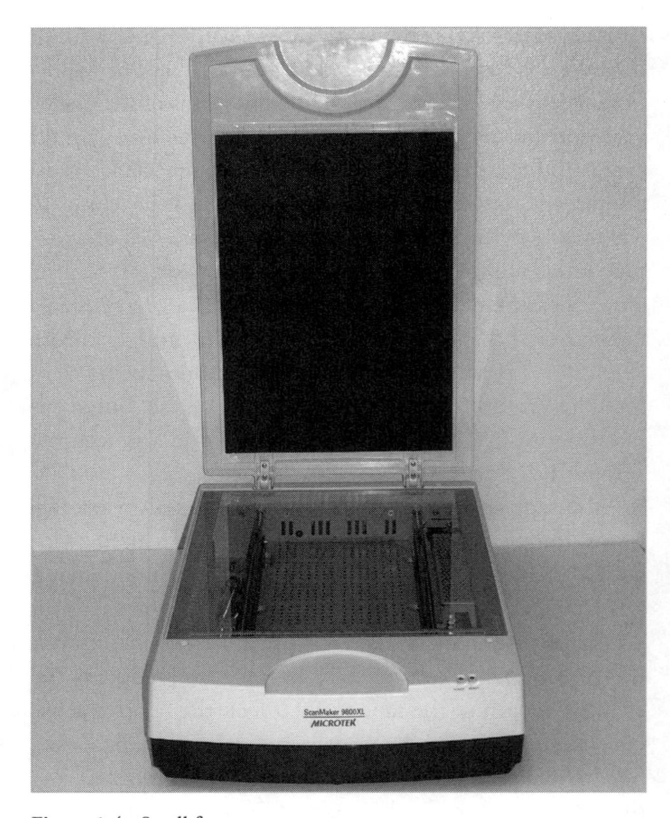

Figure 1.4 Small format scanner.

A second method of scanning involves the use of digital cameras. An array of photodetectors located within digital cameras allows you to capture and store an image. The images are saved with a raster data structure and can be transferred to a computer system and then used in a manner similar to the scanned images mentioned above. Digital cameras can be synchronized with GPS receivers so that a coordinate value and elevation are potentially associated with each image.

Remote sensing

Remote sensing involves the use of a sensor that is not in physical contact with its subject of interest (Avery & Berlin, 1992). It can include a wide variety of techniques, and in fact, capturing images with a digital camera theoretically uses remote sensing technology, since the camera is not necessarily in contact with the image being collected (the landscape). However, when discussing remote sensing technology in natural resource management, the use of satellites or cameras mounted on airplanes is frequently referenced. Remote sensing devices capture electromagnetic energy, generated by the sun or perhaps by some other device, such as a radar emitter, that is reflected off of landscape features. Most satellite sensors are designed to record the reflectance of light or heat from objects on the Earth's surface. These electromagnetic reflectances are recorded by the sensors in terms of their wavelength of energy, as described by the electromagnetic spectrum. The electromagnetic wavelengths are then converted to a digital format and transmitted back to a computer for processing and interpolation. Satellites such as the Landsat Thematic Mapper™ series can capture wide swaths of the Earth's surface (185 km, or 115 miles), and thus have the potential to record vast amounts of information over a short time period. The launching and operation of satellites for data collection has increasingly been conducted by private organizations, and as a result many different forms of highly accurate and precise data are becoming available, such as the 1 m resolution IKONOS satellite data (Land Info Worldwide Mapping, LLC, 2006). Although these advances in remote sensing technology have increased the variety of products available to consumers, the cost of acquiring and processing satellite-collected data is still prohibitive for many organizations.

Digital cameras can be mounted on airplanes (Figure 1.5), and can generally provide higher resolution images than that provided by satellites, yet this distinction is getting less clear with each passing year. It is also possible to mount digital cameras on smaller, remote controlled aircraft, and to synchronize color or infrared photography with GPS measurements. This coupling of technology may become more widespread in the future.

A relatively new technology called LiDAR (light detection and ranging) has emerged that allows for the collection of topographic or elevation data. LiDAR systems are typically mounted on an aircraft (although ground-based platforms are also used), and include a laser, an inertial navigation system, a GPS receiver, and an on-board computer for data processing. LiDAR measurement technology allows scientists to remotely sense and create digital models of landscape features such as vegetation, topography, and structures. LiDAR technology has been applied

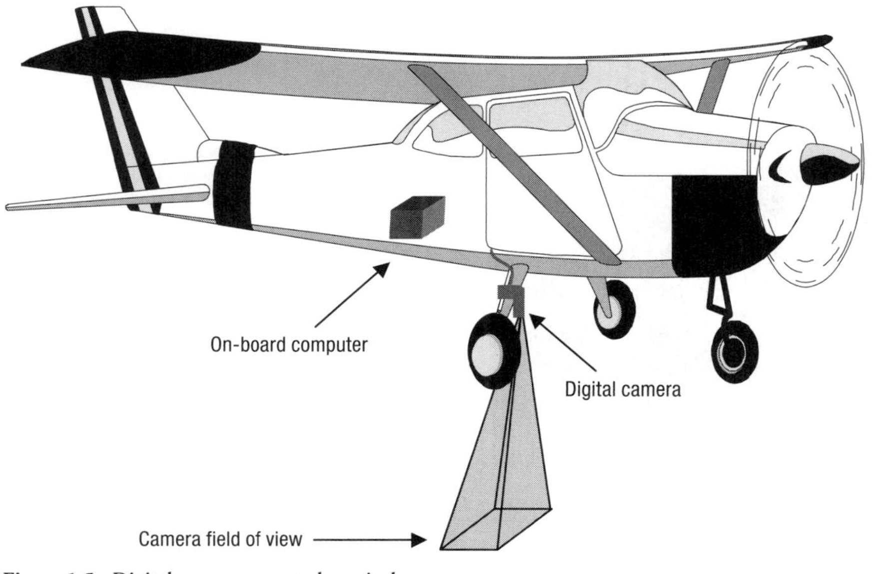

Figure 1.5 Digital camera mounted on airplane.

to natural resources to measure forest canopy structure, inventory, and biomass (Reutebuch et al., 2005). LiDAR airborne laser scanning involves directing discrete pulses of light onto a landscape in order to return the positions and dimensions of landscape features.

A LiDAR light pulse is emitted from a transmitter as the aircraft moves and travels until it reaches a solid object (Figure 1.6). Depending on the type, density, and reflectivity of an object, the light pulse is either reflected back to an airborne sensor or continues to deflect off of other objects until it reaches a solid surface, such as the ground. Typically, up to four reflected values can be returned from a single pulse. The combination of repeat returns can be fused with other multiple return pulses to create a three-dimensional visualization of landscape features. The round-trip travel time of individual light pulses is measured and stored by an airborne sensor that is coordinated with an on-board global positioning system (GPS). By comparing the return time to the speed of light, the distance to the ground or the landscape feature can be calculated. The coupling of pulse sensor and GPS measurements results in the geo-referencing of return pulses so that coordinates (longitude and latitude) and height (elevation) are associated with each returned pulse. In addition, the inertial navigation system tracks the irregularities of the aircraft's flight path and attitude (yaw, pitch, and roll) and all information is collected and processed by the on-board computer. Up to 150,000 pulses per second can be generated with contemporary LiDAR systems (Yu et al., 2006). This rapid pulse rate leads to LiDAR databases of hundreds of gigabytes for even modest sized areas (e.g. 4,000 ha). Image processing software can convert the millions of return pulses that are typical of LiDAR data projects into two and three-dimensional representations of landscape characteristics including streams, roads, and vegetation.

In addition to position and height measurements, the reflectance intensity of each LiDAR pulse is measured and stored with the geo-referenced information. Researchers have recently recognized that the strength of reflectance intensity values can potentially provide descriptive information about landscape features. Reflectance intensity is the ratio of strength of the reflected pulse to that of the emitted pulse. The reflectance intensity information is a spectral signature and can be used to determine the nature of landscape objects. There are few published studies of using LiDAR reflectance intensity values for natural resource applications but researchers have used LiDAR to investigate differences between coniferous and deciduous

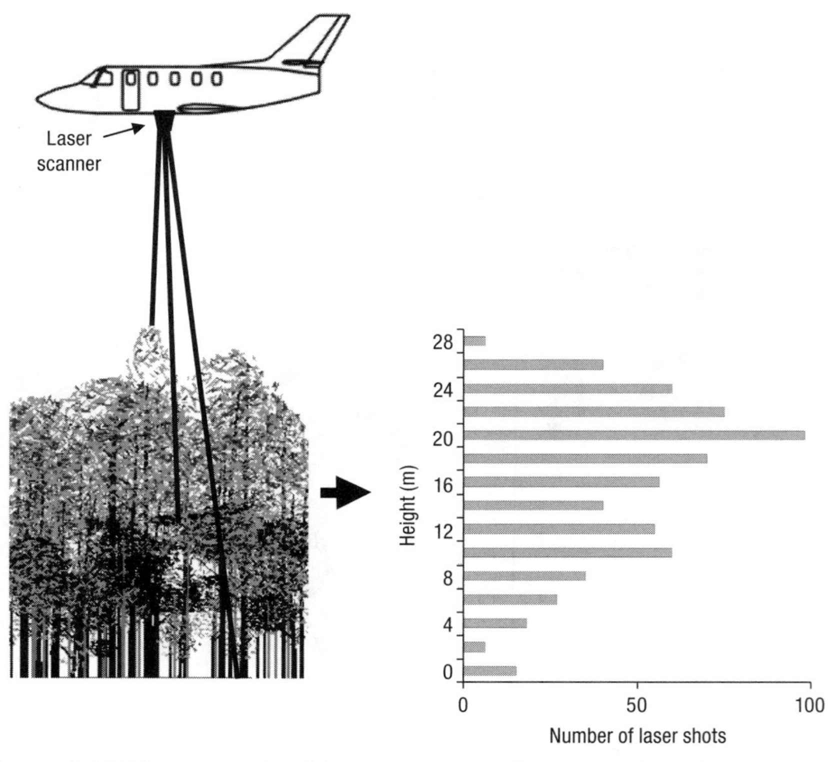

Figure 1.6 LiDAR system on aircraft (courtesy Dr Jason Drake, US Forest Service).

In Depth

In our society, electromagnetic energy is generated by a variety of sources, including sunlamps, fires, microwaves, radio towers, radar detectors, and lasers. Devices or techniques for capturing this energy can be characterized as passive or active. Passive data capture techniques, such as aerial photography or Landsat™, record electromagnetic energy that is naturally emitted or reflected. The most obvious natural producer of electromagnetic energy is the sun. The sun produces electromagnetic energy at multiple wavelengths, some of which are visible to the human eye. Devices that employ radar or laser technology transmit electromagnetic energy and record the amount of time it takes for the energy to return; these are defined as active techniques. The entire range of electromagnetic energy is known as the electromagnetic spectrum. The range of electromagnetic energy humans can see is called the visible portion of the electromagnetic spectrum, which contains wavelengths between 0.4 and 0.7 mm. Other portions of the spectrum that are not visible to humans include the cosmic, ultraviolet, infrared, microwave, and radar wavelengths. Most digital imagery developed from remote sensing devices makes use of the visible and infrared portions (0.4–0.9 mm) of the electromagnetic spectrum.

trees (Song et al., 2002) and to determine tree health (McCombs et al., 2003). While these prior research findings have had modest success in applying LiDAR reflectance intensities, the more powerful emitters and sensors that are typical of contemporary LiDAR equipment may provide descriptive information such as tree species and health, land cover type, or type of structure for mapped positions.

LiDAR has shown great potential in forestry and natural resource applications, not only in generating high-resolution digital elevation models (DEMs), but also in measuring stand structural conditions. Although the cost of acquiring LiDAR data is still prohibitive for many organizations, large areas can be flown at a cost of about $1 per acre; costs are expected to decrease in the future.

Photogrammetry

Photogrammetry is perhaps the primary method used for the creation of spatial data in forestry and natural resource management, although LiDAR acquisition is gaining steadily in its application. Within the US, many of the products produced by the US Geological Survey (USGS), including elevation surfaces and other representations of natural resources, were derived from photogrammetric techniques. **Photogrammetry** can be defined as the act of collecting measurements from the image of an object or resource. This technique dates back to the mid-nineteenth century, soon after the first photograph was created (Wolf & Dewitt, 2000). Through various techniques, photogrammetry facilitates the interpretation and measurement of features captured on photographs. Photogrammetry requires a firm understanding of photography, strong quantitative skills, and at times, creativity; successful interpretation sometimes becomes an art.

Photogrammetry offers several advantages over ground-based data collection techniques. Using aircrafts, photographs can be taken of areas at heights that might ordinarily be inaccessible by other devices. Large landscape areas can be captured, creating a permanent record of a resource at the time of data collection. Photographs can also be used for historical research because it is relatively easy to reexamine a photograph, as opposed to reviewing a field survey of a resource, which generally is difficult to reproduce. The accuracy, speed of acquisition, and cost of photogrammetric products are constantly improving, thus photogrammetry remains a popular method for collecting spatial data and for the creation of GIS databases. Digital methods of capturing images, however, are steadily replacing those that use photographic film.

The most crucial physical component of photogrammetry is the photographic system utilized. Single lens cameras are most common, and a typical frame measures 23 × 23 cm (9 × 9 in.). The camera lens is held at a fixed distance, or focal length, from the frame. Knowing this distance is critical in facilitating future measurements from photographs (Figure 1.7). The most common focal length is 152.4 mm (6 in.), but other lengths are also used (90, 210, and 305 mm). Photographic images are captured when a shutter near the lens is opened, momentarily

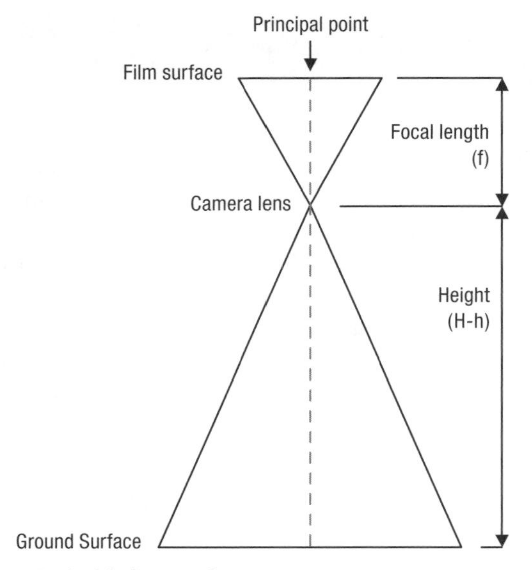

Figure 1.7 Aerial photography geometry.

allowing light to strike the film surface. Fiducial marks—usually in the form of four or eight markings located in the sides or corners of the photograph margins—are projected onto the film during the exposure of the photograph to allow you to define the geometric center of the photograph.

Aerial photographs are described by the angle in which they were captured: vertical or oblique. A vertical aerial photograph is one where the position of the camera axis is a nearly perpendicular orientation to the ground surface. An oblique aerial photograph is one where the position of the camera axis is located somewhere between a vertical and horizontal orientation to the ground. For measurement purposes, most photogrammetrists prefer vertical photographs (Figure 1.8) while oblique photographs are mostly useful for interpretive purposes.

Vertical aerial photographs usually capture images at regular intervals along a consistent heading, known as a flight line. This systematic data collection approach is followed to ensure total coverage of a resource. Flight lines are usually designed so that subsequent photographs will have an overlap of 60 per cent. In addition, photographs captured on adjacent flight lines should have an overlap of 30 per cent. An advantage of creating overlapping photographs is that you can see landscape features in stereo with a stereoscope (Figure 1.9) when simultaneously viewing photographs captured of the same area yet at dif-

Figure 1.8 Aerial photograph.

Figure 1.9 Mirror stereoscope.

ferent angles. In addition, photographic mosaics can be more easily created when using overlapping photographs.

The scale of an aerial photograph is described by the ratio of photo distances to ground distances. A scale can be calculated for any point on a photograph by using the following formula:

$$S = \frac{f}{(H - h)}$$

where S is the scale, f is the camera focal length, H is the height of the camera above a control surface, such as mean sea level (Figure 1.7), and h is the point's elevation. Variables f, H, and h, all need to be stated in the same units of measurement. If the focal length were 6″, the height of the camera 2,000', and the height of the point in question 450', the absolute scale would be 0.5 / (2000 – 450) = 0.000323. Expressed as a relative scale (the inverse of the absolute scale), these measurements represent 1:3100.

If a map is available of the photographed area, scale can be derived without using the focal length and camera height, but instead by comparing the photo distance with the map distance between two points. The following formula can then be used:

$$\text{photo scale} = \left(\frac{\text{photo distance}}{\text{map distance}}\right) \times \text{map scale.}$$

The distances used in this formula must be in the same units, and the photo scale will reflect the average elevation between the two points. Once the photo scale is known, photo distances can be converted to ground distances by multiplying the photo distance by the scale. For example, assume that the distance between two points on a photo distance was 2.5 inches, the distance between the same two points on a map was 4.5 inches, and that the map scale was 1:24,000. The scale of the photo is (2.5 / 4.5) × 24,000 = 13,333, or expressed as a ratio 1:13,333.

Analytical photogrammetry involves the use of mathematics to precisely define the locations of landscape features on stereo pairs of photographs. Stereoplotters are often used in analytical photogrammetry to register and measure photographs (Figure 1.10). There are several different types of stereoplotters; newer models interface with a computer to increase the speed of data creation and correction. Once the photographic images are placed on the stereoplotter, lights projected from different angles are directed through the photographs. The lights are adjusted so that a stereomodel is formed from the overlapping areas of the projected images on the photographs. Once the stereoplotter operator brings the stereomodel into focus, landscape features can be measured and mapped, and a potential GIS database is created. The accuracy of measurements obtained through analytical photogrammetry is usually expressed as a ratio of the camera height involved in the imaging process. Accuracy levels of around 1/12,000 of the camera height are typical. For camera heights of 12,000 ft, this translates into an accuracy of about ± 1 ft.

A relatively new product that is developed from aerial photographs is a digital orthophotograph. While an orthophotograph is derived from aerial photographs, the

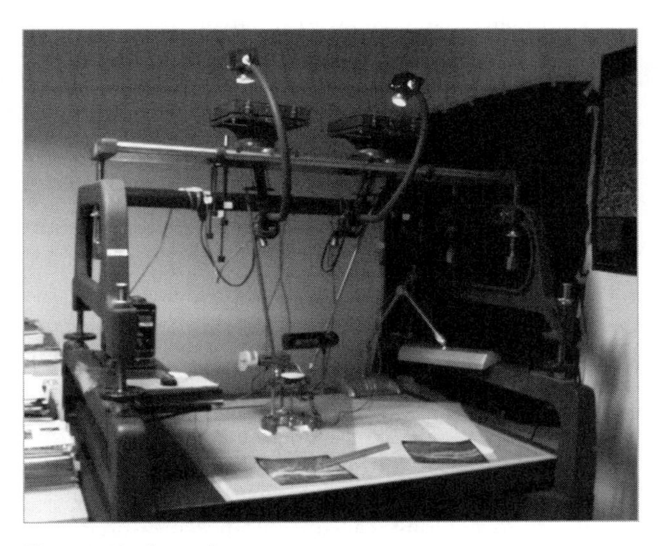

Figure 1.10 Stereoplotter.

relief displacement inherent in the photographs is minimized, and measurements of landscape features can be taken directly from the orthophotograph without the need for displacement corrections. To create a digital orthophotograph, scanned aerial photographs and digital elevation models (DEMs) are required. Orthophotographs and DEMs are discussed in more detail in chapter 2.

Field data collection

Field collection techniques for the creation of GIS databases have advanced tremendously over the past 20 years and are now fully enmeshed in the digital age. Increasingly, field data collection processes in natural resource environments are using digital data collection techniques (Wing & Kellogg, 2004). Field collection techniques were once limited to manual tools that required physical skill on the part of the operator and, depending on the instrument, technical competency equivalent to that possessed by a professional land surveyor (Kavanagh & Bird, 2000). Field crews would use metal or synthetic tapes to measure distances between objects, and clinometers or level guns to determine gradients and elevation differences. Approximate angles could be determined from compass readings, and more precise angle measurements were calculated from transits or theodolites. Measurements were recorded in field notebooks and processed in an office setting. Post-processing and adjustment of the data were almost always necessary to ensure that data collection and instrument errors were accounted for and balanced throughout the measurements. These practices are still common and appropriate today for many field crews who are involved in collecting spatial data for forestry and natural resource purposes.

Although technical competency with digital instrumentation and an understanding of measurement error and corrections are necessary skills for field crews using this technology, spatial data can be collected and processed with an efficiency and precision that far surpasses other manual field measurement techniques. Electronic distance measuring devices (EDMs) were first developed about 50 years ago and represented a major breakthrough in data collection (Wolf & Ghilani, 2002). These devices measure the amount of time it took a beam of electromagnetic energy to travel from an instrument, to a reflective surface, and back. With this information, a distance can be calculated. Current technology includes the ability to not only capture distance measurements, but also the angles between objects. In addition, the measurements are stored in a digital database. In some cases, these measurements can be highly accurate, providing positions that are within centimeters, or less, of their true locations values. Total stations and laser range finders (Figure 1.11) are examples of tools that make it possible for field crews to sight and 'shoot' distant objects. Typically, these instruments require that a reflective surface be placed on the object of interest so that a beam can be projected onto the surface and returned for measurement. Measurements include not only horizontal distances and angles, but also the elevation difference from the instrument's position. Automatically storing the measurements within the surveying instruments eliminates the potential errors that may arise when data recorded by hand on field forms is transferred to a GIS database.

Another technology that has become both more affordable and more useable in recent years is that of global positioning systems (GPS). GPS requires that a receiver, located on the Earth's surface, collect and record signals transmitted by satellites orbiting the Earth (Figure 1.12). Many natural resource professionals consider GPS receivers to be a source of frustration but recent evidence suggests that some GPS receivers are capable of reliably collecting measurements under canopy. The common limiting factor for GPS applications in natural resources has been that lines-of-sight between GPS receivers on-the-ground and space-based satellite systems have been obscured by canopy conditions, topographic barriers, or some combination thereof.

A GPS receiver calculates a position by being able to receive signals from at least four satellites, with more satel-

Figure 1.11 Laser range finder.

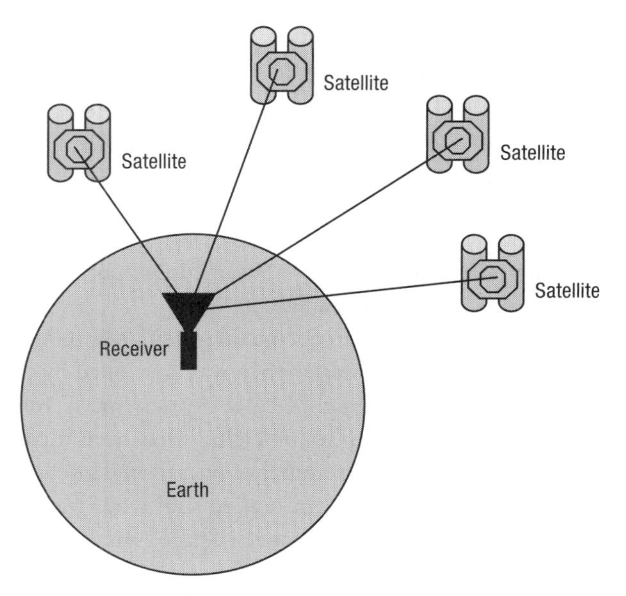

Figure 1.12 GPS schematic.

lites leading to better data collection opportunities. GPS receivers (Figure 1.13) calculate the amount of time it takes each signal to travel from the satellite. The GPS receiver uses information contained in the signals to calculate the range (distance) between the receiver and all satellites in communication. Ranges are used to estimate a position through trilateration. Satellite signal quality and reliability for measurement determination depends on satellite availability and geometry of available satellites in relation to the GPS receiver. Satellite signal quality is estimated as a Position Dilution of Precision (PDOP) statistic. Mission planning software is designed to identify the potentially best or preferred data collection times for GPS. Mission planning software can calculate an expected PDOP statistic and potentially available number of satellites for a field site. Larger values of PDOP (> 8) infer diminished satellite geometry and measurement reliability with values of 6 or below being preferred for data collection (Kennedy, 2002).

Measurement variability and error for GPS receivers can be introduced by atmospheric interference of satellite signals, timing errors between satellites and the GPS receiver, the rotation of the Earth, and satellite orbital

Figure 1.13 GPS receiver and antenna.

patterns (Leick, 2004). A portion of these errors can be estimated and removed through the process of differential correction. Differential correction uses a fixed GPS base station at a known location that continuously compares calculated GPS-derived positions to its own location. GPS base station locations are determined through repeated measurements that lead to an accurate and precise determination. Calculated differences between the known location and the GPS-derived locations serve as a correction factor that can also be applied to other GPS receivers that are collecting measurements nearby.

Another potential source of error for GPS receivers is that of multipath. Multipath errors occur when satellite signals reflect off of other objects before reaching a GPS receiver. This can introduce positional errors into the measurements (Figure 1.14). These errors generally must

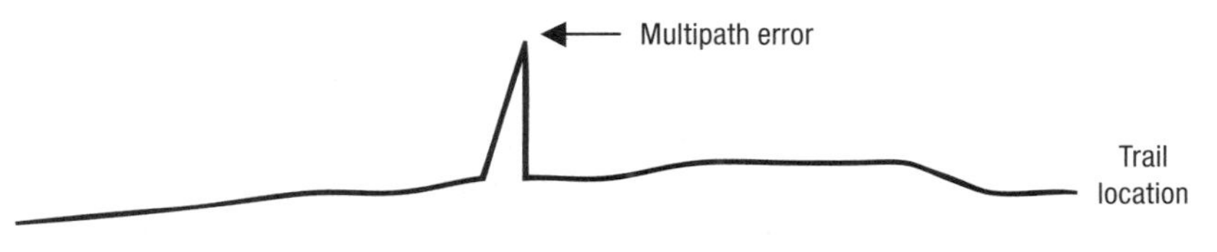

Figure 1.14 Example of multipath error in data collected through GPS.

be removed manually but some GPS manufacturers offer software routines that are designed to detect and reduce multipath errors.

Another way in which to reduce GPS receiver measurement error is to collect multiple measurements at single locations, a process known as precise point positioning. Statistical probability tells us that a coordinate determination based on the average of multiple measurements should be more reliable than that based on a single measurement.

In addition to the difficulty in receiving satellite signals in natural resource settings in forested terrain, until relatively recently there has been only one satellite system available to users worldwide. This primary satellite system is the NAVSTAR (Navigation Satellite Tracking and Ranging) system operated by the US Department of Defense (DoD). NAVSTAR became available in the early 1980s; it has a minimum of 24 operational satellites at any one time and makes satellite signals freely and continuously available to GPS users worldwide. Until 2001, the DoD, in the interest of national security, intentionally scrambled GPS satellite signals so that random errors prevented accurate location information from being collected. The scrambling process was known as selective availability (SA) and could lead to measurement errors of 100 m or more. The random errors could be removed by mapping- and survey-grade GPS receiver software through differential correction. Since 2001, selective availability has been removed and several new satellite systems have become operational. There is no guarantee, however, that selective availability will remain off in the future.

In addition to NAVSTAR, there are now Space Based Augmentation Systems (SBAS) that can provide conventional real-time differential corrections to GPS receivers as they collect data. Conventional real-time differential correction uses the more easily accessible coarse/acquisition (C/A) satellite signals rather than phase code. Although phase code signals have greater potential for accurate GPS measurements, continuous and uninterrupted satellite signals are required. Continuous and uninterrupted satellite signals are often difficult to maintain under forest canopy and in uneven terrain. SBAS derives measurement correction factors for several potential GPS error sources including atmospheric interference of signals, time sequences for satellite signal range (distance) estimates, and satellite orbital patterns. The primary SBAS for many users in North America is the US Federal Aviation Administration's Wide Area Augmentation System (WAAS). In 2008, four WAAS satellites were in orbit with at least two being operational, with more satellites expected in future years. Although a single WAAS satellite signal is required for a GPS receiver to apply real-time correction factors, reception from additional WAAS satellite signals can provide a backup should reception from one satellite become unavailable. Other SBAS include the European Geostationary Navigation Overlay System (EGNOS) and the Japanese MTSAT Satellite-based Augmentation System (MSAS). Some contemporary GPS receivers can operate with all SBAS.

Other GPS satellite systems include GLONASS (Global Navigation Satellite System), which was developed by the Russian military. Although GLONASS was made fully operational with 24 satellites in 1996, it has been inconsistent with regard to the number of operational satellites. In addition, another system, called Galileo, is under development through the European Space Agency and is expected to be operational in 2008.

The key component of a GPS from a user's perspective is that of the GPS receiver. GPS receivers can be separated by measurement accuracy and price into three broad grades or categories: survey, mapping, and consumer (Wing & Kellogg, 2004). Accuracy in this distinction is the difference between a GPS-collected measurement and the true location of the GPS receiver when it collected the measurement. The most accurate and expensive GPS receivers are survey grade and can calculate positions to within one cm of true location when used correctly. Survey-grade GPS receivers are the most full-featured of the three receiver grades and enable users to differentially correct collected data. In addition to the relatively high cost of survey-grade GPS (typically greater than $10,000) operator proficiency with the hardware and software applications is necessary. Survey-grade GPS receiver use in natural resource applications has been very limited because of the delicate nature of the equipment and the requirement for sustained satellite reception in order to derive measurements efficiently. Survey-grade GPS accuracies are likely greater than that required for many natural resource applications.

Mapping- or resource-grade GPS falls into the second level of the three GPS categories and can be purchased for $1,500–$10,000, depending on the features and the manufacturer. These GPS receivers are also sometimes called GIS-grade GPS. Many mapping-grade GPS have associated software for differential correction. Manufacturer estimates of positional accuracy are 1–5 m depending on the receiver configuration and mapping application. Accuracy estimates often reflect the best-case data collection scenarios, which may not be possible in forested environments; there have been several studies on this issue. Sigrist et al. (1999) found positional accuracies between 3.8 and 8.8 m during leaf-off and between 12.3 and 25.6 m during leaf-on conditions within a mixed-hardwood forest during selective availability. Naesset and Jonmeister (2002) reported positional errors between 0.5 and 5.6 m in sitka spruce (*Picea sitchensis*). Liu (2002) tested several mapping grade receivers under dense hardwood canopy and reported average positional errors of 4.0 m. Wing and Karsky (2005) found measurement accuracies between 1 and 4 m depending on the amount of canopy closure and the type of GPS configuration. Bolstad et al. (2005) tested a variety of mapping-grade GPS receiver configurations and found accuracies between 2.4 and 4.5 m under forest canopy in deciduous and red-pine forests. Wing et al. (in press) tested several mapping-grade GPS configurations and determined accuracies from post-processed data of 0.1 and 1.2 m in young forest and closed canopy conditions, respectively.

Consumer-grade GPS receivers are the least accurate and most affordable of the GPS grades with receivers costing between $50–$750. This price range may be attractive for many potential users but several disadvantages must be considered. Consumer-grade GPS receivers don't allow operators to set minimum thresholds for satellite signal quality through the establishment of a minimum PDOP level as a quality control. Mission planning software is usually not included with consumer GPS receivers and some do not enable users to conduct point averaging to determine a single position. While most consumer GPS afford users the ability to store measurements individually, a common storage limit of 500 points can limit the amount of time a consumer GPS receiver can be used in the field before the receiver memory is full. Differential correction capabilities through data post-processing techniques are not generally available to consumer grade GPS.

Like survey grade GPS receiver accuracy, consumer GPS receiver accuracy in forested settings has been reported in previous studies. Wing et al. (2005) tested the positional accuracies and reliability of six consumer grade GPS receivers within several different forest types and reported measurement accuracies within 10 meters of true position under dense conifer canopy and within 5 meters under partial canopy, depending on the type of consumer grade GPS receiver. Average accuracies of consumer GPS receivers between 6.5 and 7.1 m under dense primarily hardwood canopies were reported by Bolstad et al. (2005). Although the typical reported average accuracies reported by these studies (5 to 10 m) may be acceptable for many natural resource applications, consumer GPS receiver limitations, including the inability to set minimum satellite quality standards, the possibility of point averaging, and the lack of differential correction procedures, must be considered.

Data storage technology

Commonly, GIS databases consist of large quantities of data that must be stored and replicated ('backed-up') in a system that allows easy access for GIS managers and users at natural resource management organization field offices. Just a few years ago magnetic tapes and magnetic disks were commonly used to store computer files on networks and personal computers (Burrough & McDonnell, 1998) but storage technology continues to develop at a rapid pace.

Many people use optical storage devices such as compact discs (CD) and digital versatile discs (DVD) as standard storage technology. CDs generally can hold about 650 MB of data, and many personal computers now contain CD drives that allow users to both read and write to CDs. The read and write speeds of CDs are prohibitive for GIS users who work with large GIS databases. DVDs look very similar to CDs, also require a computer drive to operate, and are similar in speed. Many manufacturers now offer 'combo' drives which can read and write both CD and DVD formats. A DVD can hold approximately 4.5 GB of data, with some varieties being able to store even more. Recently, rapid read and write speeds (~10 ms) have been achieved by many brands of portable external hard drives. These external hard drives can be quickly connected to a computer's universal serial bus (USB) port. They provide fast backup and recovery, and can also serve as an additional hard drive if needed. USB keys, or what some refer to as 'thumb drives' are very popular, inexpensive solutions for transporting 'smaller' (< 2 GB) databases. These devices are comparable in size to a key-chain and less than $50 in price for a large capacity (2 GB) version. It would not seem unreasonable for a GIS workstation to include the following data disk drives: CD and DVD combo drive, large (100–200 GB) external USB drive, additional USB drives for data transfer, and a 100–200 GB internal hard drive.

Data manipulation and display

Personal computers and workstations are now nearly synonymous in meaning, although just 10 years ago the term workstation implied the use of a UNIX (uniplexed information and computing system) operating system. Personal computers are now the primary platform upon which to utilize GIS software programs and manipulate GIS databases. Even though the distinction between personal computers and workstations has become blurred, computers can be characterized by criteria that constantly change with advances in computer technology, such as speed, memory, and the operating system. Key components within a computer include random access memory (RAM), central processing units (CPUs), and storage devices (hard drives or optical discs). Windows® and Linux® operating systems have become much more common than UNIX-run computers. Most GIS software designers now focus their development efforts on Windows®-compatible systems and no longer develop UNIX-based software.

The resolution and memory of computer monitors is also important for the quick and clear display of GIS databases. Video memory cards are installed in most computers to manage the level of pixel resolution of raster GIS databases, to display a wider array of colors, and to increase the speed at which images are displayed on a computer monitor. A below-average video card can reduce the effectiveness of an otherwise fast-processing computer. Video cards are often rated by the amount of RAM installed on the card. At the time of writing of this book, 256 MB would be an adequate amount of RAM for standard GIS processes. For those working with larger GIS databases, a video card with 512 MB or more memory should be considered.

Output devices

GIS databases and the results of GIS analyses can be presented in a variety of manners, both in graphical and tabular form. For example, when examining the impacts of alternative riparian management policies on a landscape, it may be important to present data in tabular form to describe the potential economic impacts of alternatives, and in graphical form to visually display the area each policy affects. A number of common output devices that can be used to present GIS analysis results are described in the next few sections.

Printers and plotters

The most obvious output devices associated with GIS are printers and plotters. Fifteen years ago, line printers were common peripherals to computer systems. Their output quality was low and the range of symbols and colors was limited. Today, a variety of color and black and white printers are available that can produce high quality maps. Thus the cartographic creativity of GIS users is virtually unhindered. Printers are generally classified as laser or ink jet, depending on how ink is transferred to paper. Laser printers are more expensive ($100–$2,000) than ink jet printers (< $500), and they require reusable laser cartridges. However, they generally produce output products that are more stable, with regard to exposure to moisture, than ink jet printers. Ink jet printers require disposable ink cartridges and the output products produced are more easily affected by exposure to moisture.

In Depth

A GIS workstation usually includes a higher-end computer with a fast processor and large amounts of RAM and internal disk storage space. However, for the type of work performed by field foresters and natural resource managers, a typical personal computer (PC), or even a laptop computer, will generally suffice to facilitate the use of desktop GIS software such as ArcGIS or MapInfo. What are the desired characteristics of a PC computer that might allow you to use desktop GIS software? Obviously a fast processor would be beneficial: the faster the processor, the quicker a process is completed, such as buffering or overlaying. In addition, 1 GB of RAM is perhaps the minimum RAM necessary to easily handle large processing tasks. And finally, a 100–140 GB hard drive is perhaps the minimum size necessary to store GIS databases you might develop and use over the useful life of a personal computer (2–3 years). If LiDAR or other larger-size GIS databases are to be used, the minimum hard drive size might begin at 200 GB.

The main drawback with most printers is the format of the output, which is generally limited to 8½″ × 11″ or 11″ × 14″ media. Plotters allow GIS users to produce maps of a variety of sizes (Table 1.1). Plotters, however, are generally more expensive than printers, though they are now available for prices beginning at $1000; they also typically require special paper and ink cartridges. One way to categorize plotters is to use a vector/raster analogy. Vector plotters include those termed as flatbed or drum. Vector plotters draw lines using plotting pens of different colors and can produce some very precisely drawn maps. Raster plotters include those termed electrostatic, laser, ink jet, and warm wax. Electrostatic plotters use an array of electric contacts (> 100 per inch) that apply a charge to paper, which then comes in contact with negatively charged toner to produce images. The technology that is used in laser and ink jet plotters is similar to that in laser and ink jet printers—each feature on the map is drawn pixel by pixel. Warm wax plotters are similar to ink jet plotters but the resulting products have a glossy appearance. Some of these printers and plotters, in conjunction with high-quality paper, enable GIS users to produce photographic-quality output products. A thorough examination of the needs of an organization is warranted prior to making a decision regarding an investment in printers or plotters.

TABLE 1.1 Common sizes of map output from plotters

Map size	Dimensions
ANSI A	8.5″ × 11.0″ (216 mm × 279 mm)
ANSI B	11.0″ × 17.0″ (279 mm × 432 mm)
ANSI C	17.0″ × 22.0″ (432 mm × 559 mm)
ANSI D	22.0″ × 34.0″ (559 mm × 864 mm)
ANSI E	34.0″ × 44.0″ (864 mm × 1118 mm)
ANSI F	28.0″ × 40.0″ (711 mm × 1016 mm)
ISO A4	8.3″ × 11.7″ (210 mm × 297 mm)
ISO A3	11.7″ × 16.5″ (297 mm × 420 mm)
ISO A2	16.5″ × 23.4″ (420 mm × 594 mm)
ISO A1	23.4″ × 33.1″ (594 mm × 841 mm)
ISO A0	33.1″ × 46.8″ (841 mm × 1189 mm)
ISO B4	9.8″ × 13.9″ (250 mm × 353 mm)
ISO B3	13.9″ × 19.7″ (353 mm × 500 mm)
ISO B2	19.7″ × 27.8″ (500 mm × 707 mm)
ISO B1	27.8″ × 39.4″ (707 mm × 1000 mm)

ANSI = American National Standards Institute

ISO = International Standards Organization

Screen displays

A more rudimentary set of output products from GIS are those related to the image displays you can view on a computer screen. These processes consist of capturing information (data or maps) displayed on the screen of a computer and temporarily storing the information in a digital database. A number of methods are available to capture computer screen images. The resolution and detail of the resulting captured image, however, depends on the method used for image capture. Screen displays are sometimes captured and saved as image files, and at other times simply stored in the computer's 'clipboard', and thus are available for pasting into a variety of other software programs. For example, most personal computers allow users to save what may be displayed on a computer screen by pressing (all at once) the Alt and Prnt Scrn buttons on a computer keyboard. This stores the entire image displayed on the screen to the computer's clipboard, just like if you were copying text in a word processing program. Then, you can paste the captured image into either a word processing or graphics software program (Figure 1.15). One potential drawback is that screen

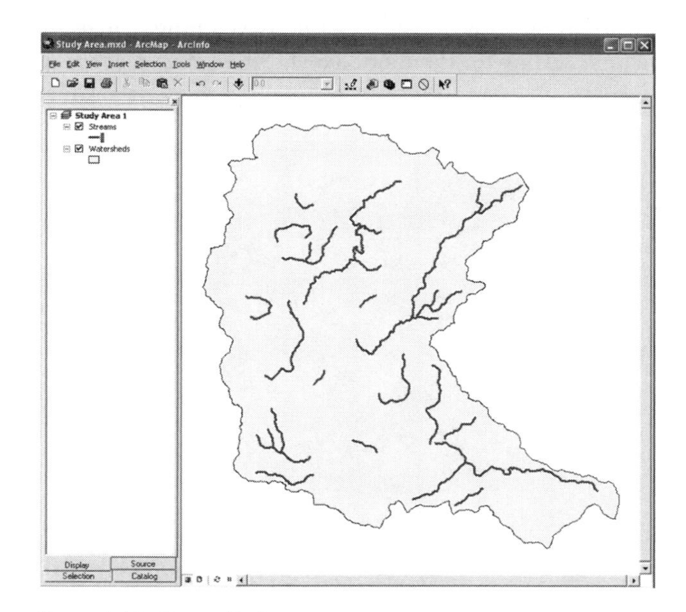

Figure 1.15 Screen display.

captures are raster images, even though you may be attempting to capture a representation of a vector GIS database displayed on the screen.

Graphic images

In addition to screen captures, most GIS software programs allow users to directly store images viewed on the screen as independent computer files. These products are also raster images, yet they are slightly different than the screen displays described above in that, generally, only the map image is captured and stored (Figure 1.16), and not everything else that may be visible on the screen. These images can be stored in a wide variety of formats (Table 1.2) depending on the availability within the GIS software program being used. However, transferring graphic images from one system (e.g., GIS) to another (graphics editing programs) or vice versa can sometimes be problematic because of format inconsistencies. In addition, the size of the resulting graphic image files will vary depending on the format used to save the image.

TABLE 1.2 Common types of graphics image output files

File extension	Description
bmp	Windows® bitmap format
cdr	Corel Draw® format
cgm	Computer graphics metafile format
dxf	AutoCad® digital exchange file format
emf	Windows® Enhanced Metafile format
eps	Encapsulated PostScript format
gif	Graphics interchange format
jpg	JPEG File interchange format
pct	Macintosh® PICT format
pcx	PC paintbrush format
png	Portable network graphics format
tga	Targa format
tif	Tagged image file format
wmf	Windows® metafile format
wpg	WordPerfect® graphics format

Tabular output

As you might assume, tabular output consists of tables or sets of data (numbers, text) derived directly from a GIS database or from the result of a GIS analysis. While maps are engaging—they draw people in and allow them to visualize the qualities and conditions of a landscape—tabular data are also important for illustrating non-spatial information. For example, while developing a map of habitat quality for the spotted owl (*Strix occidentalis*) may provide an interesting and compelling view of a landscape, decision-makers might also be interested in how much land of higher quality habitat exists. Or, as alternative riparian management policies are evaluated, the effect (e.g., the area or timber volume within the riparian management areas) will likely be of interest to decision makers. Tabular data from GIS analyses can be displayed directly on a map, or drawn together into an independent table for incorporation within a report.

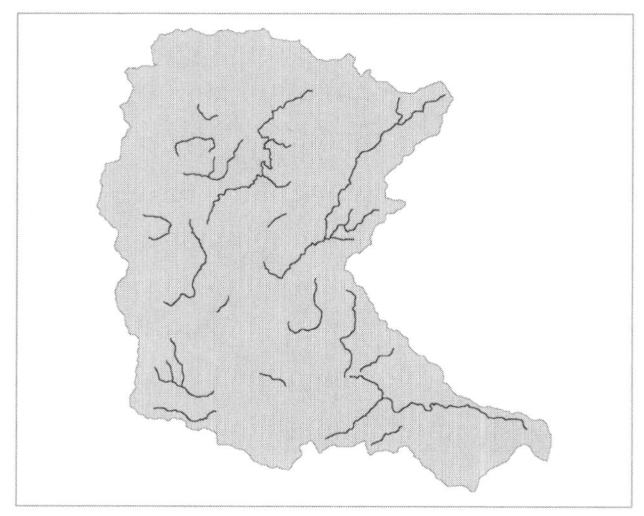

Figure 1.16 Graphic image.

GIS software programs

There are many GIS software programs available to natural resource management organizations today. The comments provided in this section are general in nature, however a list of the common GIS software programs is provided in Appendix C.

GIS software programs are categorized in a number of ways. One characterization is based on which of the two common data structures (raster or vector) is accommodated. Raster and vector data structures is discussed in more depth in chapter 2. GIS software programs have also been characterized by the operating system used. For example, GIS software programs developed for UNIX workstations were once considered 'workstation GIS software', and GIS software programs developed for PCs were considered 'PC GIS software'. This distinction has essentially disappeared as workstation GIS software is now used on computers using both Windows® and UNIX operating systems. A contemporary distinction among GIS software programs has emerged that also makes use of the term 'workstation' and contrasts it with 'desktop' systems. In this categorization, workstation GIS software programs are those that include the full range of GIS processes that allow users to create, edit, and analyze spatial data. Examples of full-featured workstation GIS software programs include ArcInfo and MGE Microstation. Desktop GIS software programs are considered to be scaled-down versions of workstation GIS software programs, possessing a portion of the tools or GIS processes found in full-featured workstation GIS software programs. Examples of desktop GIS software programs include ArcView, MapInfo, and GeoMedia. A primary distinction between workstation and desktop GIS programs (beside their associated cost) has been the ability to ensure that spatial relationships between the locations described in GIS databases (the topology) remains true or correct. Some desktop GIS software programs are limited in their ability to maintain topological relationships without the use of (and, therefore, purchase of) additional software modules.

In choosing a GIS software program, purchasers need to consider the price, the GIS databases to be managed, and the flexibility of the software to perform the likely analyses required to support management decisions. The price for most GIS software programs ranges from $500 for a raster-based desktop GIS software program to well over $15,000 for a full-featured workstation GIS software program. Those on a tight budget might consider a recent article by Bernard and Prisley (2005) titled 'Digital Mapping Alternatives: GIS for the Busy Forester'. This article compared the abilities of nine GIS software products costing less than $500.

The types of spatial databases that are accessible by a GIS software program should be carefully evaluated. For example, if primarily raster GIS databases will be used, perhaps a raster-based GIS software program is more appropriate (as opposed to a GIS software program that focuses on the development and maintenance of vector GIS databases). Finally, a consideration of the types of GIS processes likely to be performed is important in making an informed purchase decision. For example, GIS users may need to geo-reference, interpret, and classify satellite imagery, and may need to create vector GIS databases (e.g., roads or streams) from measurements collected during field data collection processes, or to perform watershed delineation and analysis processes with a DEM.

If, after pondering these issues, the choice of a GIS software program remains unclear, perhaps choosing a GIS software program that seems to have the capability to expand with the needs of an organization would be a good decision. Increasingly, GIS software programs are designed in a modular manner, which allows users to purchase separate software modules intended to work with a base GIS program. Users then purchase the base GIS program and only those modules that they deem necessary.

Maintenance charges are also becoming more typical for GIS software programs, and are often overlooked when natural resource management organizations create their budgets. Users can either purchase annual maintenance support for a GIS software program, or pay for technical support as issues arise. An annual fee is easier to predict, and to include in the budget, but a per-incident fee may result in a lower total cost (depending on the amount of support a user needs). For some GIS software programs, maintenance fees cover product upgrades, and so as new versions of the software are released, they may be available at no cost to those with annual maintenance agreements. As an alternative, online GIS user support groups are available for users with limited access to technical support from the software developer. Reliance on these groups can result in an inexpensive and often rapid response resource for those with software and hardware difficulties.

When evaluating GIS software programs, organizations should also consider the projected longevity of the use of the software. It is expensive to implement and maintain a GIS system. Once a software selection has been made and a commitment by the organization to the software has been institutionalized, changing to a different GIS software program is difficult and expensive. Organizations may not have a choice in this matter—there are many examples of the obsolescence of GIS software programs, and the subsequent elimination of support for GIS software programs (from the software developer) as new products are developed. However, organizations should consider the length of time that a GIS developer has been in business, and attempt to gauge its future prospects. GIS software programs that have been available for at least five years, demonstrate continually growing sales, and have well-established, active user support groups, might serve as criteria to consider.

Summary

This introductory chapter described the history of GIS development, why GIS is important in natural resource management, and the wide variety of input and output devices associated with GIS. In addition, a number of issues related to the selection and purchase of GIS software programs were outlined; hopefully this summary of issues will stimulate discussion among those considering the development of a GIS system within a natural resource organization. The applications of GIS to natural resource management issues will vary widely among organizations, however, understanding the capabilities and potential use of GIS is essential for natural resource management professionals. The following applications, as with those in subsequent chapters, are intended to provide students with a taste of the typical types of GIS requests posed to field foresters, biologists, and other professionals familiar with GIS, who work in natural resource management field offices.

Applications

1.1 Developing the specifications for a GIS system. You have been asked by the District Manager of your natural resource management organization, Jane Lerner, to develop the specifications for a computer to be purchased and used for, among other things, GIS analysis and map production in a natural resource management organization field office. Use resources available on the Internet to design a computer system that would be capable of running desktop GIS software in a forestry or natural resource field office.

a) What are the specifications of the computer system that you would recommend, and how much might it cost?
b) If your budget were limited to $2,500 (maximum), how might your recommendation change?
c) You have been asked to decide whether more RAM or video memory would be a better investment for a GIS computer system. What is the difference between these two types of memory and which would you select?

1.2 Terminology. The District Manager of your natural resource management organization, Steve Smith, is unfamiliar with a number of terms related to GIS. He has heard these terms distributed freely during staff meetings, and during one of your weekly reviews he asks you to help him understand what they mean. Briefly describe for him GIScience, topology, and overlay analysis.

1.3 Characterizing GIS software systems. Your supervisor, John Darling, has heard the terms 'workstation' and 'desktop' GIS software, but remains confused about how they differ. Explain to him the differences between the two types of GIS software.

1.4 History of GIS. While on vacation and visiting your relatives, you find that the conversation around the dinner table has turned to the types of work you perform in your role as a natural resource manager. Describe for your relatives, many of whom have never heard of GIS, the origin of GIS, how GIS has evolved into its current form, and how you might use GIS in natural resource management.

1.5 GIS pioneers. Identify and list the noteworthy contributions of someone who has made significant contributions to the development of GIS.

1.6 GIS data. Identify and describe one of the GIS databases described in this chapter that contained data for the resources of an entire country or for a portion of the world.

1.7 A question of scale. You measure the distance between two owl nests on a 1:24,000 scale topographic map to be 6 cm. What is the actual ground distance between the nests?

1.8 Scale revisited. You have measured the distance between two campgrounds on a topographic map to be 2 cm. From a field visit, you know that the corresponding ground distance between the campgrounds is 1 km. What is the scale of the topographic map?

1.9 GIS software. List and briefly describe three GIS software packages that were available prior to 1990.

1.10 Relative error. You measure the perimeter of a field plot with a metal tape and determine a total perimeter of 134.5 meters. Your instructor tells you that he used a total station and determined that the total perimeter is actually 136.2 meters.

a) What is the closure error between your measurement and your instructor's?
b) What is the relative precision of your metal tape measurements?

1.11 GIS data from above. You have been asked to develop a database of your county that contains elevation and landform information. What are three remote sensing data collection techniques that would be used to develop this database and what are the relative strengths and weaknesses of each?

1.12 GIS data from the ground. You have been asked to create a GIS database that contains the boundaries of a set of tree stand boundaries in a research forest. Describe three approaches to collecting this data and the relative strengths and weaknesses of each approach.

1.13 GPS considerations. A friend of yours has recently purchased a new GPS receiver from a local department store for $75 and has told you that she is excited that she will be able to collect coordinates of features that represent 'exact locations' on the Earth's surface. What advice would you offer her about the measurement accuracy of a $75 GPS receiver?

1.14 Data input devices. The Bureau of Land Management has hired you as a forestry technician. Your supervisor is aware that you have a background in GIS, and asks for your input regarding the technology that can be used to develop a vegetation GIS database. Describe three options, and their strengths and weaknesses in terms of collecting data and developing a GIS database.

1.15 Data display options. It is Friday afternoon in a natural resource organization's field office. As you are daydreaming about the forthcoming weekend's events, your supervisor enters your office and tells you that he has a meeting Monday morning with a neighboring landowner to describe the management alternatives for a portion of the forest your organization manages. A few graphics that describe the alternatives under consideration would be beneficial to the meeting, and maps are the obvious choice of output products to engage the public. However, the color plotter in your office is not working. Describe three other methods for examining output from GIS that might be useful for your supervisor's Monday meeting.

References

Avery, T.E, & Berlin, G.L. (1992). *Fundamentals of remote sensing and airphoto interpretation* (5th ed.). New York: Macmillan Publishing Company.

Bernard, A.M., & Prisley, S.P. (2005). Digital mapping alternatives: GIS for the busy forester. *Journal of Forestry, 103*(4), 163–8.

Bolstad, P., Jenks, A., Berkin, J., Horne, K., & Reading, W.H. (2005). A comparison of autonomous, WAAS, real-time, and post-processed global positioning systems (GPS) accuracies in northern forests. *Northern Journal of Applied Forestry, 22*(1), 5–11.

Brown, T.L., & Lassoie, J.P. (1998). Entry-level competency and skill requirements for foresters. *Journal of Forestry, 96*(2), 8–14.

Burrough, P. (1986). *Principles of geographical information systems for land resources assessment*. Oxford: Oxford University Press.

Burrough, P.A., & McDonnell, R.A. (1998). *Principles of geographical information systems*. Oxford: Oxford University Press.

Clarke, K.C. (2001). *Getting started with geographic information systems* (3rd ed.). New Jersey: Prentice Hall, Inc.

de Steiguer, J.E., & Giles, R.H. (1981). Introduction to computerized land-information systems. *Journal of Forestry, 79*, 734–7.

Goodchild, M.F. (1992). Geographical information science, *International Journal of Geographical Information Systems, 6*(1), 31–45.

Kavanagh, B.F., & Bird, S.J.G. (2000). *Surveying: Principles and practices* (5th ed.). New Jersey: Prentice Hall, Inc.

Kennedy, M. (2002). *The global positioning system and GIS*. London and New York: Taylor and Francis.

Land Info Worldwide Mapping, LLC. (2006). *IKONOS high-resolution satellite imagery*. Highlands Ranch, CO: Land Info Worldwide Mapping, LLC. Retrieved April 21, 2007, from http://www.landinfo.com/satprices.htm.

Leick, A. (2004). *GPS satellite surveying*. Hoboken, NJ: John Wiley & Sons.

Liu, C.J. (2002). Effects of selective availability on GPS positioning accuracy. *Southern Journal of Applied Forestry, 26*(3), 140–5.

Longley, P.A., Goodchild, M.F., Maguire, D.J., & Rhind, D.W. (2001). *Geographic information systems and science*. New York: John Wiley and Sons, Inc.

McHarg, I.L. (1969). *Design with nature*. New York: John Wiley and Sons, Inc.

McCombs, J.W., Roberts, S.D., & Evans, D.L. (2003). Influence of fusing LiDAR and multispectral imagery on remotely sensed estimates of stand density and mean tree height in a managed loblolly pine plantation. *Forest Science, 49*(3), 457–66.

Merry, K.L., Bettinger, P., Clutter, M., Hepinstall, J., & Nibbelink, N.P. (2007). An assessment of geographic information system (GIS) skills used by field-level natural resource managers. *Journal of Forestry, 105*(7), 364–70.

Naesset, E., & Jonmeister, T. (2002). Assessing point accuracy of DGPS under forest canopy before data acquisition, in the field, and after postprocessing. *Scandinavian Journal of Forest Research, 17*, 351–8.

Natural Resources Canada. (1978). *Specifications and recommendations for control surveys and survey markers*. Ottawa, ON: Canada Centre for Remote Sensing.

Peucker, T.K., & Chrisman, N. (1975). Cartographic data structures. *American Cartographer, 2*(1), 55–69.

Reutebuch, S.E., Andersen, H.-E., & McGaughey, R.J. (2005). Light detection and ranging (LIDAR): An emerging tool for multiple resource inventory. *Journal of Forestry, 103*(6), 286–92.

Sample, V.A., Ringgold, P.C., Block, N.E., & Giltmier, J.W. (1999). Forestry education: adapting to the changing demands. *Journal of Forestry, 97*(9), 4–10.

Sigrist, P., Coppin, P., & Hermy, M. (1999). Impact of forest canopy on quality and accuracy of GPS measurements. *International Journal of Remote Sensing. 20*(18), 3595–610.

Song, J.-H., Han, S.-H., Yu, K., & Kim, Y.-I. (2002). Assessing the possibility of land-cover classification using LIDAR intensity data. International Society of Photogrammetry and Remote Sensing (ISPRS), Commission III Symposium, 9–13 September, Graz, Austria, *Vol. XXXIV*, B-259–62.

Steinitz, C., Parker, P., & Jordan, L. (1976). Hand-drawn overlays: Their history and prospective uses. *Landscape Architecture, 66*(5), 444–55.

US Federal Geodetic Control Committee (FGCS). (1984). *Standards and specifications for geodetic control networks*. Rockville, MD: National Geodetic Information Branch.

Wing, M.G., & Bettinger, P. (2003). GIS: An updated primer on a powerful management tool. *Journal of Forestry, 101*(4), 4–8.

Wing, M.G., & Kellogg, L.D. (2004). Locating and mobile mapping techniques for forestry applications. *Geographic Information Sciences, 10*(2), 175–82.

Wing, M.G., Eklund, A., & Kellogg, L.D. (2005). Consumer grade global positioning system (GPS) accuracy and reliability. *Journal of Forestry, 103*(4), 169–73.

Wing, M.G., & Karsky, R. (2006). Standard and real-time accuracy and reliability of a mapping-grade GPS in a coniferous western Oregon forest. *Western Journal of Applied Forestry, 21*(4), 222–7.

Wing, M.G., & Sessions, J. (2007). Geospatial technology education. *Journal of Forestry, 105*(4), 173–8.

Wing, M.G., Eklund, A., & Karsky, R. (In press). Horizontal measurement performance of five mapping-grade GPS receiver configurations in several forested settings. *Western Journal of Applied Forestry.*

Wolf, P.R., & Dewitt, B.A. (2000). *Elements of photogrammetry: With applications in GIS* (3rd ed.). New York: McGraw-Hill.

Wolf, P.R., & Ghilani, C.D. (2002). *Elementary surveying: An introduction to geomatics* (10th ed.). Englewood Cliffs, NJ: Prentice Hall, Inc.

Yu, X., Hyyppä, J., Kukko, A., Maltamo, M., & Kaartinen, H. (2006). Change detection techniques for canopy height growth measurements using airborne laser scanner data. *Photogrammetric Engineering and Remote Sensing, 72*(12), 1339–48.

Chapter 2

GIS Databases: Map Projections, Structures, and Scale

Objectives

This chapter introduces the concepts of map projections and data structures. After completing this chapter, readers should have an understanding of the following topics related to the structure and composition of GIS databases:

1. the definition of a map projection, and the components that comprise a projection,
2. the components and characteristics of a raster data structure,
3. the components and characteristics of a vector data structure,
4. the purpose and structure of metadata,
5. the likely sources of GIS databases that describe natural resources within North America,
6. the types of information available on a typical topographic map, and
7. the definition of scale and resolution as they relate to GIS databases.

Performing GIS processes and analyses in support of natural resource management decisions requires obtaining and working with spatial databases. Many GIS users find that they spend a great deal of time and effort acquiring and modifying GIS databases to ensure that the most suitable and appropriate data is being used in subsequent analyses. One of the great challenges in working with features located on the surface of the Earth is that the Earth is very irregularly shaped, and is in a constant state of change. When you attempt to create a two-dimensional representation of the Earth (as is typically represented on maps), the Earth's irregularities must be addressed. Different map projections and projection components have been created so that data from the Earth's surface can be displayed on maps and other flat surfaces. Understanding that spatial data can be represented through any number of different map projections, and that data can be transformed from one map projection to another, is a very important component in the process of learning to manage GIS databases successfully. This chapter is intended to introduce readers to common GIS database formats, the ways in which GIS data can be structured and adjusted to represent the Earth's surface, and how GIS databases are documented and described. Some direction is also provided to allow you to begin to think about sources of GIS databases, although more detailed treatment on this subject is provided in chapter 3.

The Shape and Size of the Earth

GIS software programs are designed to work with data describing the Earth's features, to provide methods for feature measurements, and to allow comparisons of features of interest. A number of options exist by which you can collect, structure, and access GIS data. Spatial data users, however, must always be cognizant that representation of landscape features on two-dimensional surfaces, such as maps or computer monitors, are subject to distortion based on the spherical shape or the Earth. These same challenges relate to the way in which a GIS database is formatted and stored. If the Earth were perfectly flat, collecting and mapping data from its surface would be a relatively straightforward process. The Earth, however, has an uneven surface that can't be defined or modelled with complete accuracy and precision for a number of reasons. For data storage and representation, the surface of the Earth can at best be approximated through mathematical models. Depending on the source of data, data collection techniques, the geographical shape and size of a landscape, or the goals of a particular analysis, you may need to make adjustments to a GIS database to compensate for the shape of the Earth. GIS databases are generally adjusted for the shape and curvature of the Earth through application of a map projection. A map projection can be thought of as a mathematical model that attempts to associate a GIS database with its relative position on the Earth. Although some GIS software will allow GIS databases that are referenced in different map projection systems to be displayed as if they were registered to the same map projection, this capability should only be used for mapping that will serve a general purpose. When comparisons are going to be made between GIS layers in support of analytical processes, it is particularly important that each database is referenced to the same map projection. Failure to reference a GIS database to its correct map projection may also lead to inaccurate analysis results, even if only a single GIS database is involved.

The true size and shape of the Earth has been an issue of debate for millennia and, despite the recent advancements of measurement technology, data related to the size and shape of the Earth continues to be collected and analyzed. The field of collecting or calculating exact measurements of the Earth's size, shape, and gravitational forces is called **geodesy**. It is unclear who originally declared the Earth to be round (or spherical) in shape rather than flat, but Greek philosophers Pythagoras (sixth century BC) and Aristotle (fourth century BC) considered the Earth to be round. Another Greek scholar, Eratosthenes (276–194 BC) provided the first mathematical calculation of the Earth's perimeter through an ingenious method that is still used today. Eratosthenes observed that the bottom of a deep well located in Syene—today near Aswan in southern Egypt—was fully lit (without shadows from the sides of the well) only during the summer solstice (June 21). He reasoned correctly that the sun must be directly overhead at this time. Eratosthenes travelled to neighboring Alexandria, a distance that he estimated to be approximately 500 miles. Once at Alexandria, Eratosthenes measured the angle of a shadow from a vertical column of known height during the summer solstice. An angular distance of 7°12′ was recorded, and Eratosthenes subsequently applied geometric principles to determine the circumference of the Earth (Figure 2.1). By dividing the calculated angle into the total number of degrees of a circle approximating the Earth (360°), Eratosthenes reasoned that the distance between Syene and Alexandria should represent 1/50th of the Earth's circumference.

Eratosthenes took the estimated distance between Syene and Alexandria of approximately 500 miles, and multiplied this by 50 to get an estimate of the Earth's circumference (25,000 miles). Although several of the factors involved in Eratosthenes' assumptions and calculations were slightly inaccurate, the errors tended to compensate for each other. The 25,000-mile result overestimates the circumference of the Earth, but is remarkably close to today's estimate (approximately 24,900 miles) given the tools at Eratosthenes's disposal. The approach and use of geometry that Eratosthenes applied to solve a difficult question reflected a creative and very clever mind. Similar techniques are still commonly used today to solve problems of distance and angles.

Ellipsoids, Geoids, and Datums

Until the end of the seventeenth century, when Isaac Newton advanced his theory of gravity, the Earth was thought to be represented by a perfect sphere. Newton theorized that if the Earth were rotating along an axis, the shape of the Earth would tend to bulge along the equator and tend to be flattened at the poles, as a result of the centrifugal force created by the rotation. These forces would lead to a global profile that was somewhat different than

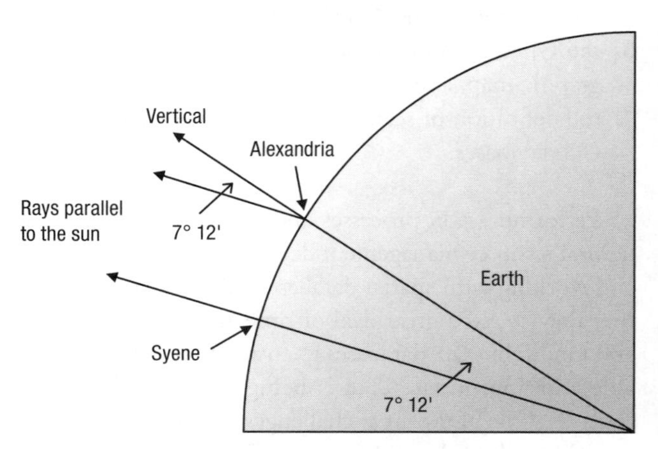

Figure 2.1 Erathosthenes's approach to determining the Earth's circumference.

that presented by a perfect circle. The theory that the profile of the Earth was not uniform was confirmed by field measurements of the Earth's surface, beginning in 1735 in Peru and Lapland, and later in other areas (Snyder, 1987).

The shape of the Earth is thus referred to as an oblate **ellipsoid** or oblate spheroid. When comparing the location of the Earth's most northern point on this spheroidal shape to where you would expect to find it on a perfect sphere, you would note a 20 km difference. A 20 km difference may have a relatively small significance when you consider mapping applications that take continents into account, however, it can be very problematic for more focused map areas. The difference of the Earth's profile from that of a perfect circle is can be expressed as a **flattening ratio** (f) and is mathematically described by the relationship $(a-b)/a$ where a is the equatorial (or semi-major) radius and b is the polar (or semi-minor) radius (Figure 2.2). The resulting flattening ratio is usually expressed as $1/f$ so the numerical results can be expressed more easily. There are over a dozen official ellipsoid models that are used throughout the world. Most of these have flattening ratios around 1/298, but some differences exist due to variations in measurement techniques that were applied during the times in which measurements were collected. The Clarke Ellipsoid of 1866 was designed to describe North America, and until very recently was commonly used by many organizations. Technological advancements and repetition in Earth measurements have lead to improved ellipsoid models, including the Geodetic Reference System of 1980 (GRS80) and the World Geodetic System of 1984 (WGS84).

A further refinement and approximation of the Earth's shape can be described with a **geoid**. A geoid attempts to reconcile Earth's local irregularities and the differing gravitational forces that are caused by varying Earth densities. The shape of a geoid is irregular and approximates Earth's mean sea level perpendicular to the forces of gravity. Within the continental US, the geoidal surface can be found on average about 30 m below the GRS80 and WGS84 ellipsoids, while usually a few meters separates its surface from the Clarke 1866 ellipsoid (Figure 2.3). Both the ellipsoid and geoid are mathematical approximations of the Earth's surface with the geoid being a more precise approximation. Once these approximations are applied to define the Earth's shape and irregularities, a measurement control system is needed upon which to base the approximate locations of landscape features on the Earth.

A **datum** is a measurement-derived representation of the Earth's surface and forms a 'control surface' upon which an ellipsoid and other location data are referenced. Datums are created from large numbers of measurements of the Earth's surface, typically assembled by land surveyors, engineers, and others involved in Earth measurements, where the location of each point has been measured using precise control surveys. From these points, a theoretical surface of the Earth is constructed. The word 'control' has a significant meaning within survey engineering applications and is used to refer to measurements that have been rigorously collected. In other words, there is assurance in the techniques that were used to establish control measurements and they can be used with relative confidence as the basis from which other measurements or applications can proceed. The greater the number of

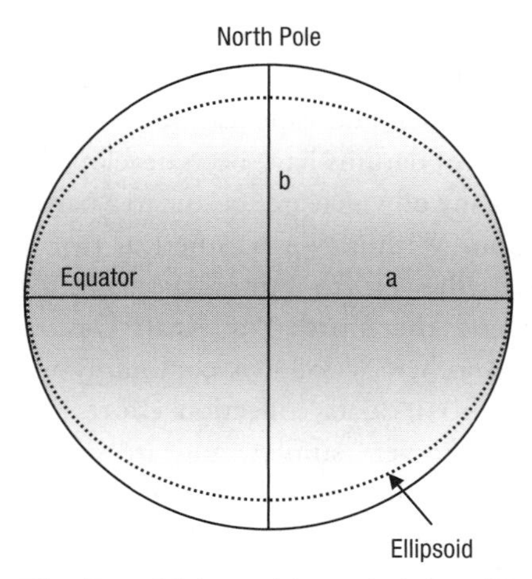

Figure 2.2 The ellipsoidal shape of the Earth deviates from a perfect circle by flattening at the poles and bulging at the equator.

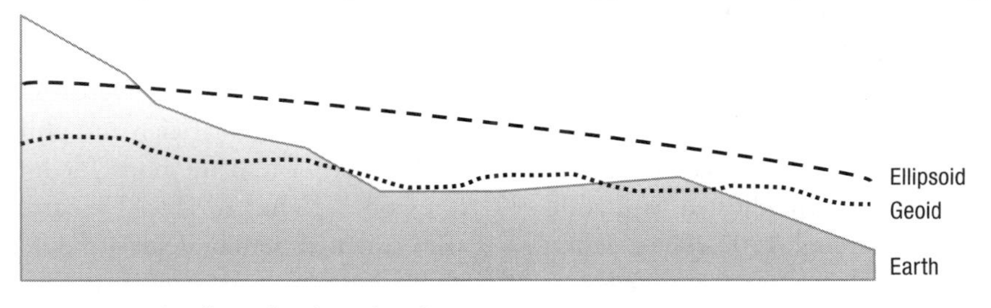

Figure 2.3 The ellipsoid and geoid surfaces.

point locations within a datum, the greater the potential of the datum to act as a reliable surface upon which you can reference other landscape features.

Hundreds of datums have been developed to describe the Earth, many of which are specific to a particular country or region. Within North America, two datums are prominent: the North American Datum of 1927 (NAD27) and the North American Datum of 1983 (NAD83). Another, WGS84, is commonly used in conjunction with GPS data collection efforts. The NAD83 and WGS84 are very similar, and are sometimes used interchangeably, although this practice may not be suitable for applications that require high data accuracy levels; the NAD83 was designed for North America whereas the WGS84 takes a global approach in representing the Earth. Differences between NAD83 and WGS84 are in the neighborhood of 1 to 2 m within the conterminous US. The primary differences between NAD27 and NAD83 datums are the number of longitude and latitude locations that were measured to create each datum, and the way in which the measured locations are referenced to the surface of the Earth. About 25,000 point locations were used to create the NAD27 datum, each of which was referenced to a central location—the Meades Cattle Ranch located in Kansas. Some 270,000 locations were used to create the NAD83 datum. Instead of referencing locations to a central location on the Earth's surface, locations are referenced to the center of the Earth's mass. The NAD83 datum has become the preferred datum for use in North America although many GIS databases continue to contain landscape features described by the NAD27 datum.

Geoids and ellipsoids are often associated with a particular datum. For instance, the Clarke Ellipsoid of 1866 was designed to describe the landscape features of North America, and is commonly used in conjunction with the NAD27 datum. Both GRS80 and WGS84 take a more global approach and are thus better suited for describing worldwide surfaces. Whereas GRS80 is commonly associated with NAD83, WGS84 can be thought of as both an ellipsoid and horizontal datum.

Datums are sometimes updated to reflect additional control measurements, shifts in the Earth's landmasses, or data corrections. When a datum is updated, the most recent year in which data were collected is often appended to the datum name. As an example, NAD83/91 would indicate that the NAD83 datum has been adjusted with additional data that were collected through 1991. These datum adjustments are often small (on the order of centimeters, or less) and are sometimes referred to as datum realizations.

Many agencies and organizations that are involved in working with spatial data in North America use NAD27, NAD83, or an adjusted NAD83 datum. A common error among users of GIS, especially those who have acquired data from a number of difference sources, is in forgetting to convert their databases to a common datum. In terms of comparison, in the US, landscape features referenced in both NAD27 and NAD83 will appear up to 40 m offset from each other in latitude and as many as 100 m off in longitude. These differences vary by region and may be hard to detect visually when using GIS to view large resource areas. This oversight can obviously lead to inaccurate analysis results.

The discussion of datums, to this point, has focused on those related to horizontal surfaces. When working with elevation data, such as a DEM, GIS users must also be aware that datums have also been developed to describe the vertical dimension. A vertical datum allows us to determine where '0' elevation begins and the heights of other objects located either above or below this point. The National Geodetic Vertical Datum of 1929 (NGVD29) was established from 26 gauging stations in the US and Canada and was a direct effort in determining the position of mean sea level. The North American Vertical Datum of 1988 (NAVD88) used additional measurements from a large number of elevation profiles to create a single sea level control surface. Between 1929 and 1988, over 600,000 km of level profiles were completed, and changes that had occurred to exiting elevation benchmarks were taken into account. These additional measurements and adjustments provided a more reliable means of establishing elevation surfaces and for this reason NAVD88 has become the preferred vertical datum. Elevation differences between NGVD29 and NAVD88 in the US could differ by as much as 1.5 m in some areas.

The Geographical Coordinate System

Now that a description of the size and shape of the Earth's surface has been presented, attention is turned to the methods by which landscape features are located on the Earth's surface. René Descartes, a seventeenth-century French mathematician and philosopher, devised one of the first written methods for locating landscape features on a planar surface. Descartes superimposed two axes, oriented perpendicular to one another, with gradations along both axes to create equal distance intervals (Figure 2.4). The horizontal axis is termed the x-axis and

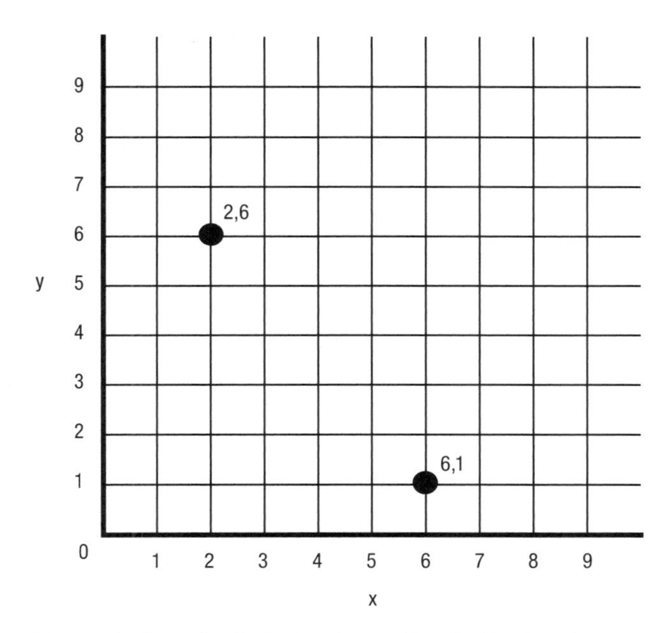

Figure 2.4 Example of point locations as identified by Cartesian coordinate geometry.

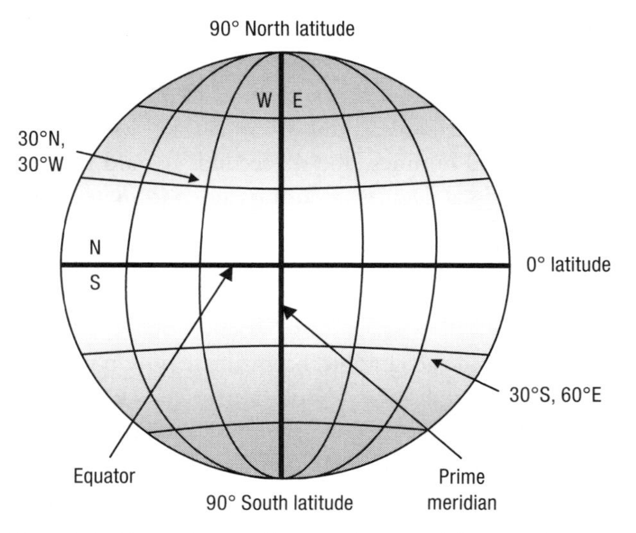

Figure 2.5 Geographic coordinates as determined from angular distance from the center of the Earth and referenced to the equator and prime meridian.

the vertical is the y-axis. The location of any point on the planar surface covered by this type of grid can be defined with respect to the interval lines that it intersects or that it most closely neighbors. This basis of determining location is known as a Cartesian coordinate system.

The most common coordinate system is the system of latitude and longitude, sometimes referred to as the geographic coordinate system. Although this system can be regarded as having x- and y-axes, angular measurements rather than distance-oriented intervals, establish the axes and intersections. The geographic coordinate system has an origin at the center of the Earth and contains a set of perpendicular lines running through the center to approximate the x- and y-axes of the Cartesian coordinate system. The orientation of the perpendicular lines is based on the rotation of the Earth. The Earth spins on an axis that, if extended, coincides very closely with the North Star (Polaris)—this axis is called the axis of rotation. This rotation axis divides the Earth in half to create a line of longitude that approximates the y-axis. A line perpendicular to the line of longitude falls along the equator (Earth's widest extent) to create a line of latitude that is conceptually similar to the x-axis. Latitudes are expressed to a maximum of 90°, in a north or south direction from the equator with the equator denoting 0° (Figure 2.5). Travelling 90° north from the equator would leave you at the most northern point of the Earth and would be noted as 90° N. Similarly a position halfway between the South Pole and the equator would be referenced as 45° S. The equator and other lines of latitude that parallel the equator are also called parallels.

Although the axis of rotation splits the Earth in half, a reference line must be established from which coordinates can start. This reference line is referred to as the prime meridian, and although there are dozens in existence, the most widely recognized prime meridian circles the globe while passing across the British Royal Observatory located in Greenwich, England. Longitude measurements are made from this reference line and are designated from 0° to 180°, in a western or eastern direction. North America is located in a region that is west of the prime meridian and is correctly described as falling into an area of negative longitude (all areas that extend within 180° west of the prime meridian), although many maps that are regional or local with usually omit the negative sign. Other lines that pass through the North and South Poles to act as guides and mark prominent longitude differences from the prime meridian are simply called meridians. The conceptual collection of meridians and parallels superimposed on the Earth's surface is known as a **graticule**.

The geographic coordinate system can be used to locate any point on the Earth's surface. To achieve a high level of precision when locating landscape features, degrees are further subdivided into minutes and seconds. There are 60 minutes (noted by ′) within each degree, and 60 seconds (noted by ″) within each minute. A location that is described as 38°30′ latitude would indicate a line between 38° and 39°. Because this measurement system does not lend itself conveniently to mathematical calculations, conversions to the decimal degree system are common. The conversion of 38°30′45″ (spoken as 38 degrees, 30 minutes, and 45 seconds) would result in 38.575° decimal degrees. By using this coordinate system and describing measurements to the nearest second, you can locate objects on maps that are within 100 ft of their true locations on the ground (Muehrcke & Muehrcke, 1998).

Although the geographical coordinate system provides a relatively straightforward solution to the complicated issue of establishing a regular system of measurements on a spherical surface, there are complications to its use. A primary problem is that the units of an arc (arc is used to describe angular distance—a sphere contains 360° of arc) are not constant throughout the system of geographical coordinates. Due to the convergence of the meridians at the Earth's polar areas, one degree of longitude ranges from 69 miles long at the equator to 0 miles long at the poles. This results in distance measurements calculated by longitudinal differences to be unreliable unless they are adjusted. Latitude measurements, in contrast, differ by minor amounts, but average 69 miles across the Earth. Field measurements of longitude are also difficult to collect without the use of GPS or other similar navigational technology. While you can calculate latitude by measuring the distance between the horizon and the North Star (in the northern hemisphere), calculating longitude involves understanding the difference between your location and the prime meridian. In addition, the calculations and conversions involved when using degrees-minutes-seconds measurements are cumbersome and time consuming.

Map Projections

Map projection can be considered a two-stage process (Robinson et al., 1995). First, measurements collected from the Earth's surface are placed on a globe that reflects the reduced scale on which you wish to visualize the measurements. This conceptual globe is called the reference globe. The second step is to take the mapped measurements from the three-dimensional reference globe and place them onto a two-dimensional flat surface. Perhaps the easiest way to develop an understanding of this concept is to picture a transparent plastic globe with the outlines of a graticule (grid of latitude and longitude) placed on its surface. If a light bulb is placed within the globe, the outline of the graticule will be projected onto any surrounding surface. To visualize the complexity of the map projection process, you might try to picture how the gracticule appears if the illuminated globe is placed immediately next to a wall or other flat surface in a darkened room so that it touches the wall or other surface. The graticule's shape would be projected onto the surface and would tend to become distorted the farther away from the contact point that the projected graticule travelled.

Three primary flat surfaces can be used to represent map projection surfaces: planes, cylinders, and cones (Dent, 1999). Map projections based on these surfaces are referred to as **azimuthal**, **cylindrical**, and **conic**, respectively (Figure 2.6). With all three surfaces, a graticule on a map will appear in a different location than it does on a globe, and a graticule will appear in a different location on each surface. In general, if you were to exam-

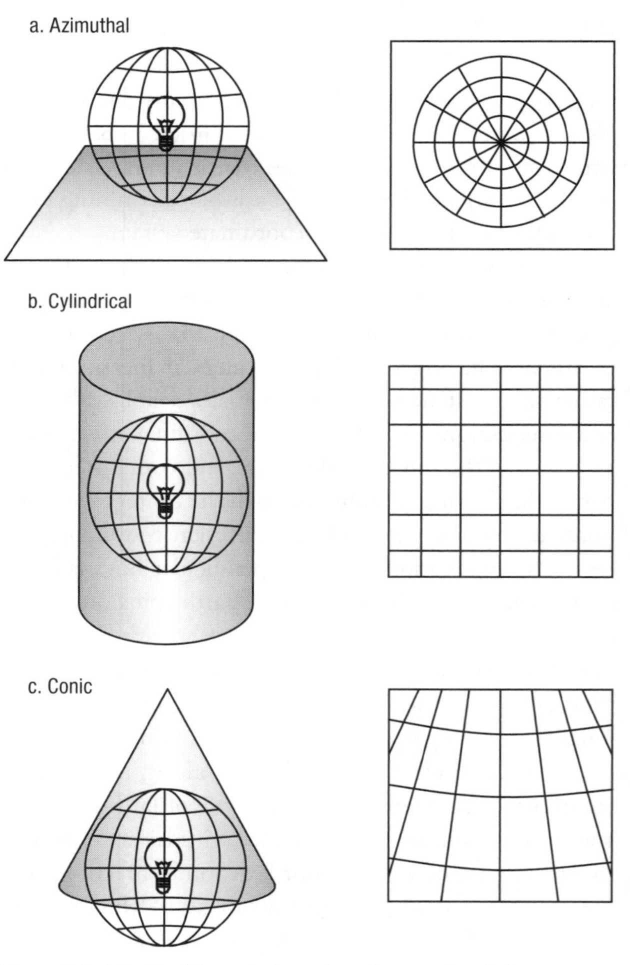

Figure 2.6 The Earth's graticule projected onto azimuthal, cylindrical, and conic surfaces.

ine a map surface after a projection has been made with one of these three surfaces, you would see that a projected graticule appears more distorted along the edges of the maps, away from the point or line(s) where a globe actually coincides with the map surface. In fact, the areas where a globe and a map surface meet are the places where distortion is minimized.

With simple or **tangent** map projection surfaces, a globe touches a map surface at one location (azimuthal). In the case of the cylindrical and conic surfaces, a globe intersects a map along one meridian, parallel, or other line that circles the perimeter of the globe. You can, however, develop a map that alters the flat surfaces so that a globe intersects them in two places rather than one (Figure 2.7). This is known as a **secant** projection surface and can increase the ability of a map projection to accurately describe landscape features. A standard parallel exists when the intersecting line is coincident with a line of latitude, otherwise the intersecting line is known as a standard line. For this reason, in the case of cylindrical and conic map surfaces, a tangent projection may have only one standard line or parallel while a secant projection surface may have two. Regardless of line location, secant projection surfaces provide additional locations where map distortion is minimized.

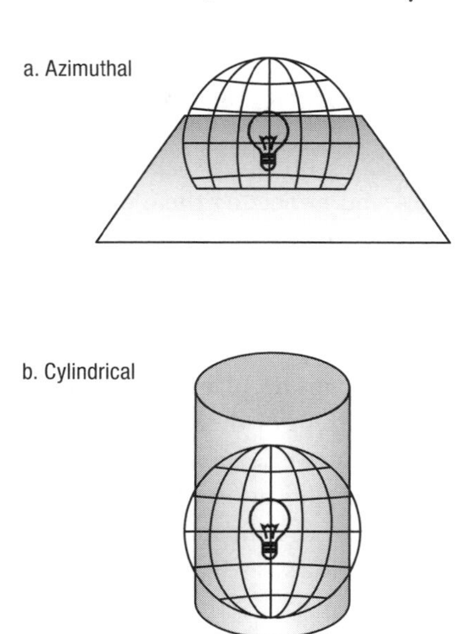

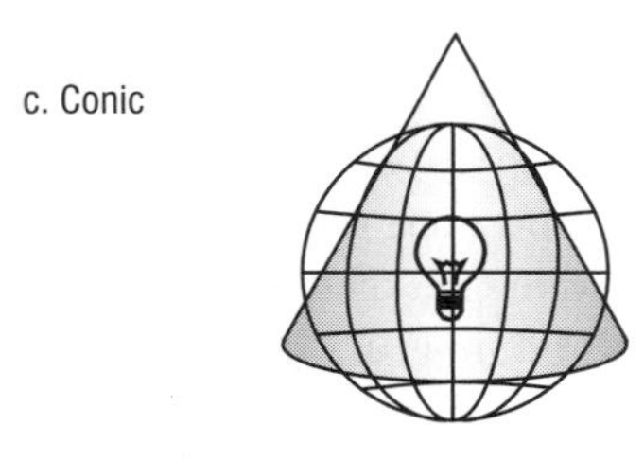

Figure 2.7 Examples of secant azimuthal, cylindrical, and conic map projections.

Common Types of Map Projections

There are hundreds of different types of map projections, and they can be classified according to how they manage the distortion related to the shape and direction of mapped landscape features. A few of the classifications include the **conformal**, the **equal area**, and the **azimuthal** projections (Robinson et al., 1995). Each of these classifications has strengths and limitations when illustrating landscape features. Users must decide which of these classifications is the most appropriate for their GIS databases.

Conformal projections are most useful where the determination of directions or angles between objects is important. Applications of conformal projections include navigation and topographic maps. Examples include the Mercator projection, the transverse Mercator projection, and the Lambert conformal conic projection. The Mercator projection is a cylindrical projection that was originally created for nautical navigation; it is probably the most widely recognized projection in the world. One useful feature of this projection system map, as it relates to navigation purposes, is that a line of constant azimuth or direction (called a rhumb line) will appear as a straight line. In contrast to the Mercator, the Transverse Mercator rotates the cylinder so that it is aligned with a parallel rather than a meridian (Figure 2.8). The Transverse Mercator projection is useful for navigation purposes in areas that have an extensive north–south orientation, but are limited in their east–west orientation. The Transverse Mercator has served as the base map projection for the USGS topographic map series, and also as the basis for the Universal Transverse Mercator coordinate system described in a forthcoming section of this chapter. The Lambert conformal conic projection is useful for mid-latitude areas of the world with an extensive east–west orientation and a limited north–south orientation. When a secant method is used for small areas, the Lambert conformal conic projection can provide a highly accurate description of directions and shapes of landscape features. Large areas of land, however, will include distorted shapes when mapped with a conformal projection. Applications of the Lambert conformal conic projection include those

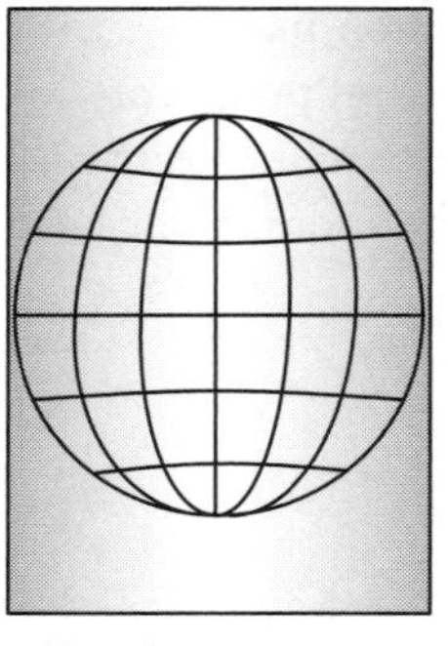

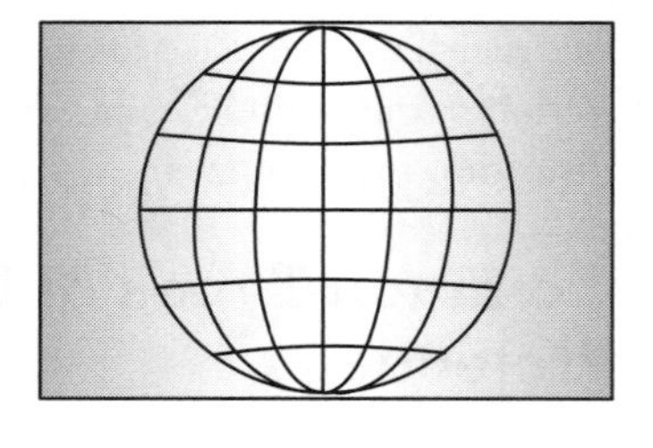

a. Mercator b. Transverse Mercator

Figure 2.8 The orientation of the Mercator and Transverse Mercator to the projection cylinder.

related to aerial navigation, meteorological uses, and topographic maps. The emphasis is usually placed on mid-latitude features of the world, such as those found in the conterminous US. Detailed applications of this projection system should focus on smaller land areas, since maintaining angular integrity across large areas is difficult.

Equal area or equivalent projections are well suited for maintaining the relative size and shape of landscape features when size comparisons are of interest. Equal area projections preserve the size and shape of landscape features but sacrifice linear or distance relationships in doing so. A tenet of map projection techniques and an important distinction between equal area and conformal projections is that areas and angles cannot be maintained simultaneously—you must decide which is more important to your work. One example of the equal area projection is the Albers' equal area projection. This projection is widely used and is typically based on a secant conic map surface. Similar to the Lambert's conformal conic projection, mid-latitude areas, which have extensive east–west orientations, are better candidates. This projection system has been selected by many US agencies as a base map projection. The Lambert equal area projection is another commonly used equal area projection, however, it is based on an azimuthal map surface.

Azimuthal projections are useful for maintaining direction on a mapped surface. Azimuthal projections can be based on one (tangent) or two (secant) points of reference. With one point of reference, distortion will occur radially from the reference point but directions near the reference point should remain true. For this reason, the azimuthal projection is appropriate for maps that have relatively the same amount of area in north–south and east–west orientations. When using two points of reference, directions emanating from either reference point should be true. The azimuthal equidistant projection offers the unique ability of maintaining uniform direction and distance from reference points. Azimuthal projections are useful for demonstrating the shortest route between two points (Robinson et al., 1995). Applications include those related to air navigation routes, radio wave ranges, and the description of celestial bodies. Azimuthal projection approaches include Lambert's equal area, stereographic, orthographic, and gnomic.

When pondering which projection system to use to describe GIS databases, you should consider the size of the area being managed, and whether maintaining direction or area is more important (Clarke, 2001). Projection distortions and the resulting analytical errors can become magnified as the size of a management area increases. A conformal or azimuthal projection should be considered when navigational or other directional properties are important. If maintaining the size, shape, and distribution of landscape features is important, an equal area projection should be employed.

Planar Coordinate Systems

Now that the process of taking shapes located on the surface of a sphere and projecting them on a flat surface has been discussed, it is time to explore the coordinate systems that are useful in order to locate landscape features on a flat surface. These systems are known as **planar coordinates**, or rectangular coordinates. Previously, the framework for examining plane coordinates was introduced with the concept of the Cartesian coordinate system. This same framework applies to planar coordinates, with a few minor modifications. For example, depending on the type of planar coordinate systems, coordinates are sometimes referred to as **eastings** or **northings**. An easting measures distance east of the coordinate system's origin while a northing measures distance north of the origin. These are usually specified by following the 'right-up' approach; eastings are numerically organized so that positive measurements begin at the origin and increase to the right (to the east) of the origin, while northings are numerically organized so that positive measurements begin at the origin and increase up (to the north) of the origin. One inconvenience of this approach is that if a coordinate system's origin is in the middle of a landscape, negative eastings and northings may occur, since some of the landscape is to the west and south of the origin. These negative coordinates might complicate the calculation of distances and areas within GIS software, and they also make manual calculations more challenging. As a remedy to these circumstances, false origins or false eastings can be constructed to prevent negative coordinates. This involves shifting the coordinate grid's numeric origin from the center of a landscape to the lower left corner (the farthest point west and south location of the landscape), or just outside the lower left corner, so that all areas of the landscape are located east and north of the origin, and can be represented by positive coordinate values.

The most common coordinate system in the US and Canada is that of the **Universal Transverse Mercator** (UTM), which has even been used to describe the surface of Mars (Clarke, 2001). The UTM system has been used for remote sensing, forestry, and topographic map applications, and it has been used in many other countries due, in part to its world-wide applicability and relative simplicity. The UTM system divides the Earth into 60 vertical zones, each zone covering 6° of longitude. The zones are numbered 1–60 starting at 180° longitude (the international date line) and proceeding eastward. The ten zones that cover the conterminous US and Canada are illustrated in Figure 2.9. The system extends northward to 84°N latitude, and southward 80°S latitude. A universal polar stereographic (UPS) grid system is used for the polar regions. Coordinates for each zone start at the equator for areas covering the northern hemisphere and at 80°S latitude for areas in the southern hemisphere. A false origin is established for each zone so that the central meridian of each zone has an easting of 500,000 meters. This arrangement ensures that all eastings are positive, and that areas of zones can overlap, if needed. As the name implies, the UTM coordinate system uses the Mercator projection to minimize distortion. The level of accuracy in the system is assumed to be one part in every 2,500 (Robinson et al., 1995). Another version of the UTM is the military grid version. The military grid version utilizes many principles of the UTM, yet divides each zone into rows, and each row covers 8° of latitude. Rows are denoted using the letters C to X with X occupying the northern latitude between 72° and 84° latitude. The military UTM can be used to further define blocks of zones into 100,000 meter squares.

The **state plane coordinate system** (SPC) was developed in the 1930s by the US Coast and Geodetic Survey (known today as the US Chart and Geodetic Survey), which created a unique set of planar coordinates for each of the 50 United States. The SPC was originally designed for land surveying purposes, so that location monuments could be permanently established. Under this system, most states are split into a smaller set of zones depending on the size and shape of the state. For instance, Florida has two zones and California has four. The SPC system uses either a Lambert's conformal conic or Transverse Mercator projection, the choice of which is usually influenced by the dimensions of the state (Lambert is used for

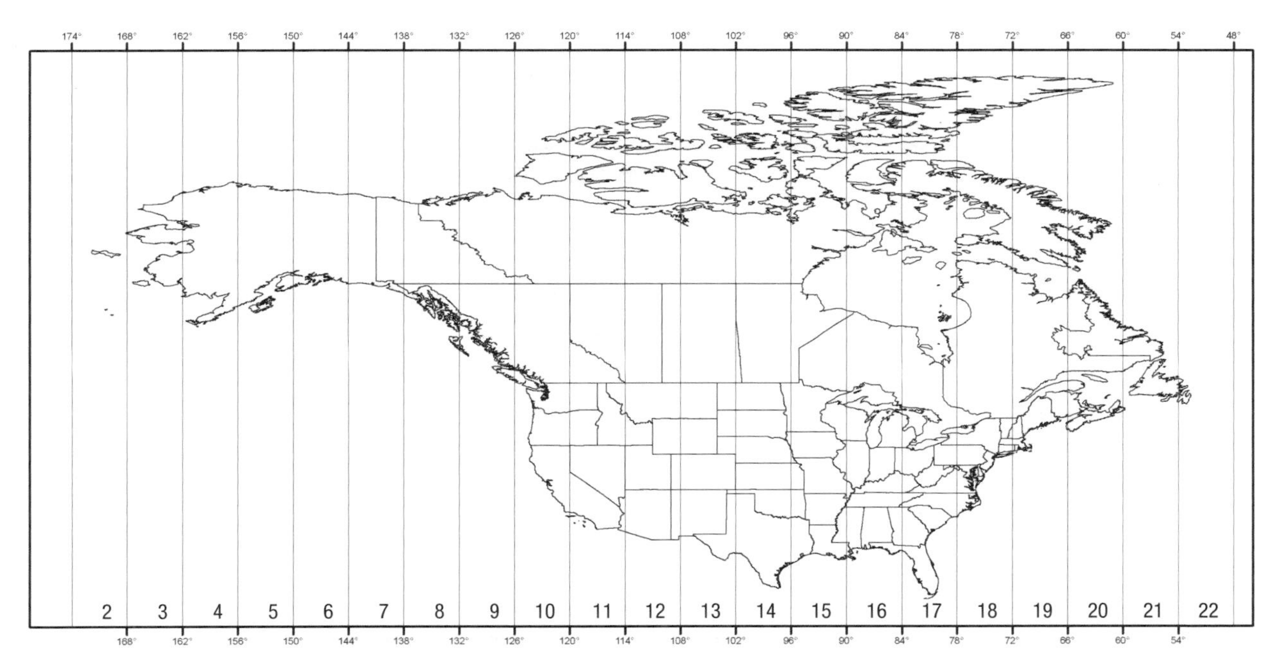

Figure 2.9 UTM zones and longitude lines for North America.

large west–east dimensions relative to north–south) and the dimensions of the zones of each state. The level of accuracy of the SPC system is approximately one part in 10,000.

Most of the original surveys in the US and Canada were described by metes (the act of metering, measuring, and assigning by measure a straight course) and bounds (a reference to general property boundaries). These were systems of describing real estate that were based on English Common Law, and involved describing the boundaries of a property by physical land features such as streams, trees, and so on. The metes and bounds system is inadequate today because of the subjective and transient nature of physical features; however, many of the original surveys were described in this manner and can be seen in property maps associated with land surveyed prior to about 1830. In some of the original US colonies, metes and bounds were used in the Headright systems that were developed to distribute land to settlers. Other more regular systems of describing land, such as the lottery system used to describe about two-thirds of the state of Georgia, followed (Cadle, 1991). Unforeseen problems with these

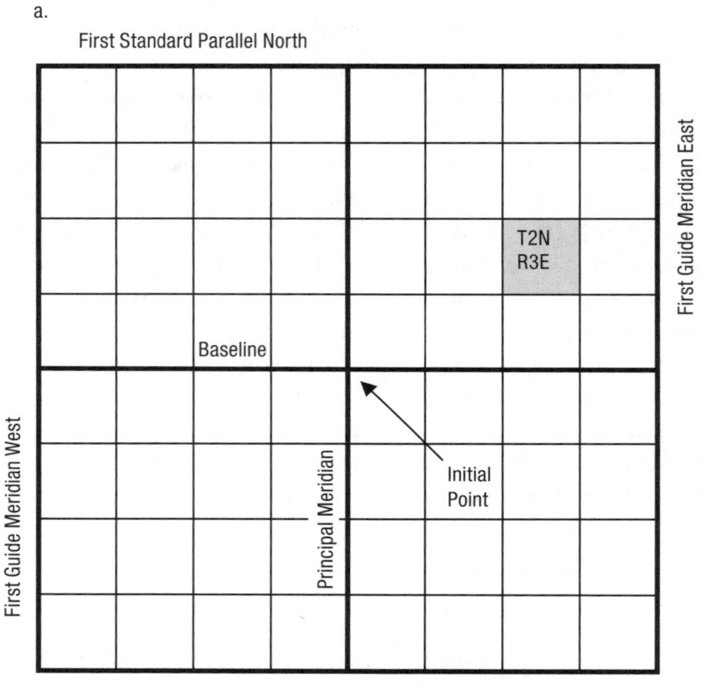

b. T2N R3E

6	5	4	3	2	1
7	8	9	10	11	12
18	17	16	15	14	13
19	20	21	22	23	24
30	29	28	27	26	25
31	32	33	34	35	36

c. NW 1/4, NE 1/4, Section 17

NW 1/4
NW 1/4 NE 1/4
NE 1/4 NE 1/4
SW 1/4 NE 1/4
SE 1/4 NE 1/4
N 1/2 SW 1/4
S 1/2 SW 1/4
W 1/2 SE 1/4
E 1/2 SE 1/4

Figure 2.10 Origin (a), township (b), and section (c) components of the Public Land Survey System.

early surveying systems included extensive fraud, surveying errors, a lack of consistency in the surveying processes, hostility from natives, and remoteness of the terrain. As a result, several rectangular systems of describing land were proposed, ranging from 6 or 7 square miles to 1,000 or 2,400 acres each (Ladell, 1993). These rectangular townships were meant to facilitate the development of communities (Stewart, 1979).

The rectangular US **Public Land Survey System** (PLSS) system was established in 1785 by the US Congress as a national system for the measurement and subdividing of public lands. Approximately 75 per cent of the US was subject to measurement by the PLSS. The original 13 colonies, which composed a significant holding, were not included in this system because they had already been inventoried through metes and bounds systems. In 1872 a similar system, the **Dominion Land Survey**, was created for administration of the Dominion Lands of western Canada. The objectives of the PLSS and Dominion Land Survey were to quantitatively measure previously non-platted land, create land portions that could be sold or distributed, and provide a means of recording ownership information. Neither system is associated with a map projection and therefore neither one can be considered a coordinate system, although you can use them to specify the general locations of property boundaries or other features. These rectangular systems begin with an initial point for a region, province, or state through which a principal meridian is astronomically derived (Figure 2.10). A baseline is also established, and it intersects a principal point at a right angle to the principal meridian. For example, 37 principal points exist within the US. In the US, at 24-mile intervals north and south of each baseline, standard parallels are established that extended east and west of the principal meridian (thus parallel to the baseline). Parallels are numbered, and are referred to as either being north or south from the baseline (e.g., 2nd parallel north or 7th parallel south of a baseline). Guide meridians were also established at 24-mile intervals east and west of the principle meridian. The guide meridians were established astronomically and are numbered relative to their position east or west of the principle meridian (e.g., 4th meridian east, 8th meridian west of the principle meridian). The grid of meridians and parallels creates blocks, each nominally 24 miles square.

Townships are created by forming range lines (running north and south) and township lines (running east and west) both at six-mile intervals. Each township is six miles square, and is further divided into sections, with each section measuring one square mile; there are 36 sections within a township. Sections are numbered 1–36 with the 36th section being at the lower right-hand corner of a township. Sections can be apportioned into smaller components such as quarter sections, half sections, or quarter quarter sections. In the naming convention, the smallest component is named first, starting with the portion of a section that a piece of land resides, then the township, range, and name of the principal meridian. An example would be the NW1/4, NE 1/4, Section 17, T2N, R3W, Mt Diablo Meridian. Use of the rectangular systems is limited within GIS to visualizing survey system themes, but it is likely, especially in natural resource applications, that you will encounter this system as your work with GIS progresses. This is particularly true for projects that involve property ownership issues, as most property locations and boundaries in areas covered by the rectangular systems are described using sections, townships, ranges, and principal meridians. At some point you may be required to re-project ownership boundaries derived from the rectangular systems so that they match the projection systems used in other GIS databases.

Mismatched map projections have been the bane of many spatial analysis efforts and doubtlessly, there are many published and reported study results that suffer from this malady. The likelihood is that there will be many future studies and written products that will also be subject to map projection problems. One of the reasons for projections problems is that many GIS users are unaware of the intent of projections and fail to realize that there are sub-components, such as coordinate systems and datums, which need to be considered when working with a projection. Another contributor to this problem is the inability of many desktop GIS software programs to manipulate spatial database projections. Although projection algorithms are becoming more common in desktop GIS software programs, they are typically more robust in a full-featured GIS software programs. As a GIS user or analyst, you must be cognizant of the projections that are associated with spatial databases. When obtaining GIS databases, either from within or from outside an organization, it is critical to obtain as much information as possible about the structure of the data, which should at least minimally include information about the map projection. Information about spatial databases can be stored in a metadata document. We'll discuss metadata in more detail later in this chapter but turn our attention now to GIS data structures.

GIS Database Structures

Like most digital files, spatial databases must be constructed so that they can be recognized and read by a GIS software program. Although GIS manufacturers have developed their own data formats, there are still two commonly used data structures for GIS data: raster and vector. Many GIS manufacturers have created their own spatial data formats but almost all make use of raster or vector format principles. The raster and vector data structures are as different as night and day, and both have strengths and weaknesses to be considered for use in various applications. Although many of the applications in this book involve vector databases, most GIS users will eventually find themselves using a combination of both raster and vector databases.

Raster data structure

Raster data structures are constructed by what can be considered **grid cells** or **pixels** (picture elements) that are organized and referenced by their row and column position in a database file. Raster data structures attempt to divide up and represent the landscape through the use of regular shapes (Wolf & Ghilani, 2002). The shape that is almost exclusively used is the square (Figure 2.11), yet other shapes can also cover the Earth completely and in a regular fashion, such as triangles and octagons. For each cell or shape in a raster database, an attribute value that stores information about the resources or characteristics is associated with the cell. These values can be numeric (e.g. temperature, elevation) or can be descriptive (e.g. forested, prairie) and are used to describe all areas represented by each cell. Some common raster GIS databases include those related to satellite imagery, digital elevation models, digital orthophotographs, and digital raster graphics. Although the particular format (how they are stored digitally) differs among raster databases and the techniques used to create raster databases may also differ, the raster approach to storing spatial information is consistent: a system of cells (usually square) that covers a landscape.

Satellite imagery

Satellite imagery is a term used to describe a wide array of products generated by remote sensors contained within satellites (Figure 2.12). Satellites are either positioned stationary above some location on the Earth, or circumnavigate the Earth using a fixed orbit. Although satellites have been sent into deep space, and have returned imagery to Earth, natural resource management is generally concerned only with imagery that provides information about planetary features. When viewing satellite imagery of the Earth, it may seem as if there is no relief associated with the landscape, since the images were collected from a very high elevation (100+ miles), however, you can associate elevation data (DEMs) with raster images, and subsequently view them in three dimensions.

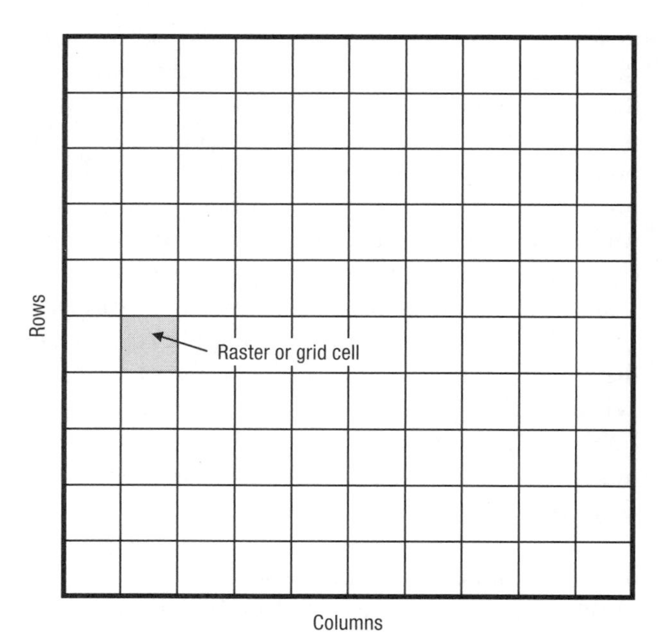

Figure 2.11 **Generic raster data structure.**

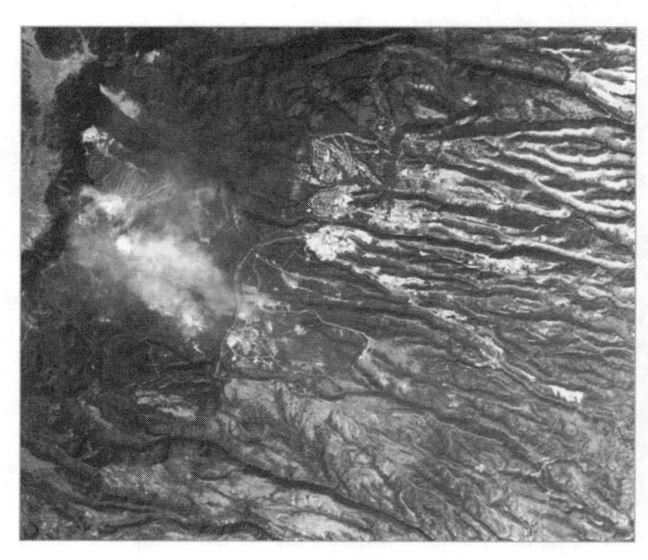

Figure 2.12 **Landsat 7 satellite image captured using the Enhanced Thematic Mapper Plus Sensor that shows the Los Alamos/Cerro Grande fire in May 2000.** This simulated natural colour composite image was created through a combination of three sensor bandwidths (3, 2, 1) operating in the visible spectrum. Image courtesy of Wayne A. Miller, USGS/EROS Data Center.

Digital elevation models

Digital elevation models (DEMs) are databases that contain information about the topography of a landscape. The grid cells in these databases contain measurements of elevation across a landscape (Figure 2.13). It is possible to derive terrain models from DEMs that represent aspect, ground slope classes, and shaded relief maps. It is also possible to perform a wide variety of terrain-based analyses, such as landscape visualization or watershed analysis. Elevation data can be collected by a variety of means, including sensors located on satellite or aerial platforms, or photogrammetric techniques that use aerial photography in conjunction with GPS data. Elevation data may also be collected from the bottom of water bodies such as oceans, lakes, or streams, through the use of sonar and acoustical sensors operated from boats or submersible watercraft. This data can be used to create cross section profiles or to support engineering projects that involve bridges or other infrastructure.

Digital orthophotographs

Digital orthophotographs are essentially digital aerial photographs (or aerial photographs that have been scanned) that have been registered to a coordinate system. The displacement common to aerial photographs is usually corrected through the use of precise positional data, DEMs, and information about the platform sensor (e.g. camera system used). The majority of the US has been represented by digital orthophotography, created through a mapping program sponsored by the US Geological Survey (USGS). Digital orthophotographs are generally made available in portions that match the extent of USGS 7.5 Minute Series Quadrangle maps, and are often referred to as digital orthophoto quadrangles (DOQs). Since the USGS Quadrangle maps cover large ground areas (7.5 minutes of longitude and latitude), digital orthophotographs have been developed to cover portions of Quadrangle maps as well. Many counties in the US have also commissioned more detailed digital orthophotographs, as have private companies. Digital orthophotographs provide a data source for those interested in obtaining a relatively fine scale image of landscape or in obtaining a base data layer for digitizing landscape features (Figure 2.14).

One of the strengths of digital orthophotographs is that each image is georeferenced to a coordinate and projection system; therefore, GIS users can use a desktop GIS software programs to digitize and create GIS databases

Figure 2.13 Digital elevation model (DEM).

Figure 2.14 Digital orthophoto quadrangle (DOQ).

using 'heads-up' digitizing. The term 'heads-up' digitizing indicates that the person doing the digitizing is looking at a computer screen (i.e., their head is up) rather than a digitizing table (which requires them to look down) when digitizing landscape features. Through heads-up digitizing, you can quickly create a GIS database from a digital orthophotograph image on a computer screen.

Digital raster graphics

Digital raster graphics (DRGs) are digitally scanned representations of the USGS topographic maps (Figure 2.15). These maps cover the entire United States, are published at several different scales, contain a wealth of information, and, very importantly for GIS projects, are available as digital databases that can be used by most GIS software programs that have raster display capabilities. Within Canada, the National Topographic Data Base (NTDB) is maintained under the administration of Natural Resources Canada. The NTDB contains vector databases that are similar to the USGS topographic maps and are available in several digital formats at scales of 1:50000 and 1:250000.

The most detailed of the maps produced by the USGS are the 7.5 Minute (7.5′) Quadrangle maps, which have a published map scale of 1:24,000. The 7.5′ refers to the total amount of latitude and longitude, in degrees, on the Earth's surface that each Quadrangle represents. In some cases, features from the Quadrangle maps are available in a vector format as a digital line graphic (DLG). Since the 7.5 Minute Series maps typically illustrate cultural resources, such as roads, large buildings, elevation contours, and natural features such as water bodies, similarly to DOQs, georeferenced Quadrangle raster databases can be used as a base layer for digitizing other landscape features of interest. Quadrangles also provide a great resource for those who want to learn more about a landscape.

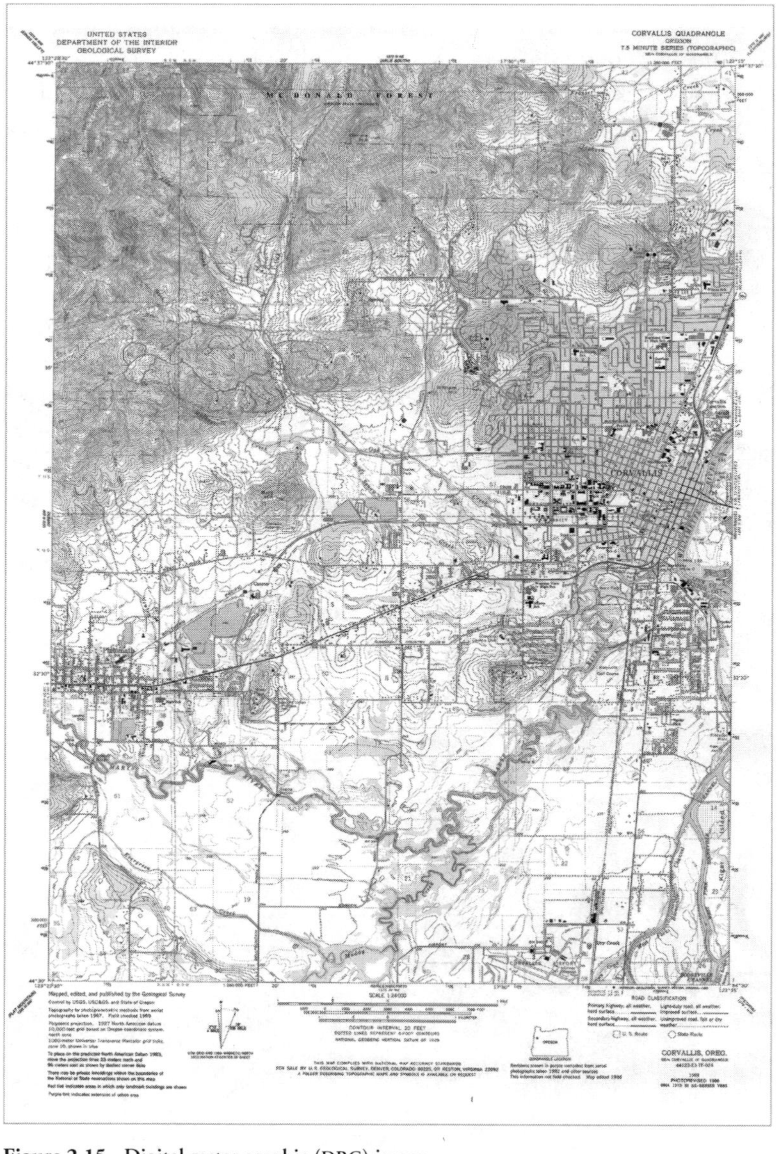

A closer look at USGS 7.5 Minute Quadrangle map Given the broad availability of the 7.5′ maps, their cartographic detail, and their popularity as a GIS database template for forestry and natural resource applications, we will closely examine one of these maps, and describe some of the more noteworthy components. Although this closer look is focused on an example using the Corvallis Quadrangle from Oregon (also the subject of the previous figures demonstrating a DEM and DOQ), the features described below should be available on most other 7.5′ maps. In particular, you should look closely at the information that appears at the bottom of the map (Figure 2.16). Quadrangle maps also provide information printed along the top margin of the map, but this is generally much more limited and a subset of what you find along the bottom margin of the map. At the time this book was being developed a digital copy of this

Figure 2.15 Digital raster graphic (DRG) image.

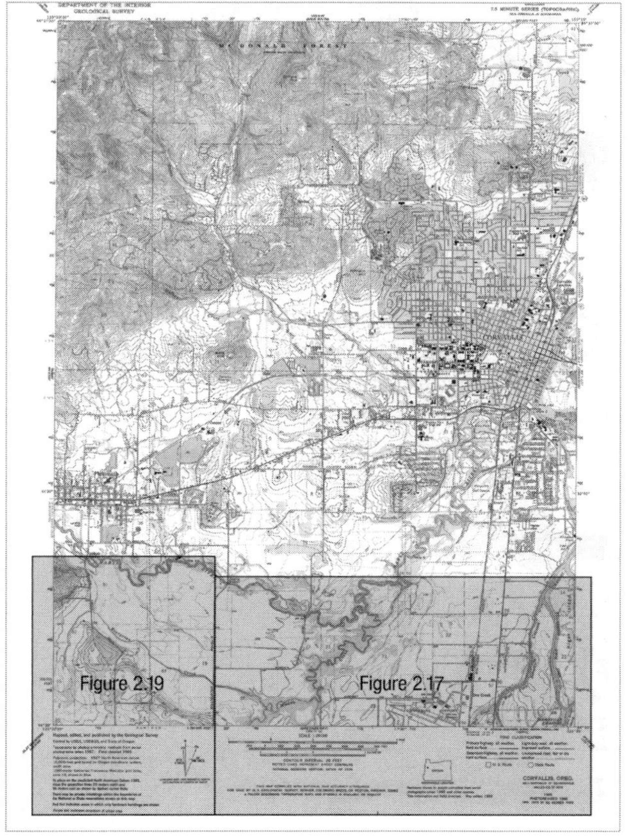

Figure 2.16 Corvallis Quadrangle with neatlines around map areas to be described in detail.

Quadrangle could be downloaded from page 8 of the list of quadrangles available at http://www.reo.gov/gis/data/drg_files/indexes/orequadindex.asp.

Lower right corner

The lower right corner of the Corvallis Quadrangle (Figure 2.17) contains a representative scale (1:24,000) and several scale bars with units expressed in miles, feet, and meters. Information is also provided for two contour intervals: the main contour interval of 20 feet (shown by solid lines on the map's surface) and a secondary contour interval of 5 feet that is represented by dashed lines. The reason for dual contour intervals is that the Corvallis Quadrangle includes a mixture of moderately-sloped, forested areas and relatively flat areas where urban and agricultural development has occurred; the dashed five-foot contours are used for the flatter areas. Below the scale bar is a statement about compliance with the National Map Accuracy Standards, a subject that will be examined in more detail shortly. Below the contour interval descriptions, there is information about where to purchase hard-copies of the Corvallis Quadrangle and the availability of topographic map and symbol descriptions.

Moving to the right, a graphic can be seen that indicates the location of the Quadrangle relative to the Oregon State border. Below this graphic, a note clarifies that the purple areas of the map were updated with aerial photography captured in 1982 (and hence edited in

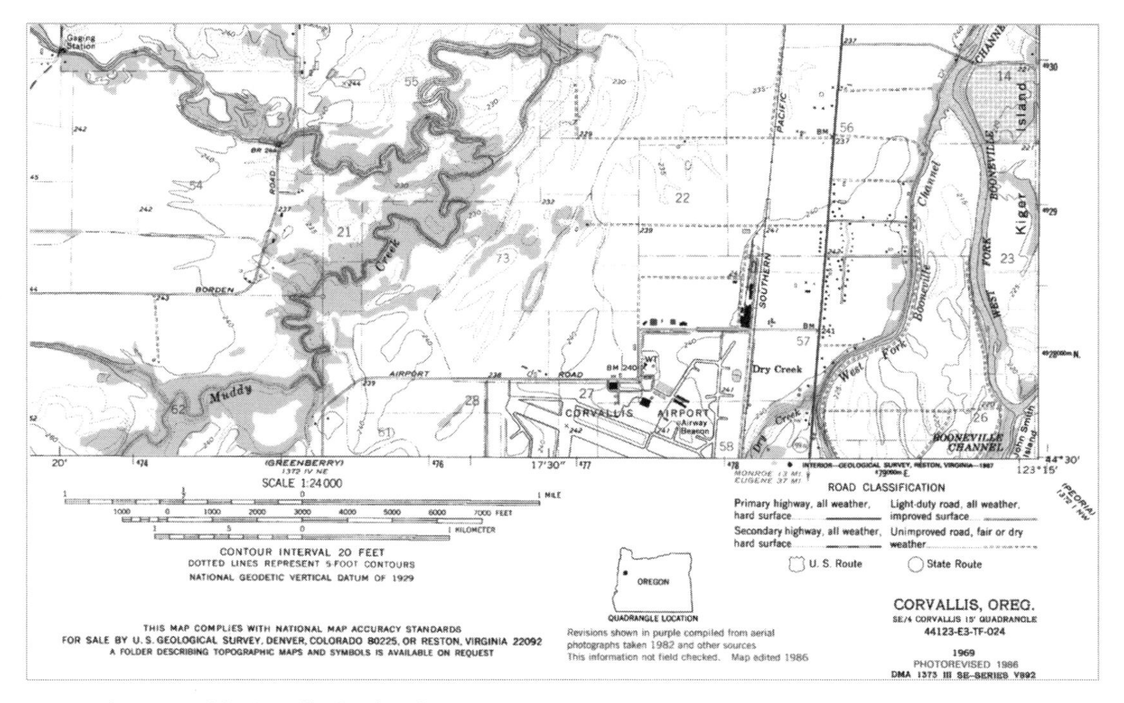

Figure 2.17 Lower right corner of the Corvallis Quadrangle.

1986). Moving to just below the bottom right corner of the map, you will find a legend for road map symbols. The name of the map is listed (CORVALLIS, OREG.) and the map is described as the bottom right corner (SE/4) of the 15′ Corvallis Quadrangle. The Ohio code description, sometimes referred to as the USGS Map Reference Code, is listed as 44123-E3. This means that the map is located in a block of geographic latitude and longitude that begins at the intersection of 44° latitude and 123° longitude (USDI US Geological Survey, 1995).

Each block of latitude and longitude can contain up to sixty-four 7.5′ Quadrangle maps that comprise an eight by eight matrix. Letters are used from A–H to denote rows starting from the corner of the latitude and longitude intersection and moving upward. Numbers are used from 1–8 moving westward to signify columns. Thus, the Corvallis Quadrangle is located in the matrix at the intersection of the 5th row and the 3rd column (Figure 2.18). Many map distributors use this Ohio code system to identify the relative location of quadrangles. The year of the original aerial photography (1969) used to create the map is listed, as is the last revision year (1986). The bottom corner of the mapped area has a set of geographic coordinates (44°30′ and 123°15′) that describe its location. Just to the east and slightly north of this corner, a cross appears. These crosses are located in similar locations relative to all four mapped area corners and demonstrate how the mapped surface corners can be adjusted to switch the datum of the map's projection from NAD27 to NAD83 (NAD is short for North American Datum). A full Universal Transverse Mercator (UTM) easting (479000 m) is found just west of this corner and a full UTM northing (4928000 m) to the north. Full UTM coordinates are also given along the opposite map corner. Hash marks that extend outward from the mapped area indicate the location on these coordinates relative to the rest of the map. These hash marks continue around the entire perimeter of the map's surface, but the last three digits of the eastings and northings that accompany the marks are not printed in order to conserve map space.

Lower left corner

The lower left corner of the map (Figure 2.19) states that the map was produced by the USGS with control or refer-

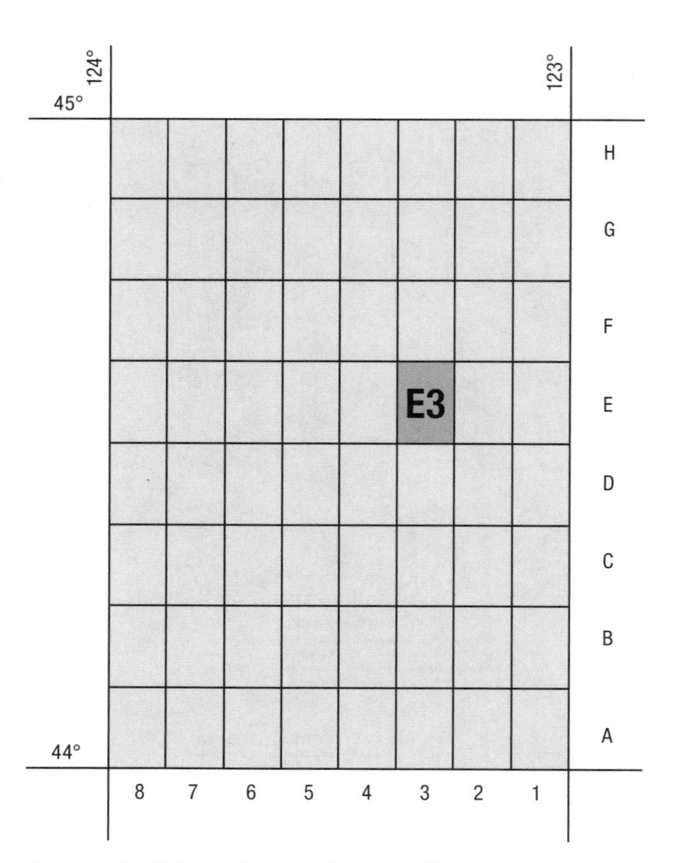

Figure 2.18 Ohio code location of the Corvallis Quadrangle.

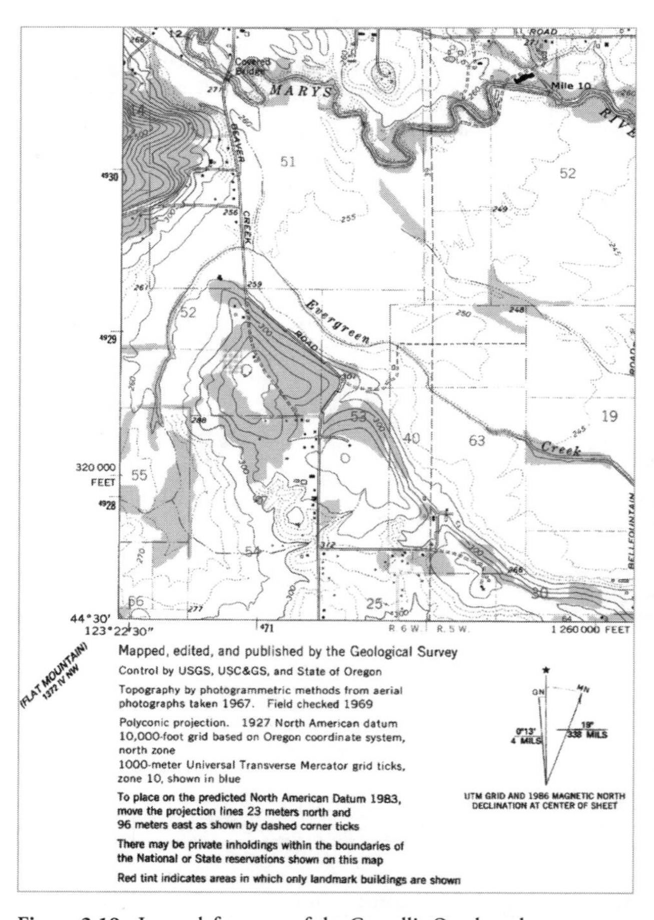

Figure 2.19 Lower left corner of the Corvallis Quadrangle.

ence points established by the USGS, US Coast and Geodetic Survey, and the State of Oregon. Mapped surfaces were taken from 1967 aerial photography and were field checked in 1969. Projection information is then listed and the base surface is described as a polyconic projection, NAD27, using the Oregon State Plane North Coordinate System. A line follows to describe the coloring of the UTM coordinates that are listed around the perimeter of the map, and the UTM zone (10) that was used. Instructions for converting the mapped surface from NAD27 to NAD83 are also given and provide a quantitative assessment of how these two datums differ in the quadrangle area. This information is useful for potential coordinate conversions in a GIS. In addition, the text states that only landmark buildings are shown in map areas that are tinted red. Landmark buildings are those that serve the public, have cultural or historical significance, or are unusually large in relation to surrounding buildings.

To the right of these statements, a graphic of the map's orientation to several definitions of north is shown and text below the graphic explains that the orientation is from the map's center. The longest north line is topped by a star symbol and refers to astronomic north. The line to the left of astronomic north refers to grid north (GN) and is oriented 0°13′ (0.22°) west of astronomic north. Grid north is the direction in which the Oregon State Plane Coordinate System is referenced. The line to the right of astronomic north shows the magnetic declination (19° east), relative to astronomic north, that existed in 1986. Since magnetic north can fluctuate from year to year (even small daily shifts are also possible), the date of the measurement is important for those who wish to convert their data to match the map's projection.

Geographic coordinates appear at the southwest corner of the mapped surface, and the UTM coordinate abbreviations are listed. A full listing of State Plane Coordinates (NAD27) are listed and also marked by hatch lines to the north (320,000 feet) and east (1,260,000 feet) of this corner. Full state plane coordinates are also listed along the upper right corner. Unmarked hash marks around the rest of the map's perimeter denote gradations of the state plane coordinates. Periodically, range and township divisions appear as longer dashed lines on the map's surface. Along the latitudinal axis of Figure 2.19, you can see R. 6 W. and R. 5 W. This signifies the division between Range 6 and 5, west of the reference meridian (Willamette) that was used for creating the PLSS for Oregon. Larger numbers appear on the mapped surface and describe the boundaries between sections and donation land claims (DLCs). As mentioned earlier in this chapter, most of the western US was originally divided according to the PLSS. The PLSS split regions of the US in a grid of townships (approximately six by six mile blocks created by the intersections of township and range lines) that were created from a reference meridian with townships further divided into sections measuring approximately one square mile. Section numbers range from 1–36 in almost all PLSS states though you might find an occasional section numbered 37 where measurement irregularities warranted adjustments to the PLSS. Some states, including Oregon, Florida, and New Mexico, had adopted less rigorous land measurement systems that superseded the PLSS. Through these systems, settlers could stake claims to lands and these systems were generally referred to as DLCs. DLC boundaries are numbered starting with 37.

In general, green shading (gray in the black and white image of Figure 2.19) is used to represent forested or natural areas, and no shading is used to represent developed areas. Roads, streams, and other cultural and natural landscape features are also visible throughout the map.

National Map Accuracy Standards

According to information illustrated in Figure 2.17, the Corvallis Quadrangle map complies with the National Map Accuracy Standards (NMAS). The US Bureau of the Budget originally developed these standards in 1941 so that guidelines would be available for the establishment of horizontal and vertical map accuracy at multiple scales. The guidelines were also intended to help protect and inform consumers about the quality of map products they acquired. The guidelines assume that organizations claiming adherence to NMAS guidelines are responsible for ensuring compliance. The NMAS was last revised in 1947 (Thompson, 1979).

The guidelines (Figure 2.20) provide horizontal accuracy standards for map scales larger than 1:20,000, and for scales at 1:20,000 or smaller. For the larger map scales, no more than 10 per cent of the points verified shall be in error by 1/30th of an inch, as measured on the map surface. For smaller scale maps, this tolerance is 1/50th of an inch. The Corvallis Quadrangle falls in the latter category. To test for NMAS compliance, locations or elevations from map points are compared to their actual measurements, where locations or elevations have been derived by highly accurate ground surveys. Within these comparisons, only 10 per cent of the points can be in error by more than the tolerance. Table 2.1 describes the tolerances in relation to some of the more common map scales. For the Corvallis Quadrangle, this threshold would be 40 feet, indicating that not more than 10 per cent of the tested

United States National Map Accuracy Standards

With a view to the utmost economy and expedition in producing maps which fulfill not only the broad needs for standard or principal maps, but also the reasonable particular needs of individual agencies, standards of accuracy for published maps are defined as follows:

1. **Horizontal accuracy.** For maps on publication scales larger than 1:20,000, not more than 10 percent of the points tested shall be in error by more than 1/30 inch, measured on the publication scale; for maps on publication scales of 1:20,000 or smaller, 1/50 inch. These limits of accuracy shall apply in all cases to positions of well-defined points only. Well-defined points are those that are easily visible or recoverable on the ground, such as the following: monuments or markers, such as bench marks, property boundary monuments; intersections of roads, railroads, etc.; corners of large buildings or structures (or center points of small buildings); etc. In general what is well defined will also be determined by what is plottable on the scale of the map within 1/100 inch. Thus while the intersection of two road or property lines meeting at right angles would come within a sensible interpretation, identification of the intersection of such lines meeting at an acute angle would obviously not be practicable within 1/100 inch. Similarly, features not identifiable upon the ground within close limits are not to be considered as test points within the limits quoted, even though their positions may be scaled closely upon the map. In this class would come timber lines, soil boundaries, etc.
2. **Vertical accuracy,** as applied to contour maps on all publication scales, shall be such that not more than 10 percent of the elevations tested shall be in error more than one-half the contour interval. In checking elevations taken from the map, the apparent vertical error may be decreased by assuming a horizontal displacement within the permissible horizontal error for a map of that scale.
3. **The accuracy of any map may be tested** by comparing the positions of points whose locations or elevations are shown upon it with corresponding positions as determined by surveys of a higher accuracy. Tests shall be made by the producing agency, which shall also determine which of its maps are to be tested, and the extent of such testing.
4. **Published maps meeting these accuracy requirements** shall note this fact on their legends, as follows: "This map complies with National Map Accuracy Standards."
5. **Published maps whose errors exceed those aforestated** shall omit from their legends all mention of standard accuracy.
6. **When a published map is a considerable enlargement** of a map drawing (manuscript) or of a published map, that fact shall be stated in the legend. For example, "This map is an enlargement of a 1:20,000-scale map drawing," or "This map is an enlargement of a 1:24,000-scale published map."
7. **To facilitate ready interchange and use of basic information for map construction** among all Federal mapmaking agencies, manuscript maps and published maps, wherever economically feasible and consistent with the uses to which the map is to be put, shall conform to latitude and longitude boundaries, being 15 minutes of latitude and longitude, or 7.5 minutes, or 3–3/4 minutes in size.

Issued June 10, 1941
Revised April 26, 1943
Revised June 17, 1947

U.S. BUREAU OF THE BUDGET

Figure 2.20 US National Map Accuracy Standards.

map points should differ from their actual locations by more than 40 feet. Vertical accuracy standards are applied by using half of the contour interval as a benchmark. For vertical accuracy of the primary contour lines (20′ intervals) of the Corvallis Quadrangle, the standard indicates that 90 per cent of the elevation points checked along the contour lines should not be in error by more than 10 feet.

Typically on USGS 7.5′ Quadrangles, 28 points are examined for accuracy verification, representing a very small portion of the population of map points. It is also worth noting that these points are not randomly selected, but represent locations that are readily visible on the photography from which the map was created, such as road intersections, bridges, and other noteworthy structures. Thus, you would expect that the points represent the mapped areas where the photogrammetric methods used to create the map were the most reliable for mapping purposes. It is also worth noting that 10 per cent of those points could be off by any distance, and the resulting map would still be NMAS compliant. Even though you are assured compliance with a published standard, potential for errors related to the accuracy of location of landscape features can still be significant.

Our discussion of raster data structures and GIS databases above provided examples of some of the more common types of raster data and a few potential applications. Raster data provide a powerful means of inventorying and analyzing Earth's features and no doubt new raster data formats and applications will arise in future years. The usefulness of raster GIS databases will depend on the needs and capabilities of each natural resource organization, but it is increasingly likely that raster data will be part of every organization's data holdings, particularly as technological advances continue to bring economies to remotely sensed landscape data. We now turn our attention to the other primary data structure for spatial databases: the vector data structure.

TABLE 2.1 Map scales and associated National Map Accuracy Standards for horizontal accuracy

Scale	Standard
1:1,200	± 3.33 feet
1:2,400	± 6.67 feet
1:4,800	± 13.33 feet
1:10,000	± 27.78 feet
1:12,000	± 33.33 feet
1:24,000	± 40.00 feet
1:63,360	± 105.60 feet
1:100,000	± 166.67 feet

Vector data structure

Vector data, as compared to raster data, is generally considered 'irregular' in its construction and appearance. This description is not a comment on the quality or usefulness of vector data structures, but just a characterization of the type of data it represents. Vector data are generally grouped into three categories: **points**, **lines**, or **polygons**. This categorization is sometimes referred to as the feature model of GIS. Almost any landscape feature on the Earth can be described using one of these three shapes, or a combination thereof (Figure 2.21). Points are

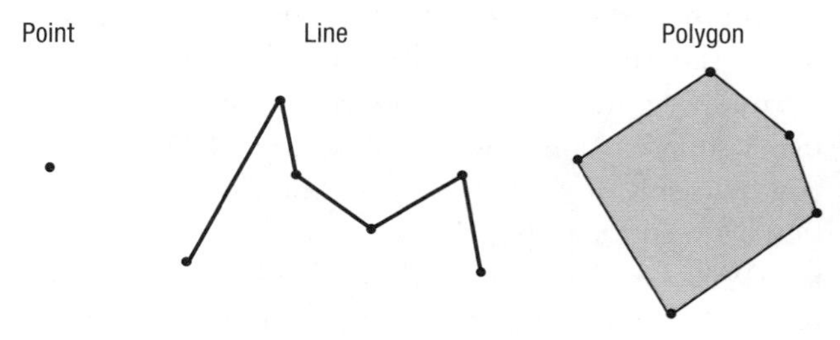

Figure 2.21 Point, line, and polygon vector shapes.

the most basic of the shapes but define the essence of all three forms. A line is a set of connected points. A polygon is a collection of lines that form a closed loop.

Point, line, and polygon vector features can be referenced by almost any coordinate system. To represent a point, a single measure from each X (east–west) and Y (north–south) axis is needed to describe the location of the point within a coordinate system. With lines and polygons, each coordinate pair is referred to as either a node or as a vertex. The coordinates of point, line, and polygon features allow calculations of distances between features, and in the case of lines and polygons, dimensions of features. Point features have no dimension, or size, because a single pair of coordinates represents them. A coordinate pair does not allow for length, area, or volume calculations. Line features are single dimension shapes in which coordinate pairs can be used to calculate a length. Polygon features are two dimensional in nature, with coordinate pairs being used to not only calculate a perimeter distance around a polygon, but also being used to calculate an area within the polygon.

Topology

Most GIS software programs store location information (X, Y coordinates that describe the position of landscape features) in separate GIS databases for each topic of interest. For example, the coordinates that are used to describe a roads GIS database are separate from the coordinates that describe a streams GIS database. Although most of the location information that defines vector features will be transparent to users of desktop GIS software programs, it is vital in establishing and maintaining **topology**. Topology describes the spatial relationships between (or among) points, lines, and polygons, and is a very important consideration when conducting spatial analyses. Topology allows you to determine such things as the distance between points, whether lines intersect, or whether a point (or line) is located within the boundary of a polygon.

Topology can be defined in a number of ways but the most common definitions involve aspects of adjacency, connectivity, and containment (Figure 2.22). Adjacency is used to describe a landscape feature's neighboring features. You might use adjacency relationships to describe polygons that share common borders (e.g., in support of green-up requirements in a forest management context), or to identify the lines that make up a polygon (area). Connectivity is typically used to describe linear networks, such as a network of culverts that might be connected by drainage ditches or a stream network. Connectivity would

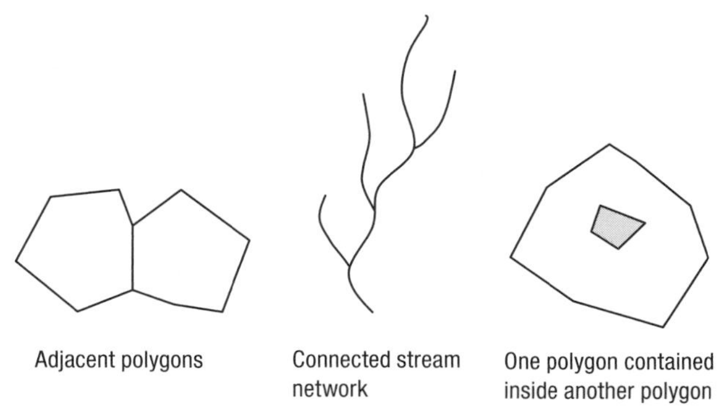

Figure 2.22 Examples of adjacency, connectivity, and containment.

allow you to trace the flow of water through the stream system. You can also incorporate direction in their description of connectivity. Based on the topography of a landscape in which a culvert system is situated, you could determine the overland flow paths of water through the system, given that water flows downhill. Containment allows you to describe which landscape features are located within, or intersect, the boundary of polygons. Containment information can be used to describe the well locations (points) or the power lines (lines) that are located within a proposed urban growth boundary, for example.

In order for topology to exist, a system of coding topology that can be understood and manipulated by a computer must also exist. With GIS databases containing point features, there is little need for anything more than a file of coordinate pairs (X, Y coordinates) since all points are ideally separate from one another, and thus there are no issues of adjacency, connectivity, and containment to resolve. However, you must be careful in describing feature locations and linkages when using GIS databases containing line and polygon features. The spatial integrity of lines and polygons is maintained by managing the nodes, vertices, and links of each feature. A **node** is the starting and ending point of a line, and also represents the intersection of two or more lines (Figure 2.23). A **vertex** is any point that is not a node but specifies a location or creates

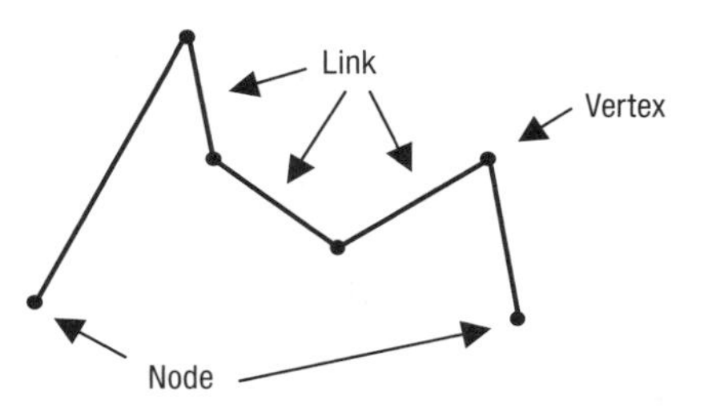

Figure 2.23 Examples of nodes, links, and vertices.

a. Network of nodes, links, and polygons

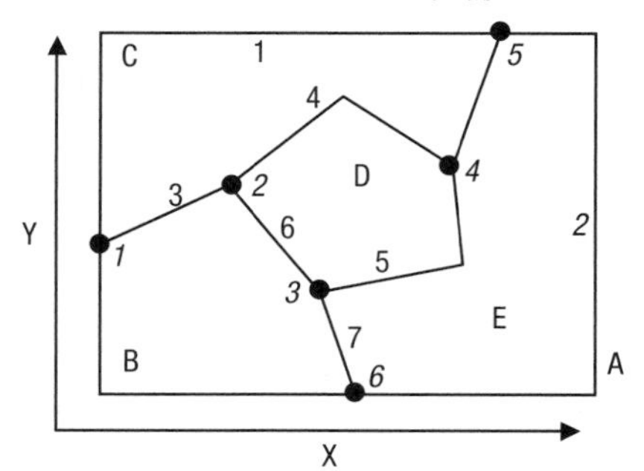

b. node coordinate file

Node	X	Y
1	0.5	2.4
2	2.1	3.1
3	3.2	1.7
4	4.7	3.3
5	5.4	5.0
6	3.6	0.5

c. topological relationship file

Link	Begin node	End node	Left polygon	Right polygon
1	1	5	A	C
2	5	6	A	E
3	1	2	C	B
4	2	4	C	D
5	4	3	E	D
6	3	2	B	D
7	3	6	E	B

Figure 2.24 Vector topological data. (a) Network of nodes, links, and polygons, (b) node coordinate file, and (c) topological relationship file.

a directional change in a line. A **link**, sometimes called an **arc**, is a line that connects points as defined by nodes and vertices. Nodes, vertices, and links are usually numbered and maintained in a GIS database file to maintain topology. In a network of lines and polygons (Figure 2.24), this would involve using numeric codes for network pieces (nodes and links) to identify the node locations, the nodes that are attached to each link, and the polygons that may form on either side of each link.

Topology also allows you to inspect the spatial integrity of lines and polygons. For instance, you can use topology information to determine whether any breaks or gaps occur in lines that are meant to represent streams. From a polygon perspective, topology would allow you to determine whether a polygon forms a closed boundary, or whether an extraneous polygon exists inside, or along the outside border, of another polygon (Figure 2.25).

One of the primary differences between full-featured GIS software programs and desktop GIS software programs is whether they can identify and correct topology problems in vector GIS database features. Many desktop GIS software programs, such as ArcView 3.2, GeoMedia, and MapInfo, may use vector data formats that are not topologically-based (Chang, 2002). Thus landscape features, such as adjacent polygons, may not be represented as sharing a common boundary line with other polygons. Most full-featured GIS software programs such as ArcGIS may also allow use of vector formats that are not typologically based, but will usually have tools or options to draw and manipulate typology. In some cases, users may be

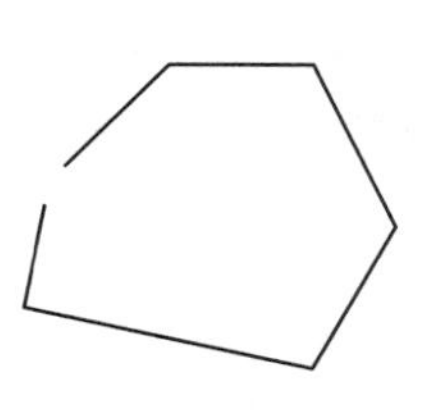

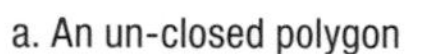
a. An un-closed polygon

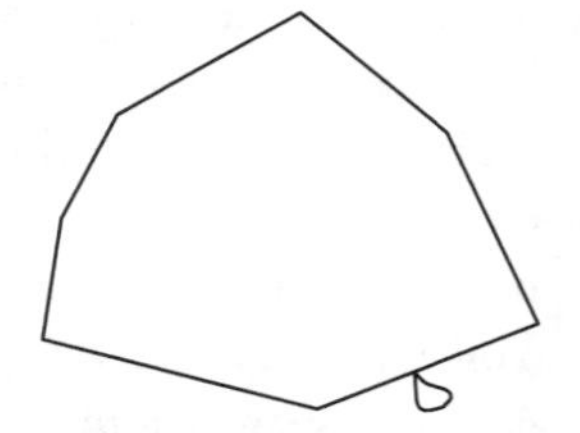
b. An extraneous polygon

Figure 2.25 Examples of topological errors. In (a), an undershoot has occurred and instead of a closed figure creating a polygon, a line has been created. In (b), a small loop has been formed extraneously adjacent to a polygon. This might represent a digitizing error or the result of a flawed overlay process.

In Depth

What is topology? Topology, or topological coding, provides the intelligence in the data structure relative to the spatial relationships among landscape features (Lillesand & Kiefer, 2000). For example, in a vector GIS database containing polygons, topological coding keeps track of each line that forms each polygon, and the common nodes each line shares with each other line. In addition, the polygons that are formed on either side of each line (since polygons may share a boundary defined by a line) are known. Thus with topology you can understand which forest stands, for example, are next to which other forest stands.

able to define their own topological rules, or whether certain topological relationships can be ignored in a specific database. The danger for users of desktop GIS software programs is that some programs will allow users to proceed, without warning, with spatial processes and analyses even though topological problems exist. This condition may lead to errors in linear and area measurement calculations, or an inability to complete certain spatial operations that rely on these dimensions for processing. Users of desktop GIS software programs may only become aware of topological problems after careful examination of GIS databases and associated analyses, or may not notice potential problems altogether.

The point, line, and polygon vector data structure provides a method to represent irregularly-shaped Earth features. More often than not, vector GIS databases do not completely cover a landscape of interest (e.g., a vegetation GIS database may only contain the vegetation located within the ownership boundary of a natural resource management organization, and not the vegetation outside of the ownership boundary—quite different from satellite imagery or digital orthophotographs), and represent landscape features that are quite diverse (e.g., polygons of different sizes and shapes, rather than a regular size and arrangement of pixels). Examples of diverse vector databases include road and stream representations. Both of these types of databases tend to have unique geographic shapes that do not completely occupy a landscape (unless, of course, the 'landscape' is the size of a pothole, or some pool in a stream). Some point databases, such as those that describe timber inventory cruise plots, come close to a regular arrangement across a landscape, yet they usually deviate from regularity as a result of the sampling method selected for each stand. Point locations of wildlife sightings are usually very irregularly distributed across a landscape. Polygons, whether they represent stands of similar trees, soils, or recreation areas, tend to be very irregular in shape, although in areas where the Public Land Survey System or other culturally-based land delineations have been implemented, some edges of forest stands now seem to have an aspect of regularity built into them.

Comparing raster and vector data structures

There are a number of ways in which the differences between vector and raster GIS data can be described (Table 2.2). Since GIS users will ultimately use databases representing both structures, and since users sometimes convert raster to vector data, and vice versa, an illustration of the differences is needed. Therefore, an examination of a generic data structure conversion process might be helpful to illustrate how the three main types of vector data (points, lines, and polygons) might be represented in a raster GIS database. The right side of Figure 2.26 also illustrates these three features but demonstrates how they might be represented in a raster database structure. The vector representation of points can yield fairly precise locations, depending on the type of coordinate system used to reference the points. This means that you could use a GIS to determine with some degree of precision where these points could be located on a map that

TABLE 2.2 Comparison of raster and vector data structures

	Raster	Vector
Structure complexity	Simple	Complex
Location specificity	Limited	Not limited
Computational efficiency	High	Low
Data volume	High	Low
Spatial resolution	Limited	Not limited
Representation of topology among features	Difficult	Not difficult

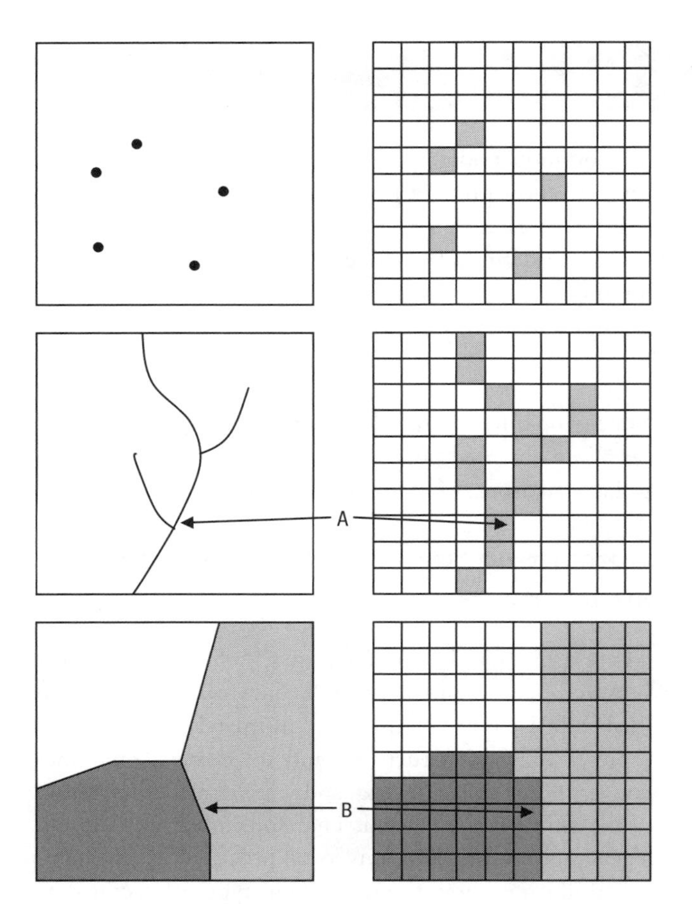

Figure 2.26 Point, line, and polygon features represented in vector and raster data structures.

includes coordinate reference marks on its axes. While the raster representation also shows these point locations, there is less precision (specificity) in their map location, which is dependent upon the raster grid cell size assumed. You would know that the point is located somewhere in the grid cell, but the precise location is elusive.

You can also see some similar relationships between the data structures when describing line and polygon features. Both the line and polygon vector features have very discrete shapes that are sometimes lost when converted to the raster data structure. This fact is perhaps most noticeable when you examine the junctions, or nodes, of the line feature (A), or those places where polygon features intersect (B).

The loss of specificity when converting a vector feature into a raster can, at least in part, be overcome by selecting a smaller raster grid cell size to represent the rasterized vector features. This choice comes with a price, however: an increasing storage size requirement for the resulting raster GIS database, and greater strain on computer processing resources. The loss of specificity when converting from vector to raster data structures may be acceptable if your goal was to simply represent the relative locations of landscape features. However, if you needed to know the precise location of a well, road junction, or property boundary, representing these landscape features with a vector data structure may be more appropriate.

In general, raster data structures may be more appropriate for representing continuous surfaces than the vector data structure. For example, if you were interested in describing precipitation, temperature, or species diversity across a landscape, raster data structures may do this more efficiently because the data may be more appropriately stored and illustrated with grid cells. Because of the regularity of features (i.e., each grid cell is the same size and shape) the computer processing requirements are lower when using raster data structures. When performing GIS processes with raster data, generally no calculation of the intersection of landscape features (lines or polygons) is needed, given the regular shape of the cells. In contrast, analysis processes that involve vector GIS data usually must deal with the potential intersection of landscape features (e.g., overlapping polygons).

Unfortunately, GIS databases stored in the raster data structure can become very large, especially when fine resolution cells are assumed. One of the hindrances to using raster data is that every cell must have a value associated with it; even cells where no landscape features of interest are present (e.g., vegetation outside of an ownership). From a computer storage perspective, this means that all raster grid cells have a value (either a valid value or a null value) that must be stored and maintained. In contrast, vector GIS data need only have points, lines, or polygon features in locations where landscape information is present.

Alternative data structures

Although points, lines, and polygons represent the most common forms of vector GIS data, several other forms of vector GIS data that may be useful in representing landscape features. These other data structures include triangular irregular networks (TINs), dynamically segmented networks, and regions. What follows is a brief discussion of each of these mysterious-sounding data structures, and a potential application that might be useful in understanding the potential uses of these alternative data structures.

Triangular Irregular Network

A **Triangular Irregular Network** (TIN), like a raster data structure, is useful for representing a continuous surface (an entire landscape). A TIN, however, addresses some of the problems that raster data structures have in accurately representing landscape features, especially those problems that result when you use regularly sized raster grid cells to describe landscape features. If the raster grid cells are small, in comparison to the size of other landscape features, you will probably have success in accurately representing those features. If the raster grid cells are large, in comparison to other landscape features, you might lose some of the integrity of the landscape features in the resulting raster GIS database. A TIN attempts to avoid this problem by using, as the name implies, a set of triangles—rather than a set of squares—to represent landscape features (DeMers, 2000). Each of the three sides of each triangle can, in fact, have a different length, making the triangles irregular in nature. Thus a TIN is composed of irregularly shaped objects, yet covers an entire landscape. In most applications, TINs are used to represent elevation models. An elevation is associated with each triangle corner, as illustrated in Figure 2.27. In landscapes that are highly irregular in terms of elevation (as are many forested landscapes, for example), the TIN may better represent topography than a raster-based data structure. Working with TINs, however, is beyond the ability of many standard desktop GIS software programs because of the complexity involved in storing and processing irregularly sized triangles and representing three-dimensional surfaces. Some software developers offer modules associated with desktop GIS software programs (at additional cost) that will allow GIS users to utilize TINs.

Dynamic segmentation of linear networks

A data structure that uses **dynamic segmentation** is based on a network of lines, and thus is a variation of the vector data structure. The dynamic segmentation data structure is designed to represent linear features, and traditional uses of this structure include modelling efforts related to river systems, utility distributions, and road networks. Dynamic segmentation allows GIS users to create routes that represent the movement or presence of an entity along a linear network. The routes are actually stored as information within a vector GIS database. Dynamic segmentation eliminates the need to create a separate GIS database for each route and facilitates advanced data handling and manipulation of GIS databases.

Figure 2.27 TIN representation of an elevation surface.

Underlying the route structure are sections and event tables. Sections are the linear components or segments that, when added together, form a route. Event themes are the data sources or attribute tables that are connected to the routes. The dynamic segmentation data model has the capability to associate information with any portion or segment of a linear feature. Event themes can be associated with each line or a single point on a line. This information can then be stored, queried, analyzed, and displayed without affecting the structure of the original vector GIS database.

Dynamic segmentation attempts to link a network of lines based on a common attribute so that the lines are grouped into categories of interest. An example of this approach might relate to a streams GIS database. A typical streams GIS database uses a series of lines to represent a stream network. Each of the lines would have a set of nodes, or beginning and ending points, and the nodes would be placed at all tributary junctions along the stream network. Depending on the size of the stream network, hundreds or thousands of lines might exist. A stream ecologist interested in analyzing the stream system, for example, could associate all lines that are used to represent the main channel of a river. Any attributes that are used to describe the main channel, such as length, depth, or temperature, can then be summarized. The stream ecologist might use this dynamic segmentation approach for the entire stream network to create a new GIS database that groups all lines based on some attribute (such as the stream name). The stream ecologist may also be interested in maintaining specific point locations along the stream database that contain features of interest, such as a smolt trap or culvert location. Point event tables of these features could be stored within the dynamically segmented stream, and show not only the locations of these features, but also distances and directions from other features within the database.

Dynamic segmentation allows GIS users to organize a

GIS database so that analysis and storage can be easier and more efficient. Dynamic segmentation can also be used to assist in scheduling management operations that involve transportation or movement within a resource area, or in planning or tracking almost any phenomenon that is associated with a linear network.

Regions

Another alternative vector data structure is called the **region**. This data structure is based on polygons or approximations of areas, such as stand boundaries or ownership parcels. One of the features of a 'topologically correct' polygon structure is that polygon features cannot be represented as overlapping areas in a topologically enforced polygon database. When two polygons meet, a new polygon is created to represent the overlapping area. As mentioned earlier, some desktop GIS software programs do not allow you to determine whether polygons are topologically correct. A region data structure will allow the existence of overlapping polygons while also maintaining topology. A forest scientist interested in capturing the locations of fallen trees within a steam channel would find regions to be useful (Figure 2.28). Polygons could be used to represent the lengths and widths of the trees, but any trees that are stacked on top of each other, like you might expect to find in a log jam, will not be accurately represented in a topologically correct polygon structure. Two fallen trees that overlap each other might result in multiple topologically-correct polygons: one or more polygons to represent the non-overlapping areas of trees, and other polygons for each of the overlapping areas of logs. For the five fallen trees displayed in Figure 2.28, enforcing correct topology for these landscape features would create a total of 11 polygons. This would result in both a loss of information and the creation of a larger set of database records than might be appropriate to describe the trees. With the use of the region data structure, you can retain individual tree data records, while also associating the overlapping trees with one another.

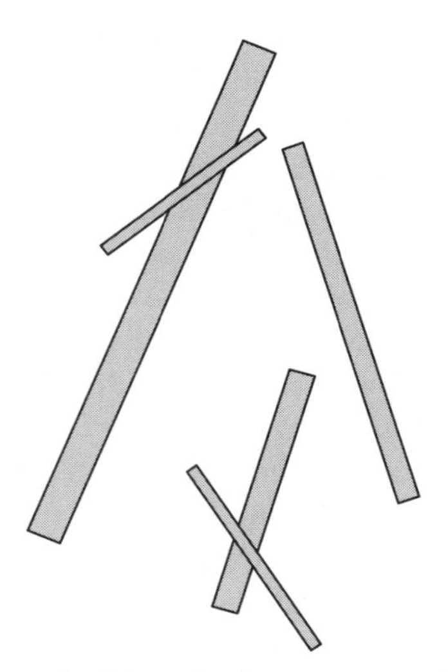

Figure 2.28 Example of the region data structure used to capture the placement of downed woody debris in a stream channel. Typical polygon topology would create 11 polygons to represent the five woody debris pieces. Regions allow for polygons to overlap, creating a five-shape database, with one complete shape for each piece.

Metadata

Metadata is 'data about data'; this is a relatively recent phenomenon in working with spatial databases. Typically, metadata is a digital document that accompanies a GIS database that describes the content and quality of the data. With recent advancements in some of the full-featured GIS software programs it is now possible to have a metadata file digitally linked to a GIS database. New software programs also make it easier to create and populate metadata fields with information about the data. Metadata is an excellent place to store and retrieve information about the characteristics of a database, including the map projection and coordinate systems used. Other useful information in a metadata file might include a description of the original data source, any editing that has been done, a list of the attributes, the intended use of the database, and information related to the database developer. The descriptions should allow users to trace the evolution of the GIS database. In the US, most federal and state agencies are required to make metadata available for any GIS database that is offered for public use. The US Federal Government has developed standards for producing and reporting metadata. Requirements for producing metadata are highly variable among private natural resource management organizations because no governing body enforces metadata compliance. GIS users should always ask for metadata whenever acquiring a GIS database.

Obtaining Spatial Data

The USGS has developed the most comprehensive collection of spatial data in the world. This collection includes DEMs, DOQs, DRGs, digital line graphs (DLGs), and many sources of raster and vector data for the US. The majority of this collection is available via the Internet, along with associated metadata. Several US and Canadian federal agencies, as well as state and provincial groups, produce and maintain spatial databases for the lands that they manage, and this information is also available to the public online. A more detailed discussion of data acquisition processes is provided in chapter 3.

Scale and Resolution of Spatial Databases

GIS databases are often characterized in terms of their scale or resolution. Scale or resolution refers to the size of landscape features represented in GIS databases. Issues of scale are usually associated with vector GIS databases, while issues of resolution are associated with raster GIS databases. Typically, the scale or resolution of a GIS database relates to the source material from which the GIS database was created. Source material, as described in chapter 1, can include aerial photographs, existing maps, satellite data, or information gathered from survey instruments such as total stations or GPS receivers. Many sources of vector data are derived from remote sensing techniques, particularly from aerial photographs. The scale that is associated with vector GIS databases typically relates to photographic scale (a function of camera height, lens length, and photo size). Scale is often expressed as a ratio, or representative fraction, such as 1:24,000 or 1:100,000 (Muehrcke & Muehrcke, 1998). The ratio expression is unitless, and implies that 1 unit of measurement on a map or photo represents 24,000 or 100,000 units on the ground. Sometimes, confusion exists as to the correct use of the terms 'large scale' or 'small scale'. The ratio 1:24,000 is a larger ratio than 1:100,000 (1 is a larger portion of 24,000 than of 100,000) and thus, 1:24,000 is a larger scale than 1:100,000. If you examine both 1:24,000 and 1:100,000 scale maps printed on the same size paper, the 1:24,000 map would show less area but greater detail than the 1:100,000 map (Figure 2.29). Scale can also be referred

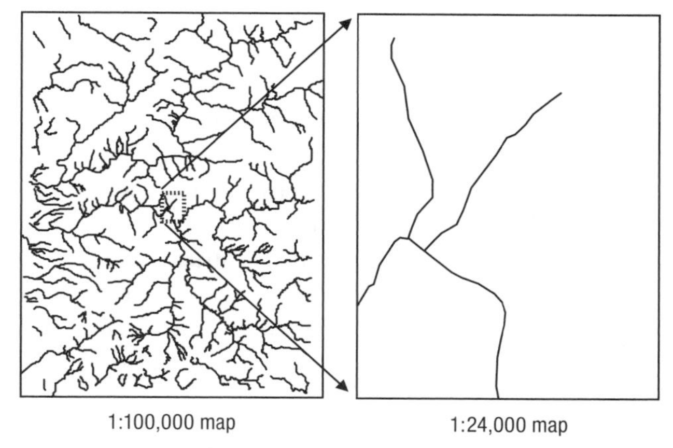

Figure 2.29 Map of stream network displayed at scales of 1:100,000 and 1:24,000.

to in terms of relative units, such as 1 cm = 1 km, or through the use of a scale that graphically illustrates approximate ground distances.

With imagery derived from satellite or aerial platforms, the ability of the electromagnetic sensor on the platform to delineate landscape features on the ground determines the resolution. A 1 m resolution image implies that the sensors used to collect the imagery captured a value for each square meter of the landscape. For raster GIS databases that were developed by scanning from maps or photographs, such as a DRG or DOQ, the size of the raster grid cell in representing landscape features determines the resolution. If each raster grid cell spans a 30 m ground distance, the raster GIS database is said to have a '30 m resolution'. This means that each raster grid cell represents 900 m^2 (30 m × 30 m) of ground area.

Although scales or resolutions are associated with spatial databases, some users mistakenly believe that they can improve the detail of GIS databases by focusing on small land areas. Users need to be cognizant that the scale or resolution of a GIS database remains static, regardless of how closely you view an area of the landscape.

Summary

This chapter discussed one of the fundamental considerations of any successful GIS program: spatial data. One of the main issues when working with GIS databases is knowing what projection system the database is set within, and how this coincides with other databases being used within an organization, or perhaps by other organizations with which data is shared. This chapter was developed to provide readers with a brief description of data projections and data structures—both raster and vector—as well as a few alternatives. In some cases, existing databases such as DOQs or DRGs may provide an excellent source of base data from which to verify other spatial databases or to create new spatial databases altogether. How GIS databases are stored and managed is a function of the decisions natural resource managers make regarding the purpose and intent of use for each GIS database. In many cases, however, you have no choice regarding the structure of GIS databases. For example, satellite imagery uses a raster data structure whereas vegetation and soils databases acquired from the US Forest Service generally use the vector data structure. In addition, most natural resource organizations use vector data structures to represent management units, roads, and other landscape features. Metadata are useful in helping you determine the characteristics of a GIS database, including the projection system used. Finally, understanding the scale or resolution of GIS databases and their associated landscape features is important, as it relates to the usefulness of a GIS database in assisting with analyses related to management decisions. Perhaps an interesting topic of conversation over a cup of coffee might be how well 30 m grid cells obtained from satellite images portray forest vegetation and help you develop management recommendations.

Applications

2.1. Projection parameters. Your supervisor, Steve Smith, has just learned that another natural resource organization that you intend to share spatial data with stores their GIS databases in a projection format that is different from yours. Steve is unfamiliar with projections, and asks you to provide some background for him.

a) What is a projection?
b) Why are projections necessary?
c) What is an ellipsoid?
d) What is a geoid?
e) What are the major projection types, what are their assumptions, and how have they been used?

2.2. Choosing a projection. You've been asked to recommend either a Lambert or Mercator projection that best fits the dimensions of the state or province that you live in.

a) What is the largest north–south dimension of your state or province in miles and kilometers?
b) What is the largest west–east dimension of your state or province in miles and kilometers?
c) Which of the two projections would you choose?
d) Defend your projection choice.

2.3. Public land survey system. With the exception of the original 13 colonies, several other states, and other designated ownerships, much of the US has been surveyed (measured) using the PLSS. Significant areas in Canada have also been surveyed using a similar approach. Given the broad use of this system and its counterparts in North America, it is important that you understand several key components.

a) How many square miles (or square kilometers) would you expect to find in a township?
b) How many acres (or square meters) would you expect to find in a section?

2.4. GIS data structures. You have been hired as a land management forester for a timber company in the southern United States. As a recent college graduate you are expected to have the most current knowledge of forest measurement and data acquisition techniques. Your supervisor, John Delaney, an older forester, is interested in GIS and is curious about database structures. Describe to him the difference between a raster and vector data structure, and give an example of a GIS database that might be designed with each structure.

2.5 Quadrangle challenge. Locate the USGS 7.5 minute Quadrangle that contains your GIS classroom or work location.

a) What is the name of the quadrangle?
b) When was the map originally compiled?
c) If the map has been updated, when was it updated?
d) What is the Ohio code description of the map?
e) When was the topography developed?
f) How much magnetic declination exists in the map?

2.6. Resolution and scale. A consultant has proposed using satellite imagery to quickly update the forest resources that your natural resource management organization manages. Some people in your organization are arguing for a complete and fresh photo interpretation of the land base to accomplish this goal, resulting in a vector GIS database of the vegetation condition of the landscape. The differences in resolution and scale are two of the hot topics when comparing these alternatives. Explain the difference between resolution and scale, and how they relate to raster and vector data structures.

2.7. Designing GIS databases. You have been hired by a natural resource consulting agency to develop and maintain a small GIS operation. While your expertise is in natural resource management, the owners of the consulting agency were intrigued by your GIS expertise, and have been interested in providing these services to their clients. You, of course, took the job because it is an opportunity to put your GIS and natural resource management skills to use. As a start, what data structure might you use to describe the landscape features in the following databases?

a) timber stands
b) streams
c) roads
d) inventory plots
e) culverts
f) logs in stream
g) precipitation
h) land ownership
i) stream buffers
j) owl locations
k) owl habitat

2.8. Map scale and ground distances. You are employed as a field forester for the Ministry of Natural Resources, and are stationed in northern Manitoba. You have been inventorying timber for the past three hours, and now it is time for lunch. Your lunch, of course, is located in your truck. The distance you measure on your map from your current position to the truck is 12 cm. If the map scale was 1:12000, how far are you from your truck?

2.9. Spatial resolution. The natural resource management agency you work for in the intermountain west is considering the purchase of 30 m resolution satellite imagery for assisting in the management of their natural resources. How much area, in acres, does a single 30 m grid cell cover? How much area would be covered by a single 100 m resolution grid cell?

2.10. Spatial scale. Your supervisor has asked that you bring a map at the largest map scale possible to your next planning meeting. You have your choice of the following map scales: 1:24000 or 1:100000. Which map do you bring with you?

References

Cadle, F.W. (1991). *Georgia land surveying history and law.* Athens, GA: The University of Georgia Press.

Chang, K. (2002). *Introduction to geographic information systems.* New York: McGraw-Hill.

Clarke, K.C. (2001). *Getting started with geographic information systems.* Englewood Cliffs, NJ: Prentice Hall, Inc.

DeMers, M.N. (2000). *Fundamentals of geographic information systems.* New York: John Wiley and Sons, Inc.

Dent, B.D. (1999). *Cartography thematic map design.* New York: McGraw-Hill.

Ladell, J.L. (1993). *They left their mark: Surveyors and their role in the settlement of Ontario.* Toronto, ON: Dundurn Press.

Lillesand, T.M., & Kiefer, R.W. (2000). *Remote sensing and image interpretation* (4th ed.). New York: John Wiley & Sons, Inc.

Muehrcke, P.C., & Muehrcke, J.O. (1998). *Map use: Reading, analysis, and interpretation* (4th ed.). Madison, WI: JP Publications.

Robinson, A.H, Morrison, J.L., Muehrcke, P.C., Kimerling, A.J., & Guptill, S.C. (1995). *Elements of cartography.* New York: John Wiley & Sons, Inc.

Snyder, J.P. (1987). *Map projections—a working manual.* Washington, DC: United States Government Printing Office.

Stewart, L.O. (1979). *Public land surveys.* New York: Arno Press.

Thompson, M.M. (1979). *Maps for America* (3rd ed.). Washington, DC: US Government Printing Office.

USDI US Geological Survey. (1995). *Geographic names information system, data users guide 6.* Reston, VA: US Geological Survey. Retrieved January 15, 2003, from from: http://mapping.usgs.gov/www/ti/GNIS/gnis_users_guide_toc.html.

Wolf, P.R., & Ghilani, C.D. (2002). *Elementary surveying: An introduction to geomatics* (10th ed.). Englewood Cliffs, NJ: Prentice Hall, Inc.

Chapter 3

Acquiring, Creating, and Editing GIS Databases

Objectives

This chapter discusses a number of topics related to acquiring, creating, and editing GIS databases. Readers should gain an understanding of the opportunities and challenges associated with the need to obtain GIS data from a variety of sources. At the conclusion of this chapter, readers should be able to understand and discuss issues related to:

1. the acquisition of GIS databases, particularly via the Internet,
2. the various methods for creating new GIS databases,
3. the processes for editing existing GIS databases, and
4. the error types and sources that are potentially associated with GIS databases.

Acquiring, creating, and editing GIS databases to address the needs of natural resource management decision-making processes is a continual process. Ideally, natural resource professionals would have a complete and robust set of GIS databases at their disposal prior to performing an analysis. However, as new and interesting opportunities to incorporate GIS analysis in decision-making processes arise, the GIS database needs change as well. Four general cases are common in natural resource organizations, as they relate to the availability of GIS databases:

1. GIS databases required for a specific analysis do not exist.
2. GIS databases exist, but they were created for other general uses and may not be quite appropriate to address the issues related to a specific analysis.
3. GIS databases exist, but they were created for other specific analyses and are not quite appropriate to address the issues related to another specific analysis.
4. GIS databases exist, and they are adequate and appropriate to address the issues of a specific analysis.

Ideally, you would hope that your organization is continually positioned near the fourth case noted above. However many GIS users find, eventually, that the first three cases are real, and that time must be spent acquiring or developing a GIS database. Several options are available to GIS users faced with having to acquire, create, or edit GIS databases. These options include having someone else create a GIS database (e.g., a GIS contractor) based on maps and other input provided to them, using GPS or some other field-based data collection method to facilitate the development of a new GIS database, modifying or editing an existing GIS database, creating a new GIS database by digitizing maps, acquiring a GIS database from the Internet (for example, the National Wetlands Inventory from the US Fish and Wildlife Service [http://wetlands.fws.gov/]), or acquiring a GIS database from other organizations. The decision to pursue one of these strategies will depend on several factors inherent to a person's job, such as the budgetary resources, time constraints, and the skills and computing resources that are available within the natural resource organization.

Acquiring GIS Databases

One of the main concerns of natural resource managers is locating and acquiring GIS data. Physically receiving the GIS databases is relatively easy; they can be sent and received as an email attachment via the Internet or by way of media such as portable USB drives, compact discs (CDs), or DVD disks. GIS databases can be acquired from a variety of clearinghouses, many of which are maintained and supported by federal, provincial, or state organizations. The US federal government, in fact, is perhaps the largest source of GIS data in the world. A variety of federal agencies in the US provide data, and the Manual of Federal Geographic Data Products (US Geological Survey, Federal Geographic Data Committee, 2002) provides a wealth of information concerning the agencies from which GIS databases can be acquired. Natural Resources Canada (Natural Resources Canada, 2007) and the Geography Network Canada (Geography Network Canada, 2007) provide access to many types of geographic data within Canada. Provincial and state agencies, such as the Washington Department of Natural Resources, also distribute a large number of GIS databases related to the resources of each state.

Acquiring GIS databases over the Internet has become a widely used practice over the past few years. The disadvantage, however, is that the format of the data is generally limited to that of the most popularly used data, and may not be directly compatible with some GIS software programs. Agencies that require GIS users to make requests for GIS databases may provide a wider variety of products to be acquired, as well as a wider variety of formats, depending on the provider. In the case of data requests, however, the agency may require that a payment accompany each request. The payment covers the cost of the media and staff time required to package and deliver the GIS databases; the rates are usually reasonable because public organizations provide the data. People requesting GIS databases are asked to provide a variety of information specific to their request (Table 3.1), so that the final product will meet (as closely as possible) their needs without the staff investing extra effort in formatting, re-projecting, or adjusting the GIS databases.

As an example of acquiring GIS databases via an Internet site, the Gifford Pinchot National Forest (Washington State) maintains a website where a number of GIS databases can be acquired (Gifford Pinchot National Forest, 2007). The Internet site is located at http://www.fs.fed.us/gpnf/forest-research/gis/. By accessing this website, you will find that all of the vector GIS databases are available in ArcInfo coverage (ESRI, 2005) export format. Databases can either be downloaded directly from this Internet site, or obtained on a CD-ROM or 8 mm tape. The cost of obtaining GIS databases on media is $40 (US currency). In addition, some metadata related to each GIS database can be accessed through the Gifford Pinchot National Forest website. The data related to the forest trail system, for example, indicates that the source scale for the GIS database was 1:24,000, that it was last updated in 1999, that the projection system is the Transverse Mercator using the Clarke 1866 spheroid, that the coordinate system is the UTM system, and that the datum is NAD27. In addition, the metadata identifies the primary contact person and alerts GIS users that some gaps in the trail system may exist so that users know to examine the data closely before using it to make management decisions.

TABLE 3.1 Typical information associated with a GIS database request

- GIS database requested
- Location (Township/Range/Section, Topographic quad index Ohio code [or name])
- File format (e.g., Spatial Data Transfer Standard [SDTS], ArcInfo export format, ESRI shapefile, Tagged Image File Format [TIF], Imagine format [IMG], ESRI geodatabase)
- Map projection, coordinate system, horizontal and vertical measurement units, and related parameters
- Metadata
- Contact person for questions related to the GIS database
- Delivery method (USB drive, DVD, CD, e-mail, FTP, etc.)
- Database compression format (Zipped, MrSID, TAR, etc.)
- Billing information
- Product license agreements

As you may have noticed from the Gifford Pinchot example, a variety of GIS databases are available in addition to the forest trails system, including hydrology, elevation, forest stands, roads, streams, and others. While these GIS database are commonly used to support resource management on the National Forest, they are mainly vector GIS databases. Raster GIS databases, such as digital orthophotographs, related to a particular landscape are perhaps more difficult to obtain. As mentioned in chapter 1, digital orthophotographs look like aerial photographs but they are stored in digital form and are registered to a coordinate and projection system. Raster GIS

In Depth

Reference is made in this book to a number of Internet sites that you can visit for additional information and resources. The Internet is in a constant state of flux, however, and organizations often change website addresses, or URLs. Since URLs change periodically, you might find that some examples provided in this book have become obsolete. Should you have a problem accessing the URLs we have provided, two strategies can be employed to reach the appropriate address.

First, verify that the website address being used *exactly* matches the URL listed in the book. Even the smallest of differences (one incorrect letter, or an extra space) can result in not reaching the appropriate website. Second, with a little ingenuity you can locate the Internet sites listed in this book by using an online search engine. For example, the URL to the Washington Department of Natural Resources Information Technology Division order form has changed several times since the first edition of this book. To locate the most current version of the order form, you could attempt an online search similar to the one outline below:

- Search the Internet for 'Washington Department of Natural Resources'
  - Result: http://www.dnr.wa.gov/
- Select 'Publications & Data' from the list of items on the DNR main page.
  - Result: http://www.dnr.wa.gov/base/publications.html
- Select 'GIS Data' from the list of items under Publications and Data sub-heading
  - Result: http://www.dnr.wa.gov/dataandmaps/index.html
- Select the 'Available GIS Data' choice from the available links on the Data page.
  - Result: http://www3.wadnr.gov/dnrapp6/dataweb/dmmatrix.html

At this point, you should be presented with a list of available databases that can be downloaded in an ESRI shapefile format (ESRI, 2005). Thus in four steps, you can reach the current website from which spatial data can be accessed from the Washington State DNR.

databases can also be acquired via the Internet. The Minnesota Planning, Land Management Information Center (2007), for example, maintains a clearinghouse (http://www.lmic.state.mn.us/chouse/metalong.html/) where GIS users can access the 'Imagery and Photographs' data category and download county-level digital aerial imagery, including orthophotographs, varying from 1 m to 10 m spatial resolution. In addition, the clearinghouse provides GIS users with some metadata related to the digital imagery. You can ascertain from viewing metadata

In Depth

How do you acquire data over the Internet? Usually, clicking on an Internet link will start the download process by presenting a dialog box that asks where the downloaded file should be stored. Another way is to download GIS database files to a computer is to use a **FTP**. FTP stands for File Transfer Protocol, and it is a widely used method of transferring data over the Internet. This method allows you to transfer computer files to other remote computers, or to download computer files from a remote computer. One advantage of this process is that PCs are able to communicate with UNIX-based machines, and GIS database file formats are automatically converted into a Windows®-compatible format during the transfer process. FTP was once only available on a PC through a DOS interface, but several low-cost Windows® applications are available that simplify this process and make additional file transfer or sharing options available. To locate such a utility, try an Internet search using the keywords 'FTP' and 'Windows' together.

that much of the imagery uses coordinates referenced within the UTM system, with an GRS80 ellipsoid and NAD83 datum. The clearinghouse also provides information on the accuracy, consistency, and completeness for some of the imagery.

With advances in technology, many private firms have emerged in recent years that develop and sell GIS databases. Most GIS and land surveying trade magazines will feature advertisements from these firms. The breadth of GIS databases and data creation services continues to improve and all indications are that growth in this area will continue. The current possibilities range from small land surveying firms who are capable of collecting highly precise and accurate vector data from relatively small land areas using either digital total stations or ground-based LiDAR, to larger organizations that offer raster imagery captured from aerial or satellite platforms that are capable of imaging large land areas (regions, nations, continents, the globe). Many of these organizations also advertise their services via the Internet.

Creating GIS Databases

If GIS databases required for an analysis do not exist in digital form in your organization, and cannot be obtained through other means, such as via the Internet or by request from a state, provincial, or federal agency, you may consider creating a new GIS database. Several factors must be considered when creating a new GIS database, including the type of information needed to adequately develop the database, the intended format of the data, the projection and coordinate system required, and the accuracy desired of the resulting GIS database. Creating new GIS databases can be a time-consuming and costly endeavour. The most common methods used to create new vector-based GIS databases include traditional **digitizing**, **heads-up digitizing**, and **scanning**. The process of creating GIS databases, either by digitizing maps, using a GPS to capture spatial coordinates that describe landscape features, or by other means, usually amounts to 70–75 per cent of the total time invested in GIS in support of a spatial analysis (DeMers, 2000). As you will find in subsequent chapters, new GIS databases can be created as a result of spatial analysis processes such as buffering, clipping, and overlay analysis. When creating new GIS databases with spatial analysis processes, concerns about the projection system, the coordinate system, the datum, and map units are lessened because the resulting GIS database is usually represented by the characteristics of the other GIS databases involved in the spatial analysis process.

If GIS databases do not exist, but maps of the landscape features of interest exist, these maps can be digitized using a manual digitizer. A series of measurement reference points (sometimes called control points or 'tics') must be available to allow you to register the map to a digitizing table. Reference points can include easily located landscape features, such as road intersections and building corners, or less easily located landscape features, such as property corners, section corners, or a systematic grid of points (Figure 3.1). Each of these reference points must be in a clearly definable location on the map, and the ground coordinates (coordinate system units) must be known. A common source for ground coordinates is USGS 7.5′ Quadrangle Maps since they usually feature coordinates along the map border in geographic, UTM, and State Plane Coordinate systems. It should be noted

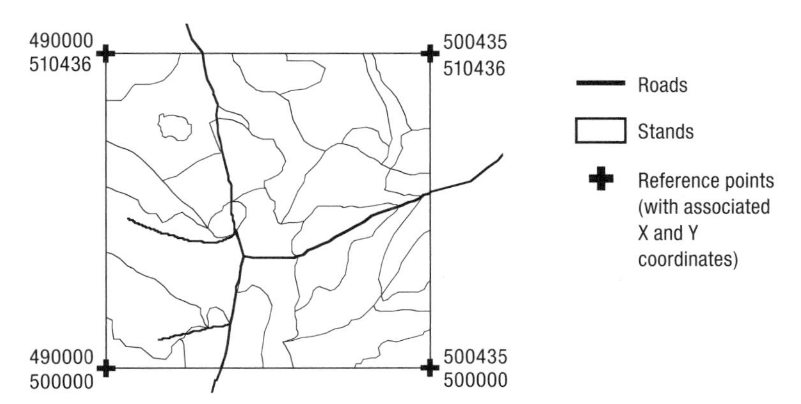

Figure 3.1 Measurement reference points for the Daniel Pickett forest to enable digitizing additional landscape features for the creation of new GIS databases.

that coordinates derived from paper Quadrangle maps would not have high accuracy since the coordinates are listed at broad unit intervals, and you must interpolate the location of landscape features. At least four registration points or tics are required to facilitate the digitizing of a map. Additional reference points, if available, will likely increase the accuracy of the spatial position of landscape features in the resulting GIS database.

At best, digitizing is an imperfect practice, and the quality of results can be dependent upon many factors. The accuracy of digitized GIS databases can be affected by the experience of the person doing the digitizing, by errors in either the location of reference marks or their associated geographic coordinates, and by imperfections in digitizing equipment and software (Keefer et al., 1988; Prisley et al., 1989). One of the often-misunderstood accuracy issues relates to the digitization of the map itself. If the map is old, or if it had been exposed to moisture (or even humidity), it may be subject to shrinkage or expansion. The shrinkage or expansion could vary across the map's surface, and thus the location (as well as the shape) of any landscape features that are digitized from it may be distorted. In addition, the methods used to delineate landscape features for digitizing could cause inaccuracies in the resulting GIS database. Regardless of how experienced a digitizing technician may be, if the map being digitized includes poorly delineated landscape features (Figure 3.2), then perhaps the 'garbage in, garbage out' principle applies. In this example, digitizing technicians may have rather precisely drawn landscape features to reference (upper left image) when creating a landslide GIS database, or some rather imprecisely drawn landscape features. When digitizing the upper right image of Figure 3.2, for example, would the landslide be defined by the outer edge of the thick line used to describe the landslide area, the inner edge of the thick line, or the center of the thick line? When using the lower left image, the landslide area is not represented by a closed polygon, thus the technician would need to use judgment or intuition in digitizing a closed polygon from the model provided.

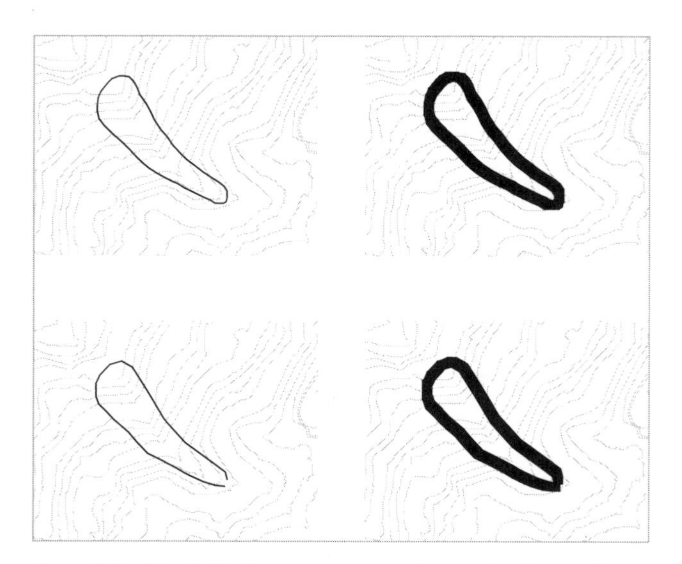

Figure 3.2 A landslide drawn on a map with a regular sharpened pencil (upper left), a marker (upper right), a sharpened pencil, yet in a sloppy manner—the landslide area is not closed (lower left), a marker, yet in a sloppy manner—the landslide area is barely closed (lower right).

Within some GIS software programs, landscape features can be digitized directly from a registered image that is displayed on a computer screen without having to establish registration points. This process is known as 'heads-up' digitizing and has increased the usefulness and popularity of geo-referenced raster products (such as the digital raster graphics and digital orthophotographs discussed in chapter 2). By not having to establish registration points, the digitizing process does not require a digitizing tablet, is faster, and offers less opportunity for error. Two advantages of heads-up digitizing are the ability to use digital imagery as an on-screen backdrop during the digitizing process, and the ability to change scales at which the digitizing takes place (by zooming in or out). Heads-up digitizing is explored in more detail in chapter 8, when the processes of updating GIS databases are described.

Scanning, as discussed in chapter 1, can also be used to convert a hardcopy map to a GIS database. Scanners sense the differences in reflection of objects on a map and encode these differences numerically in a digital file. For instance, lines and points (if identified by dark ink) would generally be distinguished from background areas (if the lines and points were drawn on white paper), and raster grid cells would be created to describe these features (Figure 3.3). The size of the grid cell (a raster data structure) will obviously be important when scanning landscape features. With increased resolution, or a greater number of grid cells per unit area, the greater the ability of the scanning process to return discrete shapes from a scanned surface will be. With larger grid cell sizes, less computer storage space is be required, but some landscape features may not be as accurately captured as desired. If needed, raster grid cells can then be converted to points, lines, or polygons (vector data structures) using raster-to-vector conversion algorithms that are common to most GIS software.

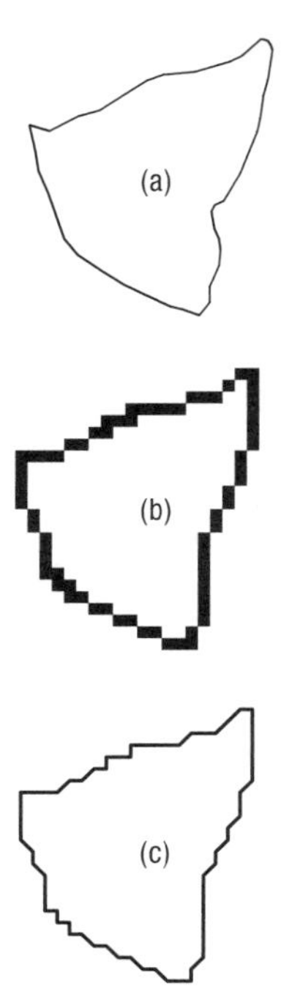

Figure 3.3 A timber stand (a) in vector format, from the Brown Tract, scanned (b) or converted to a raster format using 25 m grid cells, then converted back to vector format (c) by connecting lines to the center of each grid cell.

Editing GIS Databases

There are numerous reasons why you would edit a GIS database, such as re-projecting a GIS database to a common projection system used by a particular natural resource organization, edge-matching GIS databases describing landscape features in adjacent areas (e.g., townships, quadrangles, etc.) so that they fit together seamlessly, and other generalization and transformation processes necessary to convert a GIS database to a standard format or resolution. In addition, some GIS databases need to be continually updated and maintained, as landscape features and their attributes change over time. Processes related to updating GIS databases will be explored further in chapter 8, but as an example, most forest industry organizations update their timber stands GIS database annually to account for the changes that have occurred in the forest land base over the previous 12 months as a result of harvesting and other management activities, growth of the forest resources, and disturbance events (e.g., floods or wildfires). The data collection and reporting associated with updating processes can be labor-intensive and error-prone, indicating a need for standardized verification and editing processes.

The processes that you would use to find landscape features or attributes requiring editing can be classified as '**verification processes**'. With a verification process, the goal is to verify that a particular set of values within a database is appropriate (or reasonable, or within some standard). Verification processes should be devised so that both the landscape features and their attribute data can be assessed for completeness and consistency. These processes are probably best accomplished by involving multiple personnel so that GIS databases are checked independently, and that at least two data quality assessments are performed. The types of error that can easily be recognized include the improper location of landscape features, an improper projection system, and missing or inappropriate attribute data (data outside a reasonable range of values). These data errors could arise at any stage in the database update process, and regardless of their origin, should probably be corrected. If errors are located, or other changes need to be made, then the GIS databases need to be edited.

The following represents a short example illustrating a framework for verifying data and locating errors in GIS databases:

> Assume you work for an organization that manages a large area of forestland, over 200,000 ha. Within this organization there are a variety of people who have responsibilities for updating and managing the GIS databases that describe the forestland, from inventory foresters who collect the data, to information systems analysts who incorporate the data into standard database formats for use within the organization. A general process of updating the inventory databases (both tabular inventory and spatial landscape features) might start with the inventory forester compiling new inventory data (timber cruise reports) and maps showing changes to the forestland due to management activities over a period of time, perhaps the last year (Figure 3.4). This information is passed to the information systems analysts, who verify that the data (both maps and inventory) contain the appropriate type and format of data needed to successfully complete the updating process (verification process #1). If errors are located, some of this information is passed back to the inventory forester for clarification and editing. If the information is complete, and formatted correctly, the maps are digitized and the inventory data files are encoded (e.g., data is keypunched into

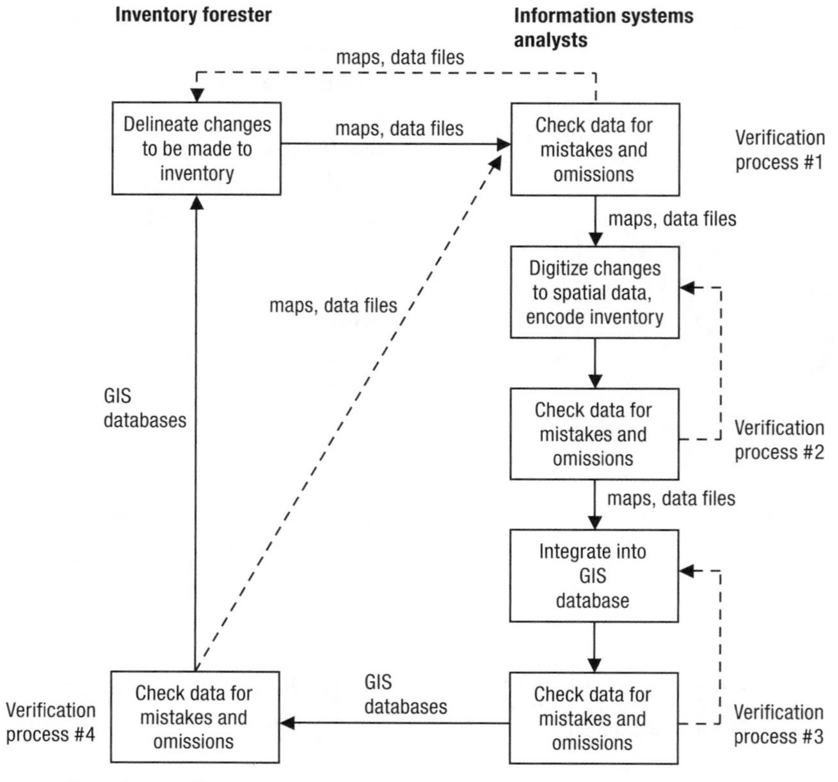

Figure 3.4 A generalized process for updating a forest inventory GIS database.

a computer file format), and a subsequent verification process (#2) is used to check whether these processes were performed successfully. The inventory data is then integrated into a standard GIS database format, and a third verification process is used to ensure that all of the changes have been incorporated into the updated GIS database. One way to do so would be to check the resulting GIS database against the information supplied by the inventory forester. Finally, the GIS database(s) are distributed back to the field office, where the inventory forester has the opportunity to verify (process #4) whether the proposed changes have been incorporated into the GIS databases.

It becomes obvious that editing processes, not just for annual inventory updates but also for periodic changes that should be made when discrepancies are found, should be considered in time and budget estimates to develop and distribute GIS databases to field offices. Some of the more common methods used to edit GIS databases include:

1. Add new landscape features (points, lines, polygons) to an existing GIS database.
2. Change the shape or position of existing landscape features.
3. Add new fields (columns) to the tabular portion of the GIS database.
4. Edit data in fields (columns and rows) in the tabular portion of a GIS database.

Editing attributes

Attributes, as described in chapter 2, are values used to characterize or describe landscape features and, thus, the qualities of the landscape. Through verification processes you can determine whether the attributes of landscape features are appropriate by assessing whether they are outside of the range of appropriate values, or missing entirely. In addition, as GIS databases are updated, you might assume that some attributes change over time. For example, as trees grow the characteristics of a forest will change. In the Daniel Pickett forest stands GIS database (Table 3.2), each stand is represented by a vegetation

TABLE 3.2 Attributes of stands in the Daniel Pickett stands GIS database

Stand	Vegetation type	Basal area[a]	Age	Original MBF[b]	Re-inventoried MBF[c]
1	A	200	50	21.2	23.2
2	C	175	40	12.9	15.3
3	A	210	55	25.8	26.8
4	A	250	65	34.2	37.0
5	C	90	20	3.1	5.6
6	A	220	55	25.7	28.2
7	C	150	35	8.7	10.5
....					
30	C	190	45	17.3	20.3
31	C	110	25	4.1	7.7

[a] square feet per acre

[b] thousand board feet per acre at the time of original inventory

[c] thousand board feet per acre after compiling new inventory

type, basal area, age, and volume (thousand board feet per acre, or MBF) at some particular time in the history of management of the forest. If you had re-inventoried the forest a few years after the stands GIS database was created, you might have found that the trees within the forest have grown, and that the stands GIS database may be in need of editing (as the age, basal area, and volume have likely changed).

Another example might include aquatic habitat inventories that attempt to monitor change in habitat conditions, such as fallen large woody debris concentrations or pool densities within a stream system. Aquatic habitat variables are constantly in flux within river systems as flow characteristics vary both by season and annually. Large woody debris is continually introduced into some stream systems as the result of natural processes or management activities, and is transported through a river system as flow regimes allow. An inventory of such streams would likely determine changing patterns of woody debris and pool concentrations on at least an annual basis. If natural resource managers wanted to track changes, new data columns could be added for each of the habitat characteristics that are of interest, and the updated inventory data could be entered into the new columns. If the position of the stream segments was also of interest, the positions of lines or areas that capture stream locations may also need to be adjusted.

Editing spatial position

Attributes of landscape features can also include the spatial coordinates of the features, and as these change, or are found to be inaccurate, they require editing. For example, as owls disperse, the spatial location of their nests may change, and the X, Y coordinates that describe owl-nesting sites require editing. Or as vegetation patterns change within a research plot, the polygon representing the plot may require splitting so that all patterns are recognized. Or as GPS data are collected and incorporated into a GIS database, multi-path error may be present, and should perhaps be either eliminated or corrected. GPS collection of data can also improve the accuracy of the location of landscape features that were previously defined and delineated with less accurate methods. For example, GPS data collection methods can improve the accuracy of a roads GIS database that was originally created from measurements collected from aerial photographs or with a staff compass and surveying tape.

Spatial position editing techniques vary widely across GIS software programs due, in part, to the different file formats that the different programs use. In general, however, editing the spatial position of a landscape feature requires that GIS users make a spatial layer 'editable'. Once a GIS database is editable, editing tools are used to move, copy, create, or delete points, lines, and polygons. While GIS databases containing point landscape features are typically easy to edit because a single coordinate pair represents them, databases that contain line and polygon landscape features usually require more care because the topological integrity of lines and polygons must be maintained. As discussed in the last chapter, topology describes the spatial structure of landscape features and is essential when comparing the positions of features to other features. While some GIS software programs have features to automatically examine and correct topological problems, others offer no tools for topological considerations and will consequently produce flawed analysis results when correct topology is not in place.

Depending on the amount of editing that is necessary, altering the spatial position of landscape features can be a demanding chore, and one that requires great attention to detail. Regardless of the GIS software program used, a good first step for any spatial editing process would be to ensure that a back-up copy of the database is made and securely stored. Once in the editing process, it may be important to remember to save the changes to GIS databases often (to avoid losing work due to power shortages or computer failure), and to periodically document the editing progress (to avoid forgetting what work has been done, and what work needs to be done next). Documentation may be particularly important if multiple people are editing (at various times) a single GIS database.

Checking for missing data

One issue that affects the accuracy of GIS databases is the omission of certain landscape features. How would you know if one, or many, landscape features have been omitted from a GIS database? Comparing GIS databases to reference maps or photographs is perhaps the simplest process. Features could be omitted because of improper map creation procedures or other blunders (for example, changes in the landscape that were not accounted for in an update process). Omissions may also occur in the attribute data associated with landscape features. For example, while a streams GIS database may contain all of the streams associated with a watershed, it may lack certain characteristics of some streams, such as the width or depth. Performing queries of landscape features to learn where attributes have been omitted will allow you to locate the landscape features that need editing. In the example illustrated in Figure 3.4, an information systems analyst could query the updated GIS database (as part of verification process #3) for stands that were altered during the update process, then examine the attributes of those stands for missing data. In doing so, the analysts could verify that all of the stands needing to be updated were, in fact, updated correctly (by comparing the updated GIS database against the maps and inventory data provided by the inventory forester).

Some GIS software programs are unable to explicitly handle missing data values for numeric attribute variables and will assume that a default value of 0 if an attribute value has not been specified. As you might imagine, this shortcoming can result in significant problems if these values are included in analysis results, as most statistical summaries or tests take sample size into account, and samples of '0' are included in the computations. One strategy for handling this problem is to assign a large negative value, such as –999 or –9999, to missing attribute values. Regardless of what value is used to signify missing data, the value should be a number that is far outside the range of acceptable data values and large enough to have a very noticeable effect in results if inadvertently used for analysis.

Checking for inconsistent data

You should expect some level of map error in each GIS database simply because all hard-copy paper maps contain errors and these errors are carried along into any digital form of the map that results from digitizing or scanning processes. With appropriate control in map creation processes, this error should be kept within desired tolerances. Map error can also result from inconsistencies in how landscape features are defined. For example some timber stands might be very finely delineated, whereas others are more coarsely delineated (Figure 3.5). In other cases, error arises because two (or more) GIS databases were created independently, using different encoding processes. This may result in features within a GIS database being represented with different precision. For example, one database may have used one process (e.g., digitizing) to create a forest stand GIS database, and another process (e.g., GPS) might have been used to independently create a roads GIS database (Figure 3.6). Upon close inspection, GIS users may find some public roads contained within timber stands, when they should more accurately be represented as being located outside of tim-

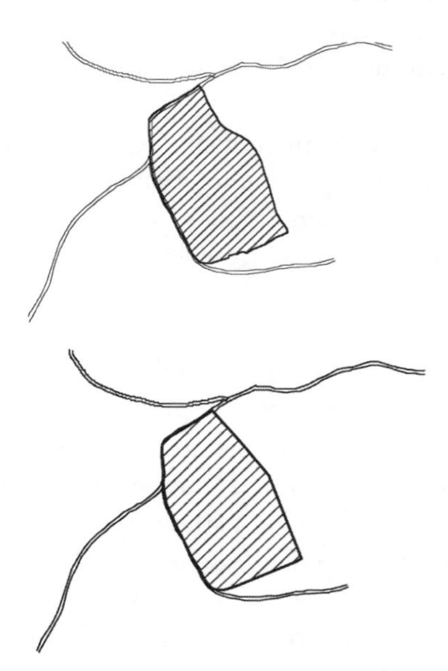

Figure 3.5 A timber stand drawn more precisely (top) and less precisely (bottom). Note that the lines on the south and eastern portion of the figures are different.

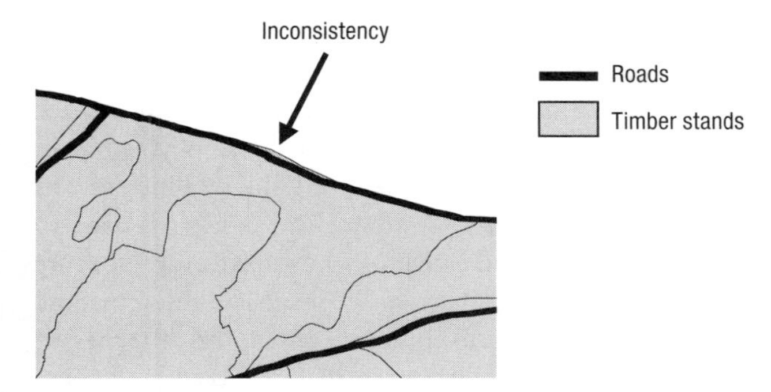

Figure 3.6 Spatial inconsistency between a timber stand GIS database created through digitizing and a roads database created through GPS measurements.

ber stands, and under the jurisdiction of some state, county, or municipality. Depending on the quality of the GPS receiver that was used, the likelihood is that the GPS database would more accurately represent the location of the road. Performing attribute queries of landscape features with values that are unusual, or outside of some logical range of data, and observing the results either through tabular or graphical means, should help identify whether problems exist and where editing processes are needed. Cressie (1991) provides some guidance with identifying spatial data outliers, and suggests verification processes such as creating histograms and distribution modelling.

Sources of Error in GIS Databases

One of the fundamental facts about computers is that they follow the instructions provided by computer users (unless they are suffering from an internal hardware problem or a computer virus). Computers have no sense of right or wrong. Therefore, assuming that the hardware is functioning correctly and you have been diligent in scanning for viruses, any errors that are found in GIS databases should be assumed to be a result of either encoding (database creation) or editing processes. Granted, some computer software programs are not perfect, and go through a number of versions to correct processing problems inherent in their computer code. However, when an error is located in a GIS database, it likely arose from a data creation or editing process. There are three sources of error commonly found in GIS databases: **systematic errors**, **gross errors**, and **random errors**.

Systematic errors, sometimes called instrumental errors, are propagated by problems in the processes and tools used to measure spatial locations or other attribute data. Systematic errors are sometimes called cumulative errors, since they tend to accumulate during data collection. They can be corrected if you can understand how each measurement is systematically affected. For example, suppose in digitizing a map that the reference points that were used to establish the initial location of the map were all erroneously shifted the same amount slightly off to the east. The landscape features that are digitized after this error is introduced will also be systematically positioned the same distance to the east of their true location. You might correct for this displacement by adding the corresponding distance, in coordinate units, in a westerly direction to the coordinates of all of the landscape features in a GIS database. Systematic errors in the collecting and processing of attribute data can also exist. For example, if you were to compute the area of a series of watersheds using acres, and then convert the area measurements to SI units (hectares) using an inappropriate conversion factor (e.g., 2.4 hectares per acre rather than 2.471), the metric areas would all be systematically incorrect. Corrections in this case can be made by a recalculation of the SI units by using the appropriate conversion factor.

Gross errors, sometimes called human errors, are blunders or other mistakes made somewhere in the data collection, map creation, or editing processes. For example, when digitizing a set of landscape features, such as landslides, a landslide that was delineated on the map being digitized might be omitted from the resulting GIS database. Or, when collecting forest stand information, an incorrect species code may be used to describe certain tree species. There may be no pattern in the occurrence of these errors, thus they may only be identified and corrected through verification processes. A thorough verification process will involve database checks by someone who did not participate in data collection or recording. Undoubtedly, human errors have the potential to exist in almost every GIS database.

Random errors are a by-product of how humans measure and describe landscape features. In contrast to machines, most humans are incapable of repeating a task over and over without any difference between repetitions. In the context of landscape measurements, it is unlikely that a person will be able to use a measurement instrument and return measurements that are consistently accurate and precise. No matter how carefully data such as tree heights from a forest are collected, there will be error in the representation of landscape features and their associated attributes. Chances are that recorded measurements will be at least slightly off from the true mark due to the limitations of vision, musculature, and instrument set-up and application. As long as consistent procedures are followed and no blunders occur, measurements will tend to be grouped closely around the actual measurement, with differences occurring slightly in all directions. These types of small errors are called random errors; they remain after all systematic errors and blunders are removed. Random error occurrence does tend to follow the laws of probability, and thus should be normally distributed in a statistical sense. The statistical distribution of repeated measurements can be estimated in order to obtain an idea of the variation expected in either the spatial location of a landscape feature or the associated attribute data. Through the process of least squares adjustments, you can attempt to remove random error from GIS databases when concern about accuracy issues is high (Ghilani & Wolf, 2006). More often however, in natural resource management it is assumed that data collection efforts were carried out diligently with respect to the accuracy and precision of measurements, and that random errors tend to cancel each other out. For this reason, random errors are sometimes termed compensating errors.

Two terms are important when assessing the usefulness of a GIS database: **logical consistency** and **completeness**. Logical consistency is the term most GIS aficionados use to describe how well the relationships of different types of data fit together within a system. In some cases this refers to the consistency of the topological relationships among GIS databases. For example, when streams are displayed in conjunction with the contour lines of an elevation GIS database, all streams should appear flow downhill (as opposed to uphill). In other cases, logical consistency refers to the attribute data of a particular GIS database. For example, the dominant tree species in one polygon may be labeled 'Loblolly', while in another polygon it may be labeled 'Loblolly Pine', and in a third, '*Pinus taeda*'. Completeness is a term used to describe the types and extent of landscape features that are included in a GIS database, and conversely, those that are omitted. For example, in some cases not all types of streams (e.g., ephemeral streams that only contain flowing water during rain events) are included in a streams GIS database. In addition, smaller streams may be omitted because they weren't apparent in the remotely-sensed data that were used to create the database. If only a portion of total number of stream types were included in a streams GIS database, you might conclude that the database is not complete.

Types of Error in GIS Databases

Since creating, editing, and acquiring GIS databases may involve many different processes, a variety of errors can obviously crop up. Some of the more common types include those related to the locational position of a landscape feature, those related to the tabular attributes of a landscape feature, and those resulting from computational problems.

Positional errors simply imply that a landscape feature is located in the wrong place in a GIS database. These errors can arise during GIS database creation processes, such as digitizing or scanning maps. As mentioned earlier, the digitizing of landscape features requires that a map be registered to a set of ground coordinates. How well the

In Depth

What is error? Error can be defined as something produced by mistake or as the difference between the true value of a feature and its observed value (Merriam-Webster, 2007). How would you know there was an error in a GIS database? Perhaps by comparing the value (or shape of a landscape feature) in a GIS database to what is known as the correct value (or shape of a feature) from a field survey, photograph, etc. When you consider GIS database errors, the goal is to understand three issues: the type of error that exists, the source of the error, and the extent of the error. All of these relate to the uncertainty associated with the landscape features contained in a spatial database. Hopefully you have a high level of confidence in the data (i.e., uncertainty regarding the location of landscape features and their attributes is minimized), so that analysis efforts can be used with confidence to positively facilitate decisions made regarding the management of natural resources.

registration is performed and how accurate the coordinates are represented on a map are both factors that contribute to errors in a resulting GIS database. Those who digitize maps also make errors—sometimes systematic errors (not using a digitizing puck correctly throughout an entire digitizing session), and sometimes gross errors (missing or displacing some objects entirely). Estimates of positional error usually indicate that some percentage of landscape features should be located within some distance either horizontally or vertically from their true position. A statement of positional accuracy, for example, might indicate that 90 per cent of the landscape features are within 150 meters of their true, or on the ground, position.

The ultimate use of a GIS database is also considered when describing the positional accuracy of the landscape features they contain. If the risk of a GIS user making a catastrophic mistake (e.g., an event that impacts human safety) when using the GIS database is high, developers of GIS databases may broaden the accuracy statement to safeguard themselves against lawsuits. An example of a situation when human safety could be affected by the accuracy of a GIS database is in databases that are used as navigational maps for maritime applications. Using an appropriate set of control standards may minimize the amount of error in a GIS database. Control standards can include using a minimum **root mean square error** (RMSE) when registering maps for digitizing processes or for establishing a set of verification processes for elimination of gross and systematic errors. In addition, when measurements are taken of a feature, be it point coordinates collected by a GPS receiver, linear features such as stream or trail lengths, or area features such as watershed boundaries, a RMSE can be used to assess and report differences between collected measurements and the true or estimated positions of features (FGDC, 1998).

While the positional accuracy of landscape features in a GIS database can be estimated, the uncertainty about what are termed 'local shapes' is more elusive (Schneider, 2001). For example, assume that a road has been digitized (Figure 3.7), and that one segment of the road has been represented by four vertices. The actual location of the road between the vertices is unknown, and can be described in a number ways other than by a direct line between each vertices. The positional accuracy of these local shapes is therefore a function of the quality of the digital encoding process associated with the landscape feature (whether manual digitizing, heads-up digitizing, or capture of data through GPS field techniques was applied).

Errors in the attributes of landscape features arise when incorrect values are assigned to features, either

In Depth

Root mean square error (RMSE) is a common term used in GIS. RMSE measures the error between a mapped point and its associated true ground position. Commonly used when digitizing a map, RMSE measures the positional error inherent in the registration points on the hardcopy map. RMSE calculates the squared differences between each reference point and its known or estimated position, sums these differences, than uses the square root of the sum to compute a measure of the positional accuracy. The formula for RMSE when location coordinates are of interest is:

$$RMSE = \sqrt{\frac{\Sigma\ [(X_{data,i} - X_{check,i})^2 + (Y_{data,i} - Y_{check,i})^2]}{n}}$$

Where:

$X_{data,i}$ = x (longitude) coordinate of each collected point i

$X_{check,i}$ = x (longitude) coordinate of true or estimated location i

$Y_{data,i}$ = y (latitude) coordinate of each collected point i

$Y_{check,i}$ = y (latitude) coordinate of true or estimated location i

n = number of observations (number of collected coordinate pairs)

A perfect RMSE is 0.00 where the reference points are located in a GIS database in exactly the same relative position as on the ground or when GPS collected data are exactly the same as control points that are used to test accuracy. This is rarely obtained when using real-world data. A sample RMSE calculation is shown in Table 3.3.

TABLE 3.3 Example of Root Mean Square Error (RMSE) calculation for GPS coordinates

Point	GPS-X (m)	Known-X (m)	GPS-Y (m)	Known-Y (m)	Squared Error (m)[a]
1	477395.3	477397.2	4934669.5	4934671.8	8.90
2	477399.7	477398.7	4934677.2	4934675.2	5.00
3	477405.5	477405.3	4934670.8	4934675.3	20.29
4	477407.5	477406.9	4934673.9	4934671.7	5.20
5	477406.9	477405.7	4934670.4	4934668.5	5.05
				Sum of squared error (m^2)[b]	44.44
				Average squared error sum (m^2)[c]	8.89
				Root Mean Squared Error (m)[d]	2.98

[a] Squared straight-line differences between GPS coordinates actual known positions: (X-direction error ^{2} + Y-direction error ^{2})

[b] Sum of the squared errors

[c] Average of the sum of the squared errors: (Sum of squared errors / number of points)

[d] RMSE: Square root of the average of squared errors: (Average squared error) $^{0.5}$

through editing processes or through spatial joins (which are discussed in chapter 9), or arise because the attribute data is outdated. Keyboard entry of attribute data can result in attribute data errors, particularly if the people performing the attribution processes are not paying close attention to the quality of their work. As we mentioned earlier, verification processes can range from an examination of printed maps and tabular databases related to the GIS database of interest, to an independent third-party examination of the landscape features and associated attributes contained in the resulting GIS database.

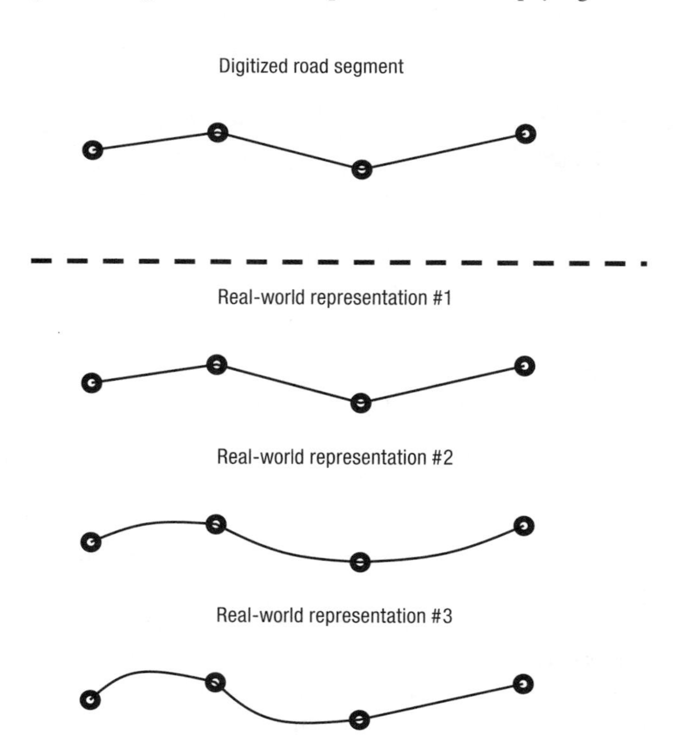

Figure 3.7 Uncertainty of the local shape of a road segment (after Schneider, 2001).

Computational processes can lead to another source of error that can be rather transparent to GIS users. Processes such as generalization, vector-to-raster conversion (or vice-versa), and interpolation cause alterations in the characterization of landscapes. In vector–raster conversion, for example, vector features are converted to raster grid cells. Obviously the size of the resulting grid cells, as discussed in chapter 2, will influence the quality of the landscape features contained in the converted GIS database. With large grid cells (e.g., 30 m or larger), the curvature of roads and streams could be lost, as well as small (and perhaps important) extensions of polygons. Users of the converted GIS databases would also need to determine whether the spatial resolution of the grid cells that result after vector–raster conversion is appropriate for the landscape features being represented. For example, are 30 m (or larger) grid cells appropriate for representing all roads and streams? Is a 10 m resolution more appropriate? The drawback of using a 10 m grid cell resolution is the relatively large amount of data required—nine times the amount of data contained in a 30 m resolution database (nine 10 m cells are contained within a single 30 m grid cell).

Ideally, in lieu of statements of error, the processes used to develop GIS databases, such as any transformations and conversions, should be documented and made known to allow users to understand the potential direction and magnitude of error associated with subsequent GIS analyses. In addition, metadata, such as projection and coordinate systems, should be made available to allow users to consider how a GIS database was developed. This is, of course, the ideal case. For example, the GIS databases used extensively in this book are void of statements of error and of metadata. They were developed as hypothetical landscapes for students in the authors' GIS applications courses.

Summary

Ultimately, most natural resource management organizations will develop needs that will outgrow the capabilities of their current collection of GIS databases. Acquiring, creating, and editing GIS databases are common processes encountered by natural resource professionals when they are in need of data or databases to assist in making management decisions or evaluating alternative management policies. When the GIS databases that are available to a natural resource management organization are not suitable to address a particular task, acquisition, creation, and editing processes must be considered. Those who simply view GIS as a system to make maps will likely underestimate the amount of work required to acquire, create, or edit the GIS databases necessary to make those maps. In addition, regardless of how GIS databases are generated, the presence and elimination of errors must be considered, and will likely require extensive verification processes to ensure that the potential errors are minimized.

Applications

3.1. Acquiring GIS data about Arizona National Forests. Assume you are interested in obtaining information about the streams related to the Prescott National Forest in Arizona. The website related to the GIS data is http://www.fs.fed.us/r3/prescott/gis/index.shtml (USDA Forest Service, 2007). Based on the information that the GIS data site provides, what are the categories that describe the available GIS databases? In reviewing the metadata for the 'Fire History' GIS database:

a) What is the purpose of the Fire History database?
b) How were the data created?
c) What datum, projection, and spheroid are used to represent fires?
d) What data structure is used to represent the fires?
e) What is the spatial extent of the fire history database in longitude and latitude?
f) Who is the primary contact should you have further questions?

Note: Please do not contact the Prescott Forest GIS coordinator for answers to these questions.

3.2. Acquiring digital orthophotographs about Massachusetts. Your position as a natural resource consultant has allowed you to become involved in a project situated in Massachusetts. As such, you are interested in obtaining digital orthophotographs about an area in Massachusetts. One source of this data might be the Massachusetts GIS website (http://www.state.ma.us/mgis/) (Commonwealth of Massachusetts, Office of Energy and Environmental Affairs, 2007). Based on what you can gather from the website:

a) What types of black and white digital orthophotographic imagery are available?
b) What spatial resolutions are available for 1:5000 color orthophotographic imagery?
c) What spatial resolutions are available for 1:5000 black and white imagery?
d) What datum and coordinate system are used to represent these images?

3.3. Data acquisition (1). You have been hired by a private consultant in Washington State to develop a GIS program. The consultant has a small office, comprised of only five employees, and is interested in developing a GIS program that will utilize desktop GIS software (ArcGIS ArcView, MapInfo, or GeoMedia) and GIS databases created by other organizations. To get started, you decide that acquiring base maps describing the counties, towns, section lines, and ownership of the state is important.

a) Is each of these databases available? List any databases that are not available.
b) What steps would you have to go through to be able to download the data?
c) What is the spatial reference of available data?
d) Who would you contact if you had problems downloading the data?

Hint: See Washington Department of Natural Resources 2007a for additional information.

3.4. Data acquisition (2). Assuming you were interested in acquiring soils GIS databases for 10 townships in Washington State from the Washington Department of Natural Resources (see Washington Department of Natural Resources, 2007b):

a) What is the format of the database?
b) What are the coordinate units?
c) Who is the contact for more information about the soils database?
d) Where did the soils information come from?

3.5. Data acquisition (3). The Washington Natural Heritage Program (Washington Department of Natural Resources, 2007c) maintains a GIS database of the high-quality terrestrial and wetland ecosystems of the State, as well as locations of rare plants of concern.

a) What groups of professionals are the intended users of the Washington Natural Heritage GIS data set?
b) How would you gain access to the Washington Natural Heritage GIS data set?
c) What data formats are available for the GIS database?
d) Are you allowed to display GIS maps of the dataset on the Internet?

3.6. Development of base map for a digitizing contractor. Assume that the standard process within your natural resource management organization for updating GIS databases is to develop hand-drawn maps of the changes desired, and then to deliver these potential changes to a contractor for digitizing work. In the center of the Brown Tract is an open area of land. Assume that the owners of the Brown Tract recently acquired this land. Develop a base map, showing the current stands and the current roads. Create a GIS database of four tics (reference marks) and place them on the map as well, along with their associated X and Y coordinates. Print the map, and then hand-draw the additional timber stands to be digitized by the contractor. Use the digital orthophotograph associated with the Brown Tract to hand-draw the new stands.

3.7. Owl locations on the Daniel Pickett forest. The wildlife biologist associated with the Daniel Pickett forest, Kim Dennis, is concerned about the quality of the owl nest location GIS database.

a) How could you determine whether the owl nest locations on the Daniel Pickett forest are represented as being in the correct place?
b) What are the coordinates associated with each owl nest location?

3.8. Errors in landscape feature and attribute data. As a user of the Brown Tract GIS databases, you are very interested in the quality of data that each provides. What might you say about the possible errors in the stands GIS database features that are located in the following places?

a) X = 1263950; Y = 371726
b) X = 1273647; Y = 363943

3.9. Verification processes. As technology progresses and the capabilities of field personnel increase (with new college courses in GIS, continuing education courses, etc.), the collection and transfer of data can occur with processes that would seemingly save effort and cost to an organization. For example, in Figure 3.4 the inventory forester would transfer cruise reports (hard copies) and maps (hand-drawn) to information system analysts in an annual update process. Several verification processes were noted in an effort to maintain the consistency and quality of data moving from the field to the information systems department and vice versa.

a) How might these processes in the flow chart change if the inventory forester were to provide spatial data that was digitized in the field office, and cruise data that was collected with a hand-held data collector?
b) Could the verification responsibilities shift under these circumstances?

3.10. Sources of error. Your supervisor, Steve Smith, is interested in understanding the types of error that may be inherent in GIS databases. Describe for Steve the differences between the following three types of error: systematic, random, and gross error.

3.11. Types of error. Given the inventory updating process noted in Figure 3.4, describe the sources of error (positional, attribute, computational) that could result at each step in the process.

3.12. Calculating Root Mean Squared Error: Woodpecker nest GIS database. Assume you are digitizing a map of red-cockaded woodpecker nest tree locations of a National Forest in Florida. You have four reference marks that you can use to reference the map to known coordinates. When registering the map you find that the X-direction and Y-direction difference between the reference marks and the actual known locations are as follows:

Registration Point	X-direction error (ft)	Y-direction error (ft)
1	3.526	0.963
2	−2.890	−2.452
3	0.985	−1.987
4	−1.598	−2.850

What is the RMSE, or positional error, associated with the resulting woodpecker nest GIS database?

3.13. Calculating Root Mean Squared Error: Trails GIS database. Assume you are employed by a National Park in Alberta, and are in the process of digitizing a map of trails that were drawn by the recreation specialist associated with the park. On the recreation specialist's map there are six reference marks that you can use to register the map to known coordinates. When registering the map you find that the X-direction and Y-direction difference between the reference marks and the actual known locations are as follows:

Registration Point	X-direction error (m)	Y-direction error (m)
1	3.125	−1.588
2	2.564	−1.992
3	2.548	−2.987
4	1.998	−2.856
5	1.268	2.857
6	1.489	3.897

What is the RMSE, or positional error, associated with the resulting trails GIS database?

3.14 Calculating Root Mean Squared Error: Tree location database. Assume you work as a natural resource manager in Oregon, and your supervisor is concerned about the quality of data that you are collecting with a GPS receiver. You supervisor has asked you to calculate the RMSE between GPS-collected coordinates and coordinates that had been collected by a digital total station. Both sets of coordinates are listed below.

Point	GPS-X	Total Station-X	GPS-Y	Total Station-Y
1	4934688.3	4934691.9	477311.7	477309.0
2	4934693.6	4934690.8	477310.5	477311.9
3	4934686.9	4934687.0	477316.1	477313.8
4	4934686.1	4934683.9	477312.7	477312.7
5	4934678.8	4934682.1	477310.2	477309.0
6	4934680.3	4934683.2	477307.6	477306.0

3.15 Calculating Root Mean Squared Error: Stream gauging stations. Assume that you are collecting data from a stream survey. A stream ecologist has given you a set of GPS coordinates that had been collected from gauging stations located next to a stream. She has also provided a set of coordinates from the same gauging stations that were collected from LiDAR data. What is the RMSE of the differences between these sets of coordinates?

Point	GPS-X	GPS-Y	LiDAR-X	LiDAR-Y
1	934681.0	477392.8	934676.6	477402.2
2	934670.8	477405.5	934675.3	477405.3
3	934673.9	477407.5	934671.7	477406.9
4	934670.4	477406.9	934668.5	477405.7
5	934665.8	477399.2	934666.9	477402.1
6	934668.1	477399.5	934668.6	477398.3

References

Commonwealth of Massachusetts, Office of Energy and Environmental Affairs. (2007). *Massachusetts geographic information system*. Retrieved April 19, 2007, from http://www.state.ma.us/mgis/.

Cressie, N. (1991). *Statistics for spatial data*. New York: John Wiley & Sons.

DeMers, M.N. (2000). *Fundamentals of geographic information systems*. New York: John Wiley and Sons, Inc.

Environmental Systems Research Institute. (2005). *GIS topology*. Retrieved April 19, 2007, from http://www.esri.com/library/whitepapers/pdfs/gis_topology.pdf.

Federal Geographic Data Committee (FGDC). (1998). *Geospatial positioning accuracy standards. Part 3: National standard for spatial data accuracy*. Reston, VA: US Geological Survey.

Geography Network Canada. (2007). *Data*. Retrieved April 27, 2007, from http://www.geographynet work.ca/data/index.html.

Ghilani, C.D., & Wolf, P.R. (2006). *Adjustment computations: Spatial data analysis* (4th ed.). New York: John Wiley and Sons, Inc.

Gifford Pinchot National Forest. (2007). *Gifford Pinchot National Forest geographic information systems, available data sets*. Retrieved April 19, 2007, from http://www.fs.fed.us/gpnf/forest-research/gis/.

Keefer, B.J., Smith, J.L., & Gregoire, T.G. (1988). Simulating manual digitizing error with statistical models. *Proceedings*, GIS/LIS '88. Falls Church, VA: American Society of Photogrammetric Engineering and Remote Sensing, American Congress on Surveying and Mapping.

Merriam-Webster. (2007). *Merrian-Webster online search*. Retrieved April 27, 2007, from http://www.m-w.com.

Minnesota Planning, Land Management Information Center. (2007). *LMIC's clearinghouse data catalog*. Retrieved April 19, 2007, from http://www.lmic.state.mn.us/chouse/.

Natural Resources Canada. (2007). *Mapping*. Retrieved April 19, 2007, from http://www.nrcan-rncan.gc.ca/com/subsuj/mapcar-eng.php.

Prisley, S.P., Gregoire, T.G., & Smith, J.L. (1989). The mean and variance of area estimates in an arc-node geographic information system. *Photogrammetric Engineering and Remote Sensing, 55*, 1601–12.

Schneider, B. (2001). On the uncertainty of local shape of lines and surfaces. *Cartography and Geographic Information Sciences, 28*, 237–47.

USDA Forest Service. (2007). *Prescott National Forest: GIS—geographic information systems*. Prescott, AZ: USDA Forest Service. Retrieved April 20, 2007, from http://www.fs.fed.us/r3/prescott/gis/index.shtml.

US Geological Survey, Federal Geographic Data Committee. (2002). *Manual of federal geographic data products*. Reston, VA: US Geological Survey.

Washington Department of Natural Resources. (2007a). *GIS data*. Olympia, WA: Washington Department of Natural Resources. Retrieved April 20, 2007, from http://www.dnr.wa.gov/dataandmaps/index.html.

Washington Department of Natural Resources. (2007b). *Available GIS data*. Olympia, WA: Washington Department of Natural Resources. Retrieved April 20, 2007, from http://www3.wadnr.gov/dnrapp6/data web/dmmatrix.html.

Washington Department of Natural Resources. (2007c). *Reference desk*. Olymipia, WA: Washington Department of Natural Resources. Retrieved April 20, 2007, from http://www.dnr.wa.gov/nhp/refdesk/gis/index.html.

Chapter 4

Map Design

Objectives

The common features of maps are described in this chapter, and emphasis is placed on developing those that field professionals can use to present results of GIS analyses or to illustrate themes of interest including forest management areas, tree species maps, harvest plans, wildlife habitat, and other natural resource management actions. At the conclusion of this chapter, readers should have acquired a firm understanding of:

1. the main components, or building blocks, of a map;
2. the qualities of a map that are important in communicating information to map users; and
3. the types of maps that can be developed to visually and quickly communicate information to an audience.

Within the various fields associated with natural resource management, we expect that maps will be available to illustrate resources and areas that we manage because of the prevalence of GIS use. Maps are amazing tools that, if constructed properly, have the ability to quickly and clearly communicate a message to an audience. Maps are an effective method of communicating spatial relationships among landscape features. Maps are also engaging—people are drawn to maps. Most GIS software programs provide users the capability to produce sophisticated maps and maps often represent the output produced by GIS analyses, thus most people mainly tend to associate GIS with map-making activities. Although this association may ignore many of the other analytical capabilities of GIS, the ability to geographically portray the results of an analysis is one of the primary distinguishing characteristics that sets GIS apart from other software programs.

Maps have been part of human civilization for millennia and have been used for many purposes, including data storage, navigation, and visualization. Maps have been used to create and sway opinion in many disciplines, including those related to the management of natural resources. In a manner similar to the use of statistics, maps can hold tremendous power over the message that is delivered to an audience and, when created skillfully, maps can be used to influence people's opinions (Monmonier, 1995, 1996). One of the great dangers presented by maps is that people assume the landscape features represented on maps are accurate portrayals of the natural resources that they claim to manage. However, maps are, at best, abstractions of the real world, and will usually possess some measure of non-conformity, be it directional or proportional or both, from a set of landscape features. A skilled map-maker will be able to choose a map projection that best preserves feature qualities (e.g., area, shape) and that best suits a map's objective. This skill might also include applying strategies for representing data characteristics or qualities through different shapes, colors, or sizes. Understanding that maps must be created and interpreted with a discerning eye is one of the first steps necessary to becoming a successful mapmaker or user.

Maps usually are two-dimensional representations of the landscape, although three-dimensional maps can be used to show volume or perspective. Symbols, colors, and text are combined to communicate information and, as with graphs, flow charts, and other diagrams, maps are graphical representations of information. Mapmakers attempt to transmit ideas in a different manner than that used by other forms of communication. The goal of the map-making process is to produce a visual display that communicates spatial information to potential map users. The human brain, with its limited capacity to store information, may be able to understand ideas more effectively when supporting concepts are presented graphically on a map (Phillips, 1989). The design of a map can affect the ability to communicate spatial information, thus a well-designed map will likely communicate ideas to an audience (e.g., co-workers and supervisors) more effectively than a poorly designed map. Poorly designed maps can lead to misinterpretations and costly or inappropriate decisions.

Cartography is the science and art of making maps, and for the better part of the twentieth century it was a skill developed through extensive experience in making maps by hand. Since the late 1980s when GIS began to be pervasive in natural resource organizations, the vast majority of maps have been made by non-professional cartographers. This shift is simply because of the accessibility and ease of use of GIS software. Some people may argue that several of the prescriptive aspects of making maps are no longer necessary (Wood, 2003). However, we would consider our effort to educate readers about the capabilities of GIS less than successful if we failed to describe the important aspects of maps, and suggest ways to make maps aesthetically appealing.

When developing maps for others to use, mapmakers should keep in mind that not all map users will be operating on the same level of competence. In fact, map users can be categorized as experienced, inexperienced, or reluctant (Franklin, 2001). The detail and clarity of mapped features will likely affect how well a map contributes to natural resource management. Within natural resource management, maps should be clear enough for users to understand two main things: (1) the land area that the map represents, and (2) the message the map intends to communicate about the land area. In order to meet these two requirements, a mapmaker needs to understand:

- the objective(s) of the map (the message),
- the people who may use the map (the audience),
- the data that will be displayed in the map (the information available),
- the use of graphics software for displaying map information, and
- the final format of the printed or digital version of the map (the product).

The size, shape, and symbology of each component of a map should reflect the likely responses to these prompts. A variety of common tools can be employed to make a map both useful for natural resource management purposes and aesthetically pleasing. The landscape features illustrated on a map should include enough landmarks to allow the users to reference themselves to the mapped area, and help navigate through a landscape. Mapmakers should keep two important aspects of maps in mind: (1) not everything known about a landscape needs to be displayed on a single map, and (2) to communicate effectively, maps should focus on displaying a limited number of landscape features. These concepts emphasize that cartographers should focus on characteristics that directly relate to the map's intended message, and thus they should ensure that other map components do not mask or cloud the message. The landscape features primarily emphasized on a map should be those associated with the main intent of the map. For example, on a map developed to illustrate stream classes, other landscape features, such as roads, timber stands, and soils, should be secondarily emphasized, or omitted from the map entirely.

Map Components

When developing a map for a natural resource management purpose, several basic components should be considered. These components include the symbols being used to describe landscape features, a north arrow, the scale, the legend, and the qualities (font, size, etc.) of the text (labels and annotation). In addition, you may find it necessary to include other components in a map, such as a description of the mapmaker, the filenames and file locations of the GIS databases used, and map quality caveats. Each of these components is briefly discussed below, and it is important to understand each element before we discuss the common types of maps that can be developed using GIS software programs.

Symbology

Symbology can be thought of as the art of expression, based on symbols (Merriam-Webster, 2007). A large suite of map symbols has been developed to identify and illustrate significant landscape features on maps. Some of these symbols (Figure 4.1) were developed as national standards for illustrating landscape features (e.g., contour lines, hydrologic symbols) found on certain widely used maps, such as the US Geological Survey topographic

Campground..............................	Campground
Gravel, sand, clay, or borrow pit......	Gravel Pit
Mine shaft................................	
Seawall....................................	SEAWALL
Shoal..	Shoal
Spot elevation............................	× 1985
State or territory........................	
Tunnel: road..............................	

Figure 4.1 A subset of USGS topographic map symbols (USDI US Geological Survey, 2003).

maps (USDI US Geological Survey, 2003) or the National Topographic System maps of Canada (Natural Resources Canada, 2006). Symbols have also been developed to represent organizational standards for identifying landscape features. For example, the US National Park Service has created a standard set of symbols for use on National Park Service maps (USDI National Park Service, 2003), which include typical symbols for roads and other landscape features as well as the highly recognizable pictographs (Figure 4.2). The International Orienteering Federation (2000) has also developed a set of standard map symbols for orienteering events. This provides a common approach to the interpretation of orienteering maps, and therefore promotes a fair competition among people involved in the sport.

Most GIS software programs provide a standard set of map symbols for map users. However, developers of maps

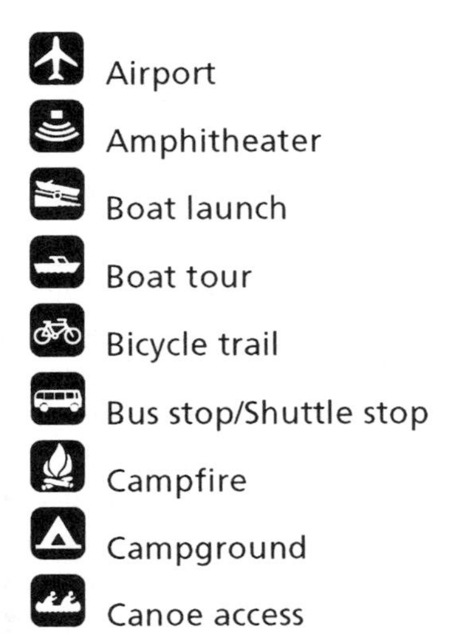

Figure 4.2 A subset of the standard National Park Service pictographs for maps (USDI National Park Service, 2003).

can easily misuse them, since documentation is usually limited within the dialog boxes provided by the GIS software. Nevertheless, a variety of symbols are available in GIS software programs that allow the mapmaker to describe landscape features. It is also possible within many GIS software programs to create a customized symbol set. In some GIS software programs symbols are merely bitmap graphic files that can be edited or created through graphic software programs. And, if the existing symbology within a GIS software program is not adequate, some GIS software programs may allow the use of customized tools or products developed by third-party software developers. A variety of free symbols sets can be obtained over the Internet (e.g., Sheahan, 2004), and you can purchase special symbols from companies such as Digital Wisdom, Inc. (2006).

Direction

Mapmakers typically use cardinal directions (north, south, east, west) to indicate map orientation. The use of a north arrow thus provides map users with a systematic method to not only help locate places on the ground, but also to understand where those places are in relation to other landscape features. While most maps are usually oriented with north at the top of the page, and south at the bottom of the page, it is appropriate to remove the uncertainty associated with orientation of a map by explicitly indicating the direction through the use of a north arrow. Omitting a north arrow from a map is considered poor cartographic practice.

A wide variety of north arrows has been developed to help map users understand direction and location, and the choice of which to use is usually determined by the mapmaker. Many of these forms of north arrows are available within GIS software programs, and a number of others can be developed by hand using lines, arrows, and text (Figure 4.3). Some organizations, such as the US National Park Service, require the use of a standard north arrow on their official maps (USDI National Park Service, 2003). Prospective cartographers should also realize that single-sided north arrows are used by some organizations to represent magnetic north. The Earth's magnetic fields are in constant flux and cause compass needles to point in alignment. Within much of North America, the magnetic variation ranges between 20° East and 20° West declination, thus creating a large angular difference between what many consider true north (which is astronomically derived) and magnetic north.

Figure 4.3 A variety of north arrow designs.

Scale

Maps are models of real landscapes, and a scale is used to indicate the ratio of the map features to the actual landscape (i.e., map distance compared to actual ground distance) to the map users. For this reason, scales are essential because they permit map users to reference map features to their actual size. A scale is almost always required on a map, and can be displayed using graphical, equivalent (verbal), or proportional scales (Figure 4.4). Graphical scales generally do not indicate the exact scale of a map (as do proportional or equivalent scales), but serve to visually associate the length of a map feature to actual ground distances. Many people relate to this form of scale more effectively than they do to the proportional or equivalent scales. Equivalent scales are those where one unit of distance on a map (usually the left-hand side of the scale) is equal to (using the equal sign) one unit of distance on the actual landscape. Each side of the scale can use a different unit of measure (e.g., 1 inch = 10 chains), which distinguishes this type of scale from the proportional scale, where each side of the scale is unitless.

Proportional scales are generally presented using a representative fraction, such as 1:24,000. With this type of scale, users should interpret 1 unit on the map as representing 24,000 *of the same units* on the ground (e.g., 1 map inch represents 24,000 ground inches, or 1 map centimeter represents 24,000 ground centimeters). Proportional and equivalent scales are also interchangeable. For example, an equivalent scale that reads 1 inch = 1 mile is the same as a proportional scale of 1:63,360 (1 inch on a map represents 63,360 inches, or 1 mile, on the ground).

Whether graphic, equivalent, or proportional scales are used, the appropriate metrics (English versus metric or SI system, feet versus miles [meters vs. kilometers]) and appropriate font sizes should be employed to avoid distracting users from the map's main message. Put another way, the scale is supplementary information on a map, and therefore it should not be so large that it attracts attention away from the map itself. Finally, the units displayed in a scale must make sense to the user of the map. For example, a proportional scale of 1:23,987, an equivalent scale of 1 cm = 2.3 km, and a graphical scale which is divided into 700-meter sections represents relatively uneven divisions. While the units displayed in scales may be automatically created in this manner in GIS, the mapmaker usually has control over them and can adjust them accordingly to provide a more logical representation of scale. In general, map users will be able to relate more easily to rounded scale figures, such as 1:24,000 or 1:100, rather than to more precise representations.

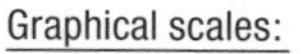

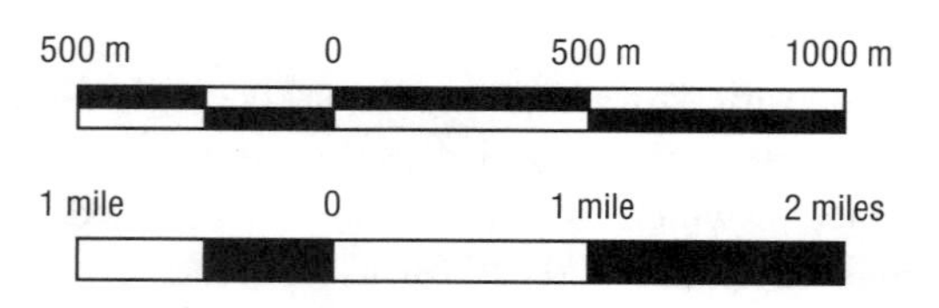

Equivalent scales:

1 inch = 1 mile
1 inch = 500 feet
1 inch = 10 chains
1 cm = 1,000 meters
1 cm = 5 kilometers

Proportional scales:

1 : 12,000
1 : 24,000
1 : 250,000

Figure 4.4 Graphical, equivalent, and proportional scales.

Legend

All of the features displayed in a map should be described in the map legend in order for users to fully interpret the map's message. Therefore, the symbology that is used to display features in a map should be replicated in the legend and associated with some text that defines the symbols (Figure 4.5). Of course, if you wanted to intentionally add mystery to a map, the legend may omit the description of certain landscape features. Some maps, such as the US Geological Survey topographic maps, may contain numerous features (and corresponding symbols). The legend that would be required for these maps would

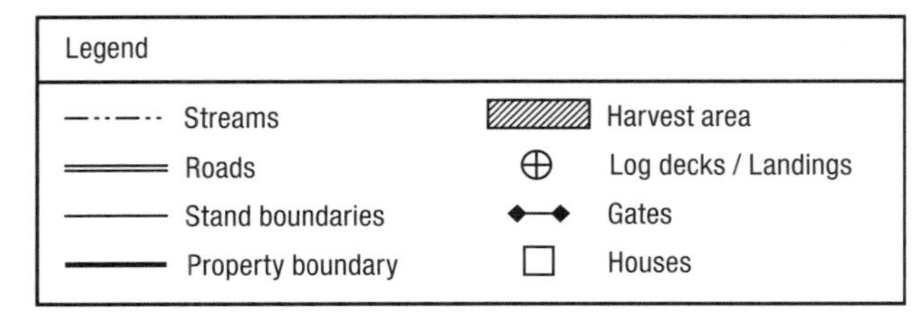

Figure 4.5 A map legend containing symbology and definitions.

overwhelm the map itself. In these cases, only a few landscape features are noted in the map's legend, and users must refer to the published standards (e.g., USDI US Geological Survey, 2003) for a full explanation of the remaining map symbols.

Legends can take many different forms and can use symbols, points, lines, polygons, colors, patterns, and text to clarify what users may see. Some legends should utilize a font size and font type that is appropriate for the map. Symbols sizes for features may also be varied to show the differences in quantities. The choices are not always obvious, but as in the case of the map's scale, the legend should not distract users from the message of a map. In addition, the appropriate descriptors for each symbol should be used. Abbreviations should be avoided if interpretation of symbols might be unclear, or if a broad audience is targeted. When data are presented and indicate quantities (such as length or area), the measurement units should be presented.

Most GIS software programs now offer tools that allow the automatic creation of map legends. These processes simply reference the GIS databases that are being used and their corresponding symbology. Typically, tools are also available in GIS software to allow you to modify automatically-created legends to suit your particular needs.

Locational inset

The approximate location of the mapped area within the context of a larger, more recognizable landscape feature (e.g., a basin, forest, county, or state or provincial boundary) can be indicated on a map using a locational inset. The locational inset may be extremely helpful for the map's audience if they are not familiar with the landscape being illustrated. The locational inset might show the location of a watershed within a drainage basin, or a property within the boundary of a county. The locational inset, however, should be a minor component of a map and it must not compete with the main feature(s) of a map for the attention of an audience. Figures 4.6 and 4.7

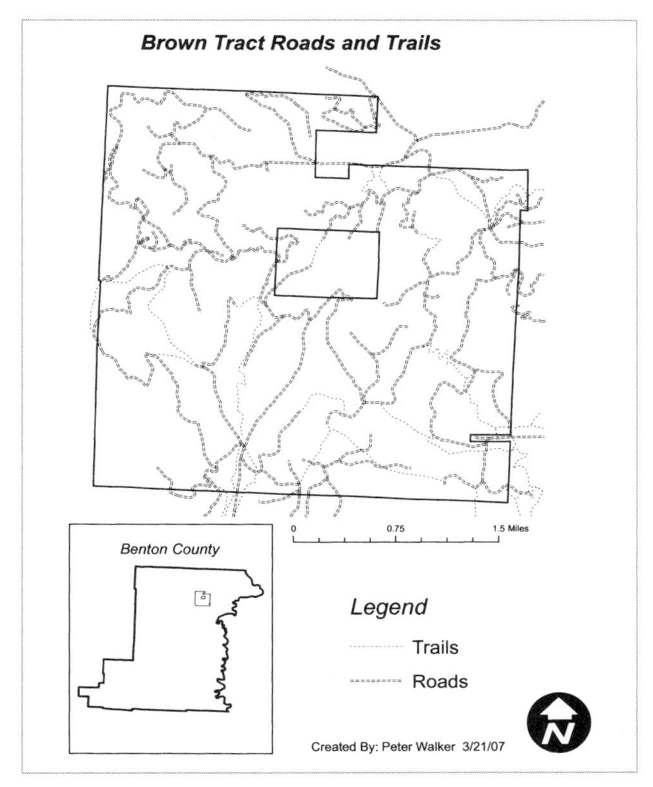

Figure 4.6 A map of the Brown Tract roads and trails containing a neatline, locational inset, title, legend, scale, and north arrow.

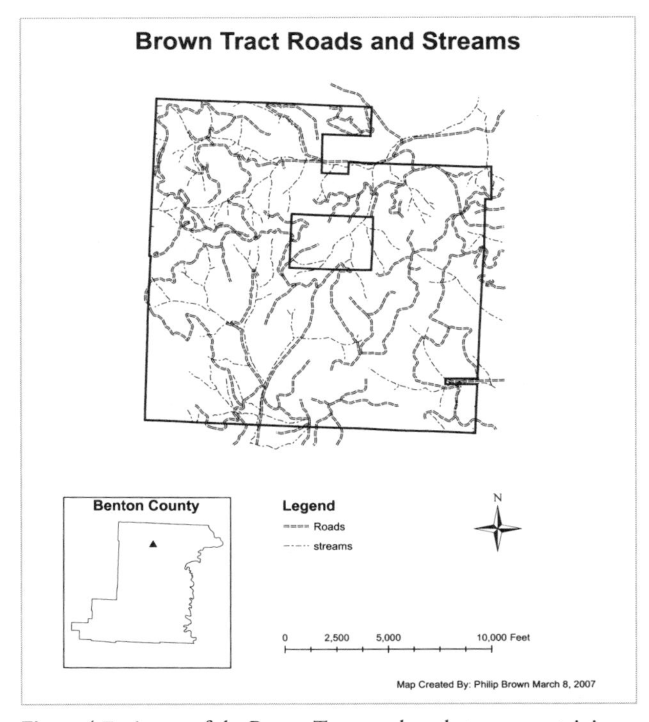

Figure 4.7 A map of the Brown Tract roads and streams containing a neatline, locational inset, title, legend, scale, and north arrow.

were developed as reference maps by students in one of our GIS courses, and although not perfect in every sense, each example provides a locational inset in the lower left corner. In one case, the locational inset contains the outline of the Brown Tract. In the other case, the location was approximated by the student using a symbol (a triangle), with the orientation and size of the symbol implying less accurate information about the location.

Three other types of insets can also be used, an enlargement inset, a related area inset, and a special subject inset. An enlargement inset can be used on a map to show more detail of a specific area located within the primary map region. A related area inset can be used on a map to illustrate features that are noncontiguous with the main map figure but are still important to display. For example, related area insets are often included in maps that illustrate the 50 United States. Here, Alaska and Hawaii are positioned in the insets, rather than in their actual geographic location (which would then include portions of Canada and vast expanses of the Pacific Ocean). A special subject inset can be used on a map to show different thematic representations of the main map area (e.g. precipitation, canopy cover). This is a popular mapping approach used in many atlases.

Neatline

A neatline is a border that surrounds all of the landscape features on a map but lies within the outside edge of the mapping medium (paper). Adding a neatline to a map is considered good cartographic practice but its presence is generally less critical than that of a scale bar or a legend. Neatlines can also be placed around other map elements to help distinguish them and keep them separate from other objects (e.g., to separate the locational inset from the legend). Usually a neatline is composed of at least one line, but multiple lines can be used to provide a more dramatic effect. Regardless of the style, neatlines can be useful tools that bring organization and distinction to mapped landscape features. Some GIS software programs include mapping tools that will not only create a neatline, but that will allow you to specify how the area contained within the neatline will be filled. For example, the background area behind the map title, landscape features, and legend, could be shaded. An automatic neatline creation tool in GIS software will allow the creation of a presentation-quality map or poster, but neatlines can also be created manually without much difficulty.

Annotation

Map annotation, or text applied directly to the map, is important in further describing landscape features beyond what can be described with a legend. In some cases, it may not be practical (or possible) to indicate all of the intended characteristics of landscape features through the use of a legend or symbology because of space limitations or because of the shape and size of the features. Therefore mapmakers may find annotation helpful in communicating additional messages to the end-users. Listed below are several examples of map annotation in natural resource management.

- Ownership: The owners of individual parcels may be displayed on a map with words, such as 'Georgia-Pacific', 'State of Alabama', or 'Province of Alberta'.
- Road numbers or names: These may be applied as labels to maps to further describe the road system.
- Surveying or other locational information: Township and range numbers (if using the Public Land Survey System or the Dominion Land System), or distances and directions from a metes and bounds survey may be illustrated on maps with annotation, as could the type of markings (e.g., pink flagging, orange paint) used to delineate treatment area boundaries.
- Areas of concern: The names of the homeowners near treatment areas may be placed on a map to allow land managers to understand who they must contact in case of a problem.
- Stand attributes: While stand attributes can be used to shade a thematic map (as we will see soon) several attributes of forested stands are commonly displayed with annotation inside individual stands. These attributes could include the stand or vegetation type, age, or area (Figure 4.8).

Typography

The content and form of text used to describe map features is an important aspect of the communicative ability of a map, and is often used informally to differentiate professional-looking maps from maps made by GIS novices. Since colors and patterns alone might not be able to fully explain the message of a map, the text used in annotation, labels, titles, and legends plays a role in the appearance and aesthetics of a map. The ability of GIS users to interpret the written information they find on maps is a function of many variables that can be described

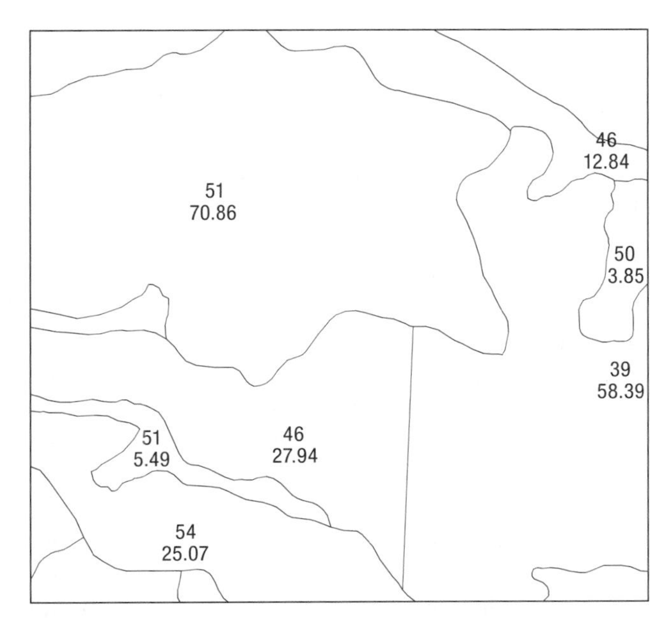

Figure 4.8 Map annotation: age (top, years) and area (bottom, hectares) of a portion of the Brown Tract stands (vegetation) GIS database.

under the broad heading of typography. Some of the most important typographical elements are typeface (font), weight (bold/normal), size (point size), and case (use of capitals) of text contained on the map itself. The font chosen will undoubtedly influence the ability of users to interpret maps, thus a normal font (Times Roman, Arial, etc.), and a normal and consistent font size is usually appropriate for most maps. Some important thoughts on map typography include the following:

- Some font types may be easier to read than others. Mixing font types on a map may create the impression that parts of a map are not clearly or logically connected.
- Use larger font sizes for map features that are relatively more important than others. However, if you were to differentiate landscape features by different sizes of labels or annotation, you must remember that only a small number of classes are discernable by most people. In addition, small font sizes may be difficult for some people to see.
- The title of a map should be displayed with a larger font size than the rest of the components of the map. Use font size judiciously to display the legend, scale, and other material not contained in the mapped area. The font size of these items should not overwhelm the information contained in the mapped area.
- In one study, Phillips et al. (1977) noted that the capability of people to search and find information on a map was enhanced when the text was displayed in a normal weight (not bold), with letters all in lower case, except an initial capital. However, capitals should be used for all letters in text when names are difficult to pronounce or need to be copied accurately.
- Text set entirely in lower case has been shown to be harder to locate on a map than text set entirely in upper case. However when the initial letter of a piece of text was slightly larger than the other letters, locating the text was quicker (Phillips, 1979).

Color and contrast

It may be difficult to believe at first but people tend to associate colors of landscape features on maps with events, emotions, and socio-economic status. Men and women respond to color with similar emotional reactions (Valdez & Mehrabian, 1994); however, people's emotional reaction to colors may vary across cultures. Listed below are general emotional reactions to various colors by south-eastern US college students, as suggested by Kaya and Epps (2004).

- Green: they felt relaxed, calm, and comforted, and associated the color with nature
- Blue: they felt relaxed, calm, and comforted, yet associated the color with sadness or loneliness
- Yellow: they felt lively and energetic, and associated the color with summertime
- Red: they associated the color with love or romance, yet also associated the color with anger
- Purple: they felt relaxed and calm, and associated the color with childhood or power
- White: they associated the color with innocence, peace, purity, or emptiness, and also associated the color with snowfall or cotton
- Black: they associated the color with sadness, depression, fear, and darkness, yet also with richness, power, and wealth
- Gray: they associated the color with negative emotions, bad weather, and foggy days

Studies of people's responses to color have indicated complex emotional relationships. For example, in the study by Kaya and Epps (2004), the color green evoked the most positive response among college students because it reminded them of nature. Yellow was a close second. Yet, students associated the color green-yellow with feelings of sickness and disgust. In an earlier study of the effects of color on people's emotions, Valdez and Mehrabian (1994) also found green-yellow to be one of the least pleasant colors, yet one of the most arousing.

In designing a map where features will be colored, two general rules should be followed:

1. Color certain landscape features with their most obvious associated color (e.g., roads–black; streams–blue).
2. When coloring polygons in a thematic map, use gradations of one or two colors to represent classes rather than a set of contrasting principal colors. The latter may confuse the ordering of importance of classes in the map user's mind.

Visual contrast is one of the most important factors in creating a map (Robinson et al., 1995). The extent of contrast employed by a mapmaker affects how well one set of information is promoted above the other information available in the map (Figure 4.9). Feature size, shape, texture, and color can all be altered to introduce contrast into a map. Contrast will help focus the attention of the audience, and will play a significant role in the ability of the audience to determine the clarity of a map. Mapmakers must also balance figure–ground relationships so that the important objects of a map are separated from those that are considered ancillary. In human perception, figures are the objects that are most strongly perceived and remembered by a map user, whereas data (text) are less distinctive and memorable (Dent, 1999). Techniques such as figure closure, contrast with other objects, and object grouping can be used to establish distinctive ground–figure relationships.

Ancillary information

In some large natural resource management organizations where multiple people share data development tasks, placing the names of people who contributed to the development of a map on a map is typically discouraged. However, in field offices of natural resource management organizations, where field personnel are generally responsible for developing maps to assist with on-the-ground decisions (and hence not specifically the development of GIS databases), it may be desirable to know both who created a map and when it was created. Since GIS databases may be modified frequently, knowing the date that a map was created might be as important as knowing the map's developer. This source information allows map users to place the content of a map in perspective with the version of the GIS database(s) used to create the map.

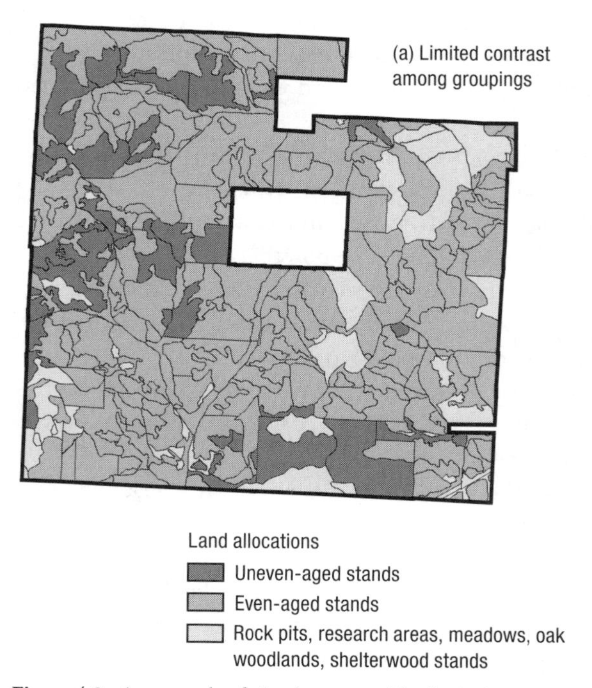

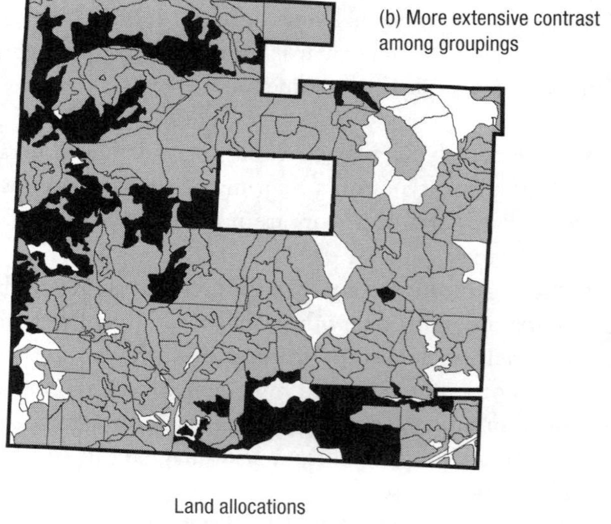

Figure 4.9 An example of visual contrast. The limited contrast among the groupings in the first map (a) does not promote the differences in stand types as strongly as the more extensive contrast of the groupings in the second map (b).

For example, suppose it is currently September 2008 and you were examining a map developed in June 2007 that represented wildlife habitat across several thousand acres of a landscape. Assume that the wildlife habitat being mapped was a function of forest stand conditions, and that the GIS database describing forest stands had been updated in December 2007. If the date were displayed on the map (June 2007), you would be able to understand that the quality of wildlife habitat illustrated in the map was estimated using an earlier version (i.e., not the current version) of the forest inventory data. Without such information, you could very likely assume (incorrectly) that the estimates of wildlife habitat are current.

It is relatively uncommon for mapmakers to provide the names of the files, projects, or computer code (e.g., macro) used in making the map on the map itself. However, providing this information would allow you to readily go back to the GIS databases or the computer code, modify some aspect related to the composition of the map, and generate a new version of the map relatively quickly. Without such guidance, you may find it difficult to remember how a map was originally constructed. The map projection might also be provided if the map perspective is affected by the projection system used. This type of ancillary information is usually placed in a subordinate position on a map, in relation to the other aspects of a map, and displayed with a relatively small font size.

Caveats and disclaimers

Increasingly, natural resource management organizations are adding more information to the maps that they produce both to clarify the accuracy of mapped landscape features and to clarify the intended uses of the maps. This information is important in helping users understand the appropriate applications of mapped information and in helping avoid damages and injuries that might result from improper map use. For example, maps have long served as vital navigation aids to mariners and pilots. The ability to safely pilot passengers depends on the quality of the mapped information used as a navigational guide. Should a landscape landmark be misplaced or unidentified, the consequences to people and vehicles that are navigating with the erroneous data could be disastrous. As you might imagine, there are other reasons why maps should contain information related to the quality of data. However, providing this information or deciding not to provide this information is, at least indirectly, a function of the litigious nature of today's society.

A map disclaimer is a statement that embodies the legal position of the mapmaker with respect to map users. In many cases, the mapmaker uses a disclaimer to distance himself or herself from any legal responsibility for damages that could result from use of his or her map. Caveats, similarly, warn others of certain facts in order to prevent misinterpretation of maps. In general, caveats are less sweeping than disclaimers, and may only address certain portions or aspects of a map. Warranties, on the other hand, are usually written guarantees of the integrity of a map, and of the mapmaker's responsibility for the repair or replacement of incorrect maps. In practice, disclaimers and caveats are regularly used, and warranties are rarely (if ever) used in association with maps and GIS databases. Quite often organizations add disclaimers or caveats to their maps in an attempt to warn users of the limitations of the map content. In some cases, disclaimers are noted directly on a map, and in other cases disclaimers are provided on websites devoted to the distribution of maps. Pima County, Arizona, for example, provides a very thorough disclaimer about its products on a website (Pima County [Arizona] Department of Transportation, 2003). The main ideas found in caveats and disclaimers include:

- rights reserved via copyright and permission requirements for modification to maps;
- degree of error found on the maps;
- suitability for use;
- liability, or responsibility, for errors or omissions (e.g., organizations usually assume no responsibility for misuse of their maps and subsequent losses); and
- contact information (e.g., addresses, phone numbers, e-mail addresses).

Caveats and disclaimers vary in form and content from organization to organization. Listed below are four examples.

- Indiana Geological Survey (2007): 'The maps on this web site were compiled by Indiana University, Indiana Geological Survey, using data believed to be accurate; however, a degree of error is inherent in all maps. The maps are distributed "AS-IS" without warranties of any kind, either expressed or implied, including but not limited to warranties of suitability to a particular purpose or use. No attempt has been made in either the design or production of the maps to define the limits or jurisdiction of any federal, state, or local government. The maps are intended for use only at the published scale. Detailed on-the-ground surveys and historical analyses of sites may differ from the maps.'

- USDI Bureau of Land Management (2001), Glennallen, AK: 'Information displayed on our maps was derived from multiple sources. Our maps are only for graphic display and general planning purposes. Inquiries concerning information displayed on our maps, their sources, and intended uses should be directed to: . . .'
- Orange County (Florida) Property Appraiser (2002): 'The Orange County Property Appraiser makes every effort to produce and publish the most current and accurate information possible. No warranties, expressed or implied, are provided for the data herein, its use, or its interpretation. The assessed values are NOT certified values and therefore are subject to change before being finalized for *ad valorem* tax purposes. OCPA's on-line (cadastral) maps are produced for property appraisal purposes, and are NOT surveys. No warranties, expressed or implied, are provided for the data therein, its use, or its interpretation.'
- Town of Blacksburg (Virginia) (2007), on the Blacksburg WebGIS site: 'DISCLAIMER: The information contained on this page is NOT to be construed or used as a "legal description". Map information is believed to be accurate but accuracy is not guaranteed. Any errors or omissions should be reported to the Town of Blacksburg Geographic Information Systems Office. In no event will the Town of Blacksburg be liable for any damages, including loss of data, lost profits, business interruption, loss of business information or other pecuniary loss that might arise from the use of this map or the information it contains.'

Map Types

The type of map that you develop should be a function of: (1) the type of data (i.e., point, line, polygon, raster) that is contained in GIS databases, and (2) the message(s) that you wish to communicate to an audience. For example, if the main GIS database used to create a map contains point features, and you wish to illustrate differences between the point values, you may want to show the points as dots or graduated symbols (different sizes of points based on the point attribute values). If the main GIS database used to create a map contains line features, you may want to illustrate the differences in the lines with different line types. If features of interest represent areas, GIS databases containing polygons can be displayed as thematic maps, qualitative area maps, and others. Volumetric databases, such as digital elevation models, can be displayed as gridded fishnet maps, as shaded relief maps, or simply as images with different shades or colors assigned to individual pixels.

If you were interested in illustrating features of a landscape across time, you can develop maps with multiple panels, where each panel would contain a view of the landscape during a different time period. In some cases, maps can be animated, as in a short movie. They can also contain a 'fly through' ability when viewed on a computer screen. The next few sections of this chapter describe the most common types of maps developed for natural resource management purposes in more detail.

Reference maps

Reference maps are those that illustrate a number of different landscape features, and that provide users with a broad perspective of the landscape. Road maps are one example of reference maps, and may contain not only road types, but also the locations of towns, major rivers, and political boundaries (provinces, states, counties, etc.) that are set among the road system. Maps that display stream systems or watersheds are another example of reference maps. Here, you would be able to place a watershed or stream system within the context of a larger geographic area, and thus these maps may also contain the locations of towns and political boundaries. Land ownership maps are a third example, and these may contain roads, streams, towns, and other features necessary to place the land you manages within a larger landscape context. Reference maps are commonly made when you develop management-related activity maps, such as those for tree planting or locations for new features such as trails or roads.

The characteristics of reference maps will vary depending on the audience. For example, some reference maps might display unique landscape features that are essential for high quality recreational experiences. Edwards (2001) describes the desirable content and features of fishing maps developed for anglers. It is suggested that these types of reference maps include the complete road system surrounding a fishing area, water depths, access points to water, names of local features, locations of off-limit fishing areas, locations of fishing lodges and places to buy fishing permits, places to park vehicles, and, interestingly, local pubs. In the case of fishing maps, Edwards (2001) suggests that they be developed in such a way that the important information is easily accessible to the eye, and that they be represented with easily understandable cartography.

Thematic maps

Thematic maps use colors or symbols to describe the spatial variation of one or more landscape features. Map features displayed with a combination of color and texture have been shown to be easier to find on maps than features displayed with variations on texture alone (Phillips & Noyes, 1982). Several types of thematic maps are common. Perhaps the most common is the choropleth map on which a range of appropriate values (Figure 4.10) or gradations of a color illustrate the relative magnitude of attributes of landscape features. Color schemes generally range from an empty shaded fill (for lowest valued attributes) to a full shaded fill (for highest valued attributes), with various shades of color for intermediate classes.

The legend is critical when developing choropleth maps because the colors related to the values must be explicitly described in order for map users to interpret the values effectively. Several design aspects must be

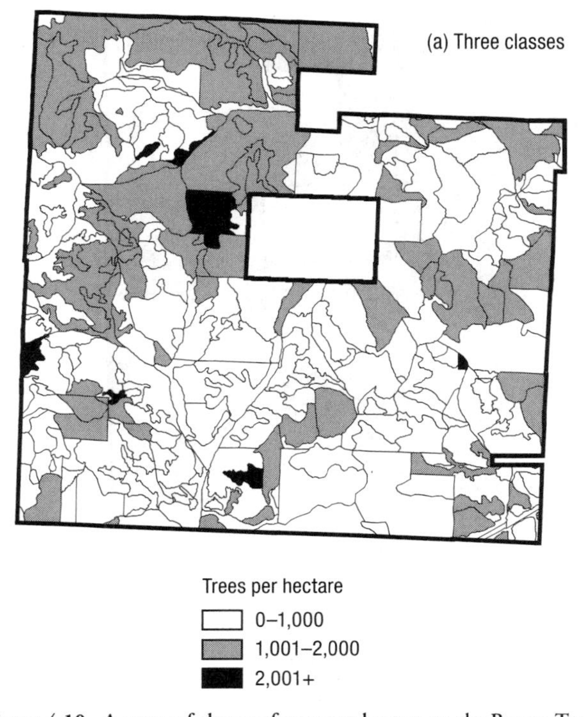

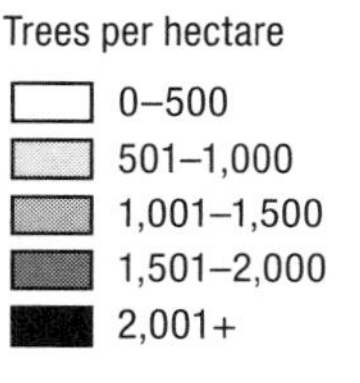

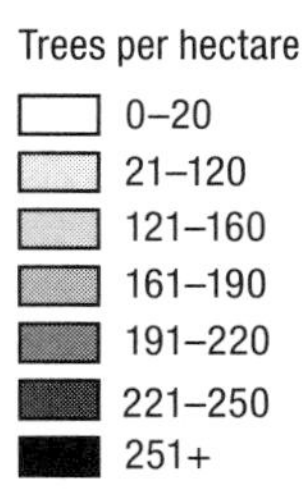

Figure 4.10 A range of classes of trees per hectare on the Brown Tract illustrated in a choropleth map.

addressed when creating a legend. The number of legend classes, or categories, is important. Too few classes may not contain enough information, while too many classes may present too much detail or result in an overly 'busy' map. Sometimes it may be necessary to experiment with several choices to determine which might work best (see Figure 4.10). Dent (1999) provides some guidelines related to this topic. Humans have difficulty differentiating more than 11 gray tones, so as a general rule, a minimum or four, and not more than six, classifications should be used on a map.

The size or ranges of the intervals for each class also has a significant impact on a map's message. Most GIS software programs offer processes to help create legends, and they may take into account the distribution of the data that are being mapped. An equal interval legend, for example, would take into account the range of data values and create intervals (classes) that share an equal distribution of the range (Figure 4.11). A quantile distribution would put an equal number of observations (e.g., polygons) into each interval (class). Intervals might also be created based on how many standard deviations an observation is from the mean, or be created using natural break points in the distribution of observations. While these automated processes can save time when compared to manual methods of creating classification ranges, it is advisable for mapmakers to visually examine the distribution of the data they are mapping to better understand its character, and then to decide what legend type might be useful. Again, an important concept when utilizing GIS software is that the software will usually do what you ask. The responsibility falls on the mapmaker to determine the appropriate classification for a legend.

The distribution of data, such as the range of basal area within the stands of a property you manages, can take on many different shapes, but the most common (for map-making purposes) are normal, random, and even distributions (Madej, 2001). Normal distributions follow a statistically-based representation of values that you would expect to see from most populations or population samples. As a result, a standard deviation legend classification, with its emphasis based on statistical variation from an average, works well. Random distributions are present in data where you cannot discern a regular pattern in the occurrence of data vales. Natural breakpoints can be established by locating sub-groupings of random data, thus you would manually create divisions between subgroups. Even distributions include data where the values do not appear to change very much. For example, if you managed 1,000 hectares of land, you might expect to find

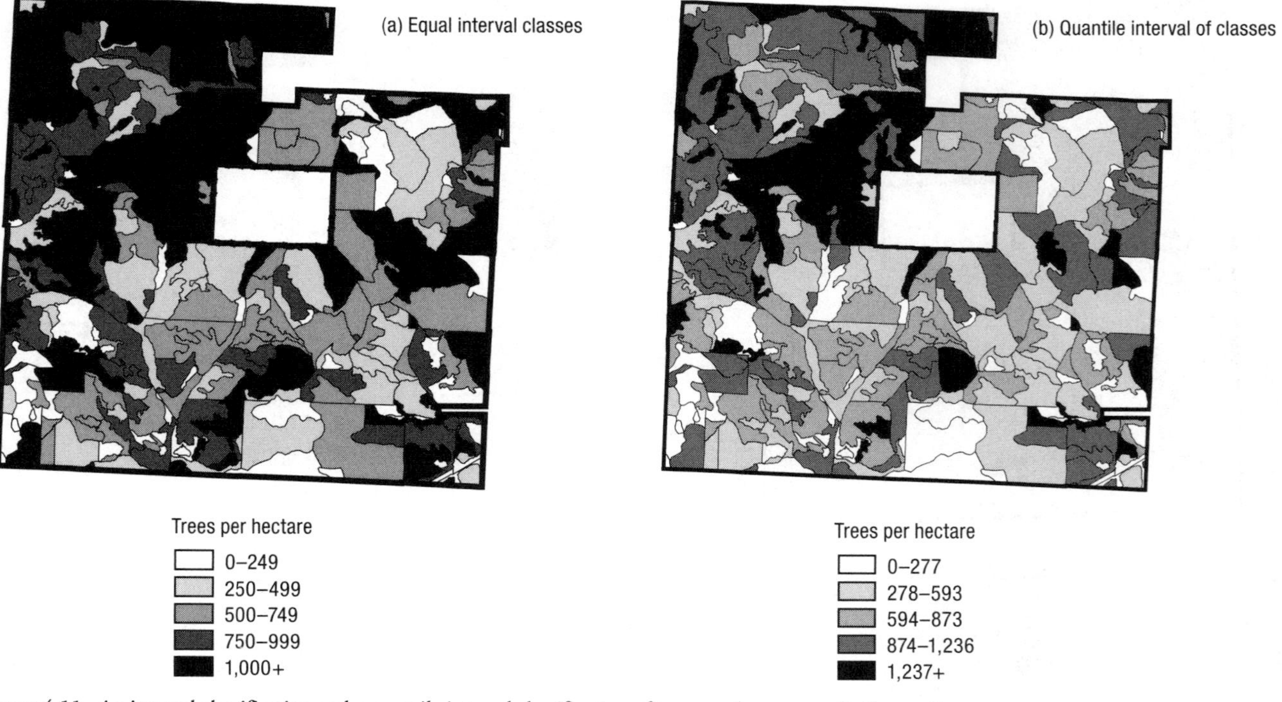

Figure 4.11 An interval classification and a quantile interval classification of trees per hectare on the Brown Tract.

an equal number of hectares in each 10 m^2ha^{-1} basal area class, and thus you would expect to find a relatively small standard deviation in the data values here. A quantile classification works well for even or uniform distributions because an equal number of observations are placed in each category.

Regardless of how thematic classes are created, mapmakers need to ensure that the intervals are consistent and that they make sense from an interpretation point of view. This also includes verifying that legend classifications do not overlap and do not inadvertently omit data ranges (Figure 4.12). If a characteristic of a landscape feature (e.g., basal area per hectare) can be placed into more than one class, then the classes overlap. Alternatively, if a characteristic of a landscape feature cannot be placed into any class, a data range has been omitted and the feature falls into a classification level 'gap'. Either way, the potential problems with the classification must be addressed.

With the increasing capabilities and affordability of color printers and plotters, the use of color to graphically portray different classifications in thematic maps has become standard. In general, tonal progressions of a single color are useful for illustrating magnitudes of change, with the lighter tones of colors indicating a lesser quantity (or quality) of an attribute value than the darker tone colors. A similar effect can be generated with gray tones, as demonstrated with some of the figures in this chapter. Single color progressions are particularly helpful for continuous numeric variables. For nominal data classifications, such as ownership or land use, or numeric data with only a few categories, distinctly different colors or patterns can be used to make certain categories stand out on a map.

Contour maps (Figure 4.13) are also a type of thematic map, and are sometimes called isoline or isarithmic maps. Here, lines or collections of similar features are used to emphasize gradients or distributions, such as elevations or precipitation levels across a landscape (Star & Estes, 1990). The contour interval is the distance between adjacent contour lines. The choice of an interval is important when creating contour maps: tight intervals may result in a cluttered map while wide intervals might misrepresent landscape variation. To reduce clutter on maps, not every contour interval is described with a data value, only those representing significant changes in elevation—usually denoted themselves by the elevation interval. For example, while the contour interval between adjacent contour lines may be 10 meters, the only contour lines represented with data values may be those that represent every 50-meter change in elevation.

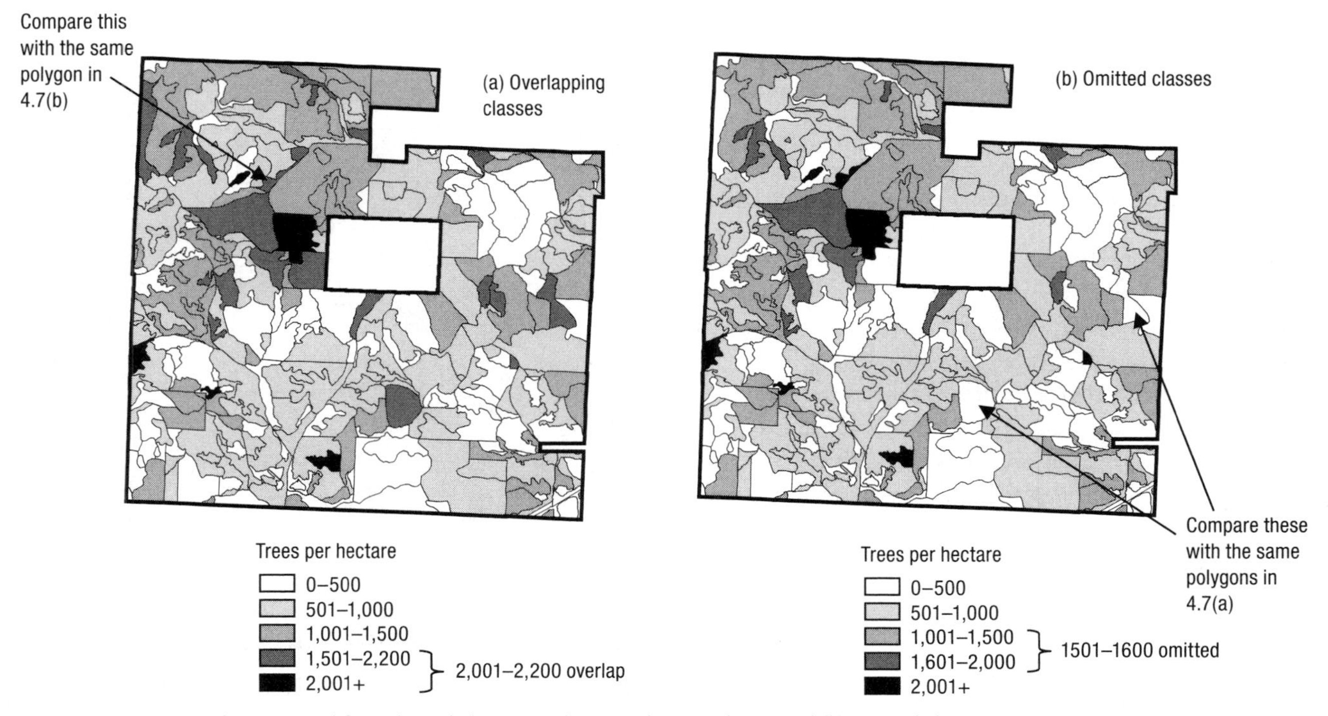

Figure 4.12 A range of criteria used for a choropleth map, with (a) overlapping classes, and (b) omitted classes.

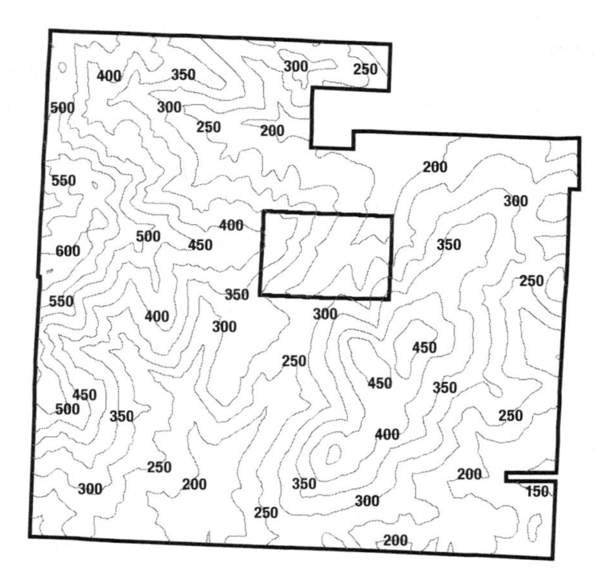

Figure 4.13 A contour map of the Brown Tract (elevation in meters above sea level).

Until now, the discussion has focused on maps created using vector GIS databases, but raster-based maps are also important for displaying data stored in the raster GIS data structure. These maps are very similar to the choropleth maps noted above: mapmakers must classify the values held by the raster pixels and then display them on a map using a legend. Raster-based maps, unless created by rasterizing a vector GIS database, usually have more of a fuzzy appearance than vector-based maps, and therefore generally represent more heterogeneity across a landscape (Figure 4.14).

Other types of maps

Dot density maps (Figure 4.15) were once commonly used in natural resource management, although less so now. Each dot in a dot density map represents a given value of an attribute. Thus areas with a greater density of dots are meant to represent areas with greater values of some particular attribute. Graduated circle maps place circles on top of landscape features and scale their diameter proportionate to one of the feature's attribute values (e.g., population of towns) to demonstrate differences among the landscape features. Cartograms (Figure 4.16) are another type of maps in which more than one attribute of landscape features can be viewed. These types of maps are also fairly uncommon because of the large amount of clutter that can be generated in a map with a large number of landscape areas, or in maps with a high density of landscape features within a given area (e.g., a multitude of small polygons).

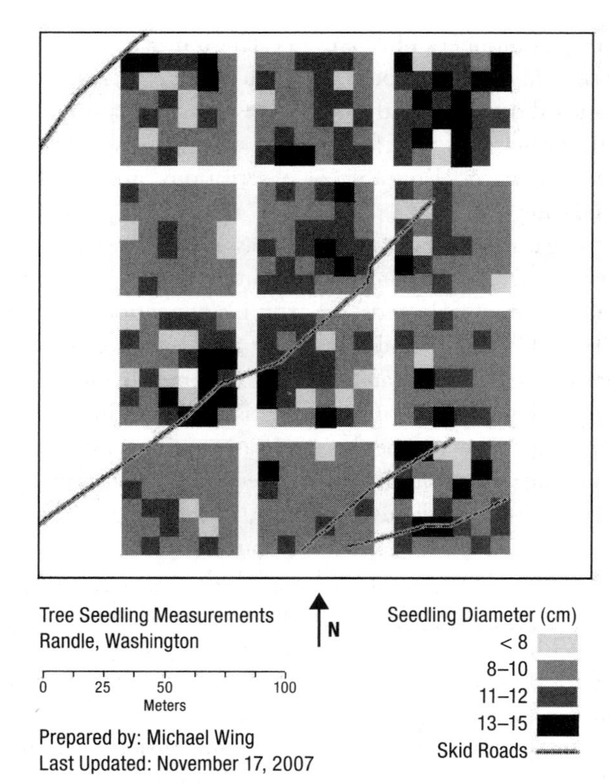

Figure 4.14 A raster map of tree seedling measurements.

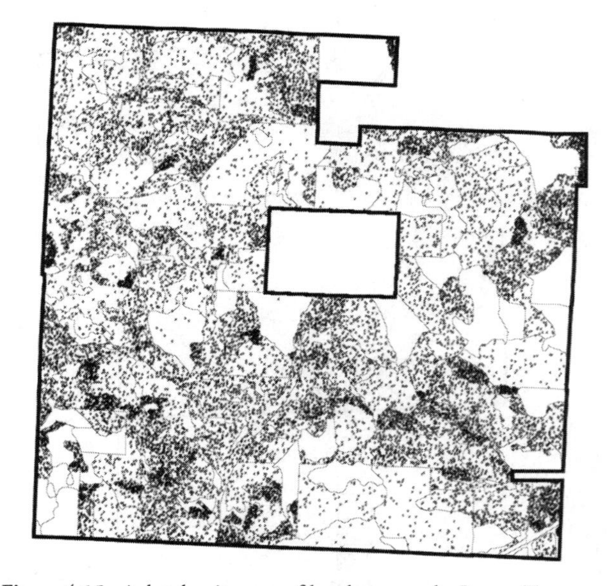

Figure 4.15 A dot density map of basal area on the Brown Tract.

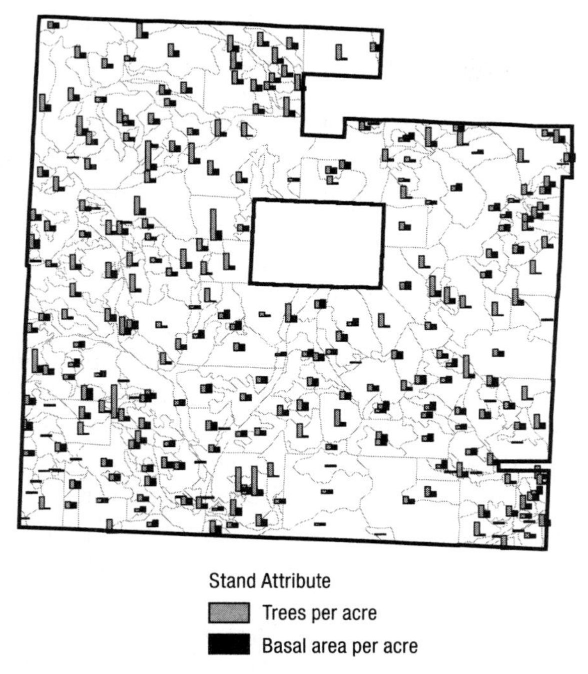

Figure 4.16 A cartogram map illustrating two measures of forest density for each stand—trees per acre and basal area per acre on the Brown Tract.

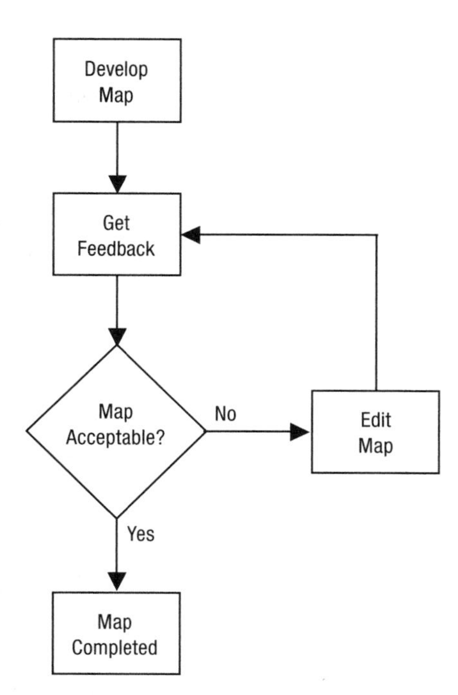

Figure 4.17 A basic design loop for making maps.

The Design Loop

Novice mapmakers, especially students, assume that their first attempt at a new map will be sufficient to effectively communicate information to their audience (or to complete an assignment). However, maps usually should go through more than one version before they are delivered to a customer, whether changes are needed based on the mapmaker's visual assessment of the map, or based on a customer's suggestions (Figure 4.17). Each iteration in the development of a map could be considered one iteration in the design loop. Feedback from supervisors and co-workers will allow you to fine-tune maps that are made for reports or management activity plans. Besides the aesthetic concerns presented in this chapter (map type, number of classes shown, etc.), mapmakers should strive to achieve visual balance within their map products. This concern is one of the reasons why maps may need to be edited numerous times. Visual balance is affected by the size of text (title, legend, and ancillary information) and the location of map components (north arrow, scale, and location inset) in relation to the visual center of the map. One key to identifying an unbalanced map is the presence of a large, empty space in some portion of the map.

The number of iterations of a design loop will be a function of your ability to address a range of visual concerns (visual contrasts, visual balance, legend, etc.), your ability to address a range of illustrative concerns (showing the correct information), and your time constraints (the time remaining before a deadline). It might be advisable to start the map-making process by first developing some hand-written notes that contain the main ideas about the intended map message(s) or purpose and the type of audience that is likely to view the map. An outline format with a primary objective and sub-headings that address other intended map purposes might be worth considering. Once these concepts have been identified, it might also be worthwhile to create a hand-drawn sketch of the basic map components and how they fit together. At this point, the development of the map within GIS can begin. A well-developed, visually-centered map will be a reflection of one's professionalism as a natural resource manager.

Common Map Problems

Mistakes (typographical errors), oversights (wrong color or symbology used), and omissions (missing information) can impair a map's ability to deliver the intended message to an audience. Probably the most important aspect of creating a map is to focus on the audience that will be viewing the map. If your co-workers (other foresters, biologists, etc.) will be the primary audience, then perhaps the basic map elements (e.g., a north arrow, bar scale, or locational inset) may not be as important to include; this audience may expect to see instead more of the technical information, such as annotation or other descriptive information, that is related to the landscape features illustrated. They may also expect to see a more extensive variation of feature classes across the landscape. However, if the audience is composed of the general public—those not familiar with the landscape resources displayed on a map—then it would seem important to include the basic map elements (e.g., title, credits, north arrow, scale bar, and inset) that help the audience orient themselves to the map. Less variation (fewer mapped classes of information) would reduce the confusion associated with the map. An audience composed of academics or scientists may want to see more detail in the data values.

Map problems are usually the result of leaving a key element, like a scale or another detail important to the audience, off of a map. Sometimes, these problems result from misjudging the audience, and other times it is simply a matter of oversight on the part of the mapmaker. For example, since automatic spell-checking processes are available in most word processing software, many mapmakers sometimes forget that spell checking is usually a manual operation within most GIS software. In addition, some place names (names associated with cities, regions, provinces, etc.) have unusual spellings and, if spelled incorrectly, may not be detected even if a spell-checker is used. This requires that mapmakers carefully read all map text before presenting the final product to their customers.

Excessive detail, or clutter, can also detract from a map's message and intent. Most GIS software programs now feature impressive arrays of map-making symbols and tools that are capable of producing any number of cartographic symbols and other aids. While many of these tools can be quite helpful—not to mention interesting to experiment with—too many objects on a map produce clutter and hinder a map's intended message. Excessive detail, especially in maps that are intended general information displays, can also result when mapmakers insert too much text (annotation or labels) onto a map.

It is also common that output devices (e.g., printer or plotter) produce maps with colors that appear slightly different from what is viewed on a computer screen. These problems can be very frustrating, especially after you have painstakingly put together a colored legend scheme that differentiates between different colors or values. The subsequent adjustments can be frustrating. Sometimes the easiest solution to a color mismatch problem is to simply choose colors from a palette file that is compliant with the output device. Other solutions include using only the additive primary (red, blue, green) or subtractive primary colors (magenta, cyan, yellow), or creating color schemes that take plotter translations of monitor colors closely into account.

Summary

Maps are used to convey information about the spatial location and characteristics of natural and man-made resources. When they are well developed, maps are an effective way of communicating ideas to an audience. This chapter described a variety of map components that can be used to help develop effective maps. These included legends, scales, annotation, symbology, and typography. The representation of these components on a map may be important to inevitable customers of the map. For example, there are certain components of maps that are required when developing maps for co-workers or others internal to a natural resource management organization, certain components that are required when developing maps for external clients (e.g., when submitting a harvest plan to a state or provincial agency), and certain components are required when developing a map for personal use. In addition, some organizations require that all maps produced by their employees contain an organizational logo, use a standard layout format, and use other features that reflect the organization and the map's intended purpose. There is no singular, correct format that fits all organizations, and you should balance your creativity with the advice provided in this chapter. You should, however, have an understanding of the map type options, and develop the appropriate map for the intended audience. Keep in mind that every map is potentially a reflection of your reputation as a natural resource professional.

Applications

4.1. Age class distribution map for the Brown Tract. The manager of the Brown Tract, Becky Blaylock, would like you to produce a map that illustrates the age class distribution of the forest. To complete this exercise, develop a thematic map showing 11 age classes: 0–10 years old, 11–20, ..., 91–100, and 100+ years old.

4.2. Tree density map for the Brown Tract. After reviewing your previous work, Becky Blaylock would like a map that illustrates the trees per hectare for vegetation stands contained within the Brown Tract; she needs the map for an annual report that she is developing. Develop a thematic map that classifies the stands by trees per hectare, using five logical classes.

4.3. Owl locations on the Daniel Pickett forest. The wildlife biologist associated with the Daniel Pickett forest needs a map that illustrates the historical spotted owl (*Strix occidentalis*) nest locations that were known to have been used in the forest. Develop a reference map illustrating the two nest locations, and annotate the map with the date of the last known sighting of the owls.

4.4. Stream types of the Daniel Pickett forest. The hydrologist associated with the Daniel Pickett forest needs a map illustrating the different stream types, in order to direct a summer crew to the locations he would like to survey for fish species and habitat conditions. Develop a reference map that illustrates the different stream types associated with the Daniel Pickett forest.

4.5. Potential harvest unit. The land manager of the Daniel Pickett forest is considering a timber sale in unit number 13 on the Daniel Pickett forest. He would like you to produce a management map indicating that unit 13 is a proposed harvest area, and to display the road and stream systems in relation to the unit.

4.6. Brown Tract hiking map. Becky Blaylock, manager of the Brown Tract, would like you to develop a map illustrating the trail system, highlighting both the authorized and unauthorized trails. Recreationists who visit the forest will likely use this map. Develop a reference map that includes the road system and the contour lines (with associated elevations) as supplementary information.

4.7. Culvert installation dates. The road engineer associated with the Daniel Pickett forest, Bob Packard, is in the process of developing a culvert replacement plan. He would like you to develop a map illustrating the road system, the culverts, and the culvert installation dates for the Daniel Pickett property.

4.8. Disclaimers, caveats, and warranties. Imagine that you work for an agency that has developed a statewide streams GIS database. At one of your regular staff meetings, the discussion shifts to this GIS database and the need to add or associate some sort of disclaimer, caveat, or warranty with the GIS database. During this conversation you conclude that the group is very confused about the use of the terms 'disclaimer', 'caveat', and 'warranty'. What guidance can you provide to help the staff understand the differences between the terms, and how the terms might be used in relation to the streams GIS database?

4.9. Map scales. One of your colleagues, Mike Marshall, does not like the type of scale that you commonly incorporate into your management maps. He prefers to use another type of scale, and insists that you use it as well. Identify, define, and describe the possible advantages and disadvantages of three types of approaches for creating map scales.

4.10. Map development. A small consulting firm in British Columbia has recently hired you, and one of your first assignments is to make management maps for various planned activities on the land managed by your firm. Identify five items or objects that should be placed on almost every map.

4.11. Map legends. The manager of the Brown Tract desires a map illustrating the trees per hectare for each stand on the property. There are different approaches for organizing and displaying spatial data into a map legend.

a) What is a general guideline for choosing the number of categories in a legend that uses gray tones?
b) What is an equal interval legend, and how does it display numeric data?
c) What is a quantile distribution legend, and how does it display numeric data?
d) What is a standard deviation legend, and how does it display numeric data?

References

Dent, B.D. (1999). *Cartography thematic map design.* New York: McGraw-Hill.

Digital Wisdom, Inc. (2006). *Cartographic symbols & map symbols library.* Tappahannock, VA: Digital Wisdom, Inc. Retrieved March 28, 2007, from http://www .map-symbol.com/sym_lib.htm.

Edwards, D. (2001). Maps for anglers. *The Cartographic Journal, 38*(1), 103–6.

Franklin, P. (2001). Maps for the reluctant. *The Cartographic Journal, 38*(1), 87–90.

Indiana Geological Survey. (2007). *Copyright, map disclaimer, and limitation of warranties and liabilities.* Bloomington, IN: Indiana Geological Survey. Retrieved March 3, 2007, from http://igs.indiana.edu/disclaimer.cfm.

International Orienteering Federation. (2000). *International specification for orienteering maps.* Radio-katu, Finland: International Orienteering Federation.

Kaya, N., & Epps, H.H. (2004). Relationship between color and emotion: A study of college students. *College Student Journal, 38*, 396–405.

Madej, J. (2001). *Cartographic design using ArcView GIS.* Albany, NY: OnWord Press.

Merriam-Webster. (2007). *Merriam-Webster online search.* Retrieved March 28, 2007, from http://www.m-w.com/cgi-bin/dictionary.

Monmonier, M.S. (1995). *Drawing the line: Tales of maps and cartocontroversy.* New York: Henry Holt and Company.

Monmonier, M.S. (1996). *How to lie with maps* (2nd ed.). Chicago, IL: University of Chicago Press.

Natural Resources Canada. (2006). *Topographic map symbols—introduction.* Ottawa, ON: Natural Resources Canada, Earth Sciences Sector, Mapping Services Branch. Retrieved March 23, 2007, from http://maps.nrcan.gc.ca/topo101/symbols_e.php.

Orange County (Florida) Property Appraiser. (2002). *OCPA map record inquiry system.* Retrieved March 29, 2007, from http://www.ocpafl.org/docs/disclaimer_map.html.

Phillips, R.J. (1979). Why is lower case better? Some data from a search task. *Applied Ergonomics, 10*, 211–14.

Phillips, R.J. (1989). Are maps different from other kinds of graphic information? *Cartographic Journal, 26*, 24–5.

Phillips, R.J., & Noyes, L. (1982). An investigation of visual clutter in the topographic base of a geological map. *Cartographic Journal, 19*, 122–32.

Phillips, R.J., Noyes, L., & Audley, R.J. (1977). The legibility of type on maps. *Ergonomics, 20*, 671–82.

Pima County (Arizona) Department of Transportation. (2003). *Department disclaimer and use restrictions.* Tucson, AZ: Pima County Department of Transportation, Geographic Information Services Division. Retrieved March 28, 2007, from http://www.dot.co.pima.az.us/mapdis.htm.

Robinson, A.H, Morrison, J.L., Muehrcke, P.C., Kimerling, A.J., & Guptill, S.C. (1995). *Elements of cartography.* New York: John Wiley & Sons, Inc.

Sheahan, B.T. (2004). *The unofficial Arc/Info and ArcView symbol page.* Victoria, BC: Spatial Solutions, Inc. Retrieved March 28, 2007, from http://www.mapsymbols.com/.

Star, J., & Estes, J. (1990). *Geographic information systems: An introduction.* Englewood Cliffs, NJ: Prentice-Hall, Inc.

Town of Blacksburg (Virginia). (2007). *Blacksburg WebGIS site.* Retrieved March 28, 2007, from http://arcims2.webgis.net/blacksburg/default.asp.

USDI Bureau of Land Management. (2001). *Map disclaimer.* Glennallen, AK: Bureau of Land Management. Retrieved March 28, 2007, from http://www.blm.gov/ak/gdo/documents/map_disclaimer.doc.

USDI US Geological Survey. (2003). *Part 6 publication symbols. Standards for 1:24,000- and 1:25,000-scale quadrangle maps.* Reston, VA: US Geological Survey. Retrieved March 23, 2007, from http://rockyweb.cr.usgs.gov/nmpstds/acrodocs/qmaps/6psym403.pdf.

USDI National Park Service. (2003). *NPS map symbols: Updated March 4, 2003.* Retrieved March 27, 2007, from http://www.nps.gov/carto/PDF/symbolsmap1.pdf.

Valdez, P., & Mehrabian, A. (1994). Effects of color on emotions. *Journal of Experimental Psychology: General, 123*, 394–409.

Wood, D. (2003). Cartography is dead (thank God!). *Cartographic Perspectives, 45*, 4–7.

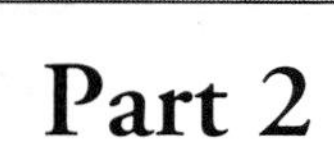

Applying GIS to Natural Resource Management

Part 2 of this book focuses on the GIS applications common to natural resource management organizations. The majority of data supporting daily GIS use in natural resource management is stored in vector GIS databases, although the integration of vector and raster GIS databases is becoming increasingly common. For example, the development of resource management plans may require the use of GIS databases representing forest stands (polygons), roads and streams (lines), water sources and wildlife observations (points), and topography (raster). The chapters included in this part of the book present a number of GIS processes such as querying, buffering, clipping, raster analysis, and simultaneous integration of vector and raster GIS databases in an analysis. The topics increase in complexity with each passing chapter, and the applications associated with each topic integrate processes presented in previous chapters. Some advanced topics are also presented, such as delineating the land classifications of a landscape and delineating the recreation opportunity spectrum classes of a landscape. In particular, the raster-oriented chapters provide examples of advanced analysis techniques and possibilities.

A variety of processing pathways can be used within GIS to address a single management concern and it is important to note that there may be more than one processing solution for a particular challenge. While we provide direction for the most straightforward processes, students should feel free to be creative in their approaches to addressing each application. Finally, some of the applications require students to think more broadly (beyond GIS) about how the results of an analysis may affect natural resource management opportunities within a given landscape.

Chapter 5

Selecting Landscape Features

Objectives

A variety of methods can be used to locate and select landscape features based on either their attributes or their proximity to other features. Spatial and attribute reference queries can be used to select features based on the information stored in attribute tables (layer database) or the spatial location of landscape features. At the conclusion of this chapter readers should be familiar with, and have a working understanding of:

1. the variety of methods that can be used to select landscape features from a GIS database;
2. the meaning of the term 'query', when applied spatially or referentially; and
3. the methods you can use to develop a description of the resources located on a landscape.

In a recent annual report to their stockholders, one of the largest timber companies in the United States said the following about the use of GIS: 'Our foresters use advanced Geographic Information System (GIS) models to get a picture of timber species, sizes, and age classes—along with a multitude of environmental details, from streams and fish to wildlife populations and habitat' (Plum Creek Timber Co., 2001). This statement recognizes the extent to which GIS is viewed as a valuable management tool and acknowledges the power of GIS to assist in the management of resources. Other natural resource management organizations also rely on GIS to store location information about the natural resources that they manage (as well as those that they might not manage but that are important in the management of their property). These organizations include public agencies such as the US Forest Service, the Canadian Forest Service, the US Park Service, the US Bureau of Land Management, state and provincial organizations, private industrial landowners, and those organizations that manage smaller tracts, such as forestry consultants, and university research forests. Each organization has a different mission, yet each relies on similar methods for storing and organizing their geographic databases. The methods used to extract information from these databases are also very similar, and help provide a picture of the natural resources that they manage. The chapter emphasizes one of the most common methods of data extraction from a database: the use of queries.

Besides asking questions like 'what is here?' or 'what type of feature is that?'—the types of questions which hold our attention in the first part of this chapter—there are at least four types of information acquisition processes you can use when asking questions of a GIS database. The four processes are outlined briefly below, and when we arrive at the section entitled 'Selecting features based on some database criteria', we will become engaged in one or more of these processes. The applications provided at the end of the chapter will further reinforce the need for these acquisition processes in natural resource management.

1. *Obtaining specific facts*
   Is there spotted owl habitat within the property that is being managed?
   Are there any steep hiking trails within the property being managed?

Is there any old-growth forest within the property being managed?

2. *Obtaining extended information*
   Besides spotted owl habitat, what other types of important wildlife habitat are found within the property being managed?
   Besides steep hiking trails, what other types of trails are contained within the property being managed?
   Besides old-growth forests, what other classifications of forest are contained within the property being managed?

3. *Obtaining broader information based on complex queries*
   Are there any areas that meet all of the habitat requirements for all of the species of interest within the property being managed?
   Are there any trails that, when combined, provide a relatively easy hiking experience?
   Are there any forest stands that can be managed to provide both wildlife habitat and timber resources?

4. *Obtaining information on resources in limited supply*
   Are there any areas of habitat that appear in small quantities across the area being managed?
   Are there any trails that provide experiences that occur rarely across the area being managed?
   Are there any forest stands that represent unique forest types within the area being managed?

Selecting Landscape Features from a GIS Database

As we noted in the introduction to chapter 1, natural resource managers are consistently called upon to describe the condition of a landscape using GIS software programs and GIS databases. Generally speaking, natural resource managers are concerned with understanding where landscape features are located and what characteristic(s) they might have now, have had in the past, or will have in the future. There are at least eight processes you can use to select landscape features from GIS databases:

1. Select one feature (manually).
2. Select many features (manually).
3. Select all features (manually or automatically).
4. Select no features (manually or automatically).
5. Select features based on some criteria.
6. Select features from a previously selected set of features.
7. Switching (inverting) a set of selected features so that all unselected items become selected.
8. Select features within some proximity of other features.

Selecting one feature manually

The ability to select landscape features using a computer mouse or digitizing puck is an essential tool for examining or editing individual landscape features in GIS, and for editing tabular data contained in an attribute table. As we have just implied, within GIS software programs landscape features can be selected either from the window that presents the spatial display of landscape features, or from the window that presents the associated attribute table. GIS software programs are either designed to allow users to select landscape features by default (e.g., MapInfo) or to provide a well-positioned, easily accessible function (i.e., the 'select features' tool in ArcMap or the 'select features' button in ArcView 3.3) that allows users to manually select individual landscape features. A careful positioning of the computer mouse's cursor over a landscape feature (either in the display of spatial landscape features or in the tabular database) and a simple click of the mouse will usually do the job. Selected landscape features will normally be colored or shaded differently (in both the spatial display of landscape features and tabular database) from other landscape features in the GIS database of interest, allowing users to visually verify what has been selected. GIS software programs use standard characteristics for displaying selected features, such as the light blue color used by ArcMap or the yellow color used by ArcView 3.3 for displaying a selected spatial feature or its attribute record. Some GIS software programs allow users to define the characteristics (i.e., color) of selected spatial features or attributes, and most GIS software will allow users to specify which GIS layers can be 'selectable'. As a result, either those layers that are visible in a window are selectable, or those that have been chosen from a list are selectable. This ability can make feature selections more efficient and also prevent analysis errors.

Selecting many features manually

There are times when selecting many landscape features manually will also be of value in assisting the development of information (map or data) to facilitate a natural resource management analysis. The process for selecting multiple landscape features is similar to the process for selecting single landscape features manually, with the computer mouse playing the central role. The selection of multiple landscape features can generally be accomplished in one of two ways:

1. Using the select tool associated with a GIS software program, draw a rectangular box in the window that displays the spatial features. Here, you would specify the location (by clicking and holding down a mouse button) for one corner of the box. Then, drag the mouse to specify the diagonal corner, and release the mouse button.
2. Using the select tool associated with a GIS software program, select features individually while depressing a key on the computer keyboard (e.g., the 'shift' key when using ArcMap). This process can be used in the window that displays the spatial features or in the attribute table.

Each GIS software program may involve a variation (or two) on these techniques. More than likely an alternative exists that allows users to more efficiently select multiple landscape features manually. For example, assume you were interested in selecting vegetation polygons that contained trees that were, on average, over 100 years of age. Within an attribute table you could sort the vegetation polygon records by their age, either using an ascending (youngest to oldest) or descending (oldest to youngest) sort. This will re-position the vegetation polygon records such that the ones that contain the oldest trees are grouped together, thus facilitating a more efficient manual selection of multiple features. Had you not sorted the records, you would have needed to scroll through the attribute table to locate records with average tree age values over 100; this likely would have lead to errors of omission (missed records) and it would not have been an efficient use of your time.

Selecting all of the features in a GIS database

The ability to select all landscape features from a GIS database is a standard process among GIS software programs. In addition, most GIS software programs generally have specific functions to allow users to select all landscape features with just one (or a few) click(s) of a mouse rather than having to select all the landscape features manually. This ability is useful when summarizing data about all of the landscape features in a database or when you are considering spatial analysis processes. Some users of GIS prefer to select the landscape features to which the GIS processes would be applied (all landscape features in this case), rather than rely on the common default (if no landscape features are selected, an analysis applies to all landscape features). Another opportunity to select all the features of a GIS database is if you are calculating values for fields in an attribute table. Calculations are usually performed on selected records in a tabular database, so, to perform a calculation on all of the records, such as the calculation of some portion of a habitat suitability score for vegetation polygons, you may need to select all of the records as a preliminary step in the process.

Selecting none of the features in a GIS database

The ability to select none of the landscape features in a GIS database (or to 'un-select' or 'clear' the selection) is also a standard process associated with GIS software programs. Most GIS software programs have specific functions—either menu items or buttons—that allow users to un-select all landscape features with just one click of a computer mouse (rather than having to un-select all landscape features manually). There are many reasons for wanting to perform this action; one of the most common reasons is to clear previously selected landscape features from a GIS database before performing a spatial process such as buffering. In a buffering process, when no spatial features are selected, generally all of the features are used to develop buffers. If one or more features are selected, only the selected features are buffered. It is usually after viewing the results of a spatial process that you realize you have forgotten to un-select landscape features before performing the operation.

Selecting features based on some database criteria

By now it may be evident that there must be a faster way to select a subset of the entire set of landscape features rather than having to select them manually. Within all GIS software programs, users have the ability to ask questions, that is, to develop queries, about the landscape features

In Depth

What is an **attribute**? It is defined as a characteristic or quality of an object, or in our case, a characteristic or quality of some natural resource. Within GIS, you are usually interested in a characteristic or quality of some feature found on the landscape, such as the vegetation, the soil, the water, or the land. More specifically, a forester might be interested in the basal area, timber volume, or habitat quality of the stands that are delineated on a property. An attribute of a landscape feature can be extended, however, to include its spatial position, size, perimeter (or length), and even the type of data it represents (e.g., point, line, polygon, raster grid cell).

contained in a GIS database. A **query** is simply a question, or set of questions, used to request information about some resource contained (or described) within a database. Imagine a co-worker asking, 'Please help me find all of the forested areas that might require a pre-commercial thinning treatment.' To locate the potential pre-commercial thinning areas, the request should be refined, then re-directed by asking questions about the information contained in one or more GIS databases, in this case, perhaps a forest stand GIS database. You might ask the GIS database 'Where are all of the young, overstocked, conifer stands?' This would seem to be a good start, yet a GIS software program would need more specific, quantitative instructions, detailing the attributes to use in the search for the appropriate landscape features, and detailing the bounds of the values of the attributes. For example, to find the young stands in a forest stands GIS database, you might search an 'age' attribute field for those forest stand polygons that have values between 10 and 20 (10 and 20 years old). The more refined query then becomes 'find all of the forest stands where age $\geq$ 10 and age $\leq$ 20.'

You might ask, 'Why would we need to perform queries?' One reason was illustrated with the need to identify the potential pre-commercial thinning areas—there are times when natural resource managers need to know where certain resources are located to help facilitate making management decisions regarding the resources. Queries can range from the rather simplistic (finding the most appropriate hunting area) to the rather complex (finding the most appropriate areas to commercially thin trees over the next two years). Suppose a management decision needed to be made regarding locating the most appropriate area to develop a new hiking trail. You might first locate and describe all of the characteristics of a landscape (current hiking trails, timber stands of various characteristics, etc.) that would influence the placement of a new trail. A query of these GIS databases could facilitate the selection of resources of interest across a landscape, and help a management team focus on the more suitable areas.

In developing GIS queries, you must build a set of criteria to enable a search of database, and subsequently to enable the selection of appropriate landscape features. For those more attuned to visualizing processes from a computer programming perspective, queries are similar to the development of *if statements*. As the attributes of each landscape are examined, *if* some *conditions* (criteria) of the landscape features are *true* (or conform to the criteria), the landscape feature will be placed into a 'selected' set. The following examples of queries relate to the data presented in Table 5.1. Answers are provided to allow students to work through the queries on their own, and to understand how (and why) the results were obtained.

TABLE 5.1 A timber stand database

Stand	Acres	Hectares	MBF[a]	Age	TPA[b]	TPH[c]
1	100	40.5	12	25	200	494
2	70	28.3	20	45	150	371
3	250	101.2	13	26	200	494
4	80	32.4	6	18	300	741
5	60	24.3	2	12	575	1,421
6	120	48.6	10	23	200	494
7	40	16.2	7	20	400	988
8	60	24.3	14	28	150	371
9	75	30.4	3	15	550	1,359
10	95	38.4	1	10	600	1,483

[a] Thousand board feet per acre
[b] Trees per acre
[c] Trees per hectare

Single criterion queries

Single criterion queries examine a single attribute (also called a variable, or field) of each landscape feature, use a single relational operator, and include a single threshold value.

a) How many timber stands are 20 years old?
   Attribute: age
   Relational operator: = (equals)
   Threshold value: 20
   Query: age = 20
   Answer: 1 stand (7)

b) How many timber stands are greater than or equal to 25 years old?
   Attribute: age
   Relational operator: ≥ (greater than or equal to)
   Threshold value: 25
   Query: age ≥ 25
   Answer: 4 stands (1,2,3,and 8)

c) How many timber stands are less than or equal to 20 years old?
   Attribute: age
   Relational operator: ≤ (less than or equal to)
   Threshold value: 20
   Query: age ≤ 20
   Answer: 5 stands (4,5,7,9, and 10)

d) How many timber stands contain at least 15 thousand board feet (MBF) per acre of timber volume?
   Attribute: MBF
   Relational operator: ≥ (greater than or equal to)
   Threshold value: 15
   Query: MBF ≥ 15
   Answer: 1 stand (2)

e) How many timber stands have more than 700 trees per hectare (TPH)?
   Attribute: TPH
   Relational operator: > (greater than)
   Threshold value: 700
   Query: TPH > 700
   Answer: 5 stands (4,5,7,9, and 10)

f) How many timber stands have at least 950 trees per hectare?
   Attribute: TPH
   Relational operator: ≥ (greater than or equal to)
   Threshold value: 950
   Query: TPH ≥ 950
   Answer: 4 stands (5,7,9, and 10)

g) How many timber stands are larger than 100 hectares in size?
   Attribute: hectares
   Relational operator: > (greater than)
   Threshold value: 100
   Query: hectares > 100
   Answer: 1 stand (3)

Multiple criteria queries

Multiple criteria queries are combinations of single criterion queries, held together by **logical operators** (*and*, *or*, *not*). They allow you to develop a complex query without having to perform several single criterion queries in sequence. Below are several multiple criteria queries that relate to the data found in Table 5.1.

a) How many timber stands are less than or equal to 20 years of age, and contain more than 950 trees per hectare (TPH)?
   Attributes: age, TPH
   Relational operators:
      Age: ≤ (less than or equal to)
      TPH: > (greater than)
   Threshold values:
      Age: 20
      TPH: 950
   Logical operator: and
   Query: (age ≤ 20) and (TPH > 950)
   Answer: 4 stands (5,7,9, and 10)

b) How many timber stands are at least 25 years old, or contain at least 10 thousand board feet (10 MBF) per acre of timber volume?
   Attributes: age, MBF
   Relational operators:
      Age: ≥ (greater than or equal to)
      MBF: ≥ (greater than or equal to)
   Threshold values:
      Age: 25
      MBF: 10
   Logical operator: or
   Query: (age ≥ 25) or (MBF ≥ 10)
   Answer: 4 stands (1,2,3, and 8)

c) How many timber stands are at least 20 years old, and are no older than 30 years old, and contain more than 500 trees per hectare?
   Attributes: age, age, TPH
   Relational operators:
      Age: ≥ (greater than or equal to)
      Age: ≤ (less than or equal to)
      TPH: > (greater than)
   Threshold values:
      Age: 20
      Age: 30
      TPH: 500
   Logical operators: and, and
      Query: (age ≥ 20) and (age ≤ 30) and (TPH > 500)
      Answer: 1 stand (7)

To illustrate the use of a complex query, we will ask a few questions regarding the polygons contained in the Brown Tract stands GIS database. First, assume that the managers of the Brown Tract are interested in managing the forest for timber production, and maximizing the growth potential of the trees in the forest. One way to achieve this goal may be to use precommercial thinning. As a result, they need to understand whether any potential commercial thinning opportunities exist. Assume that the criteria developed by the managers of the Brown Tract to assist in the analysis was based on four ideas:

1. Thinning should occur about 10 to 15 years prior to the final harvest age assumed by the organization (45–50 years).
2. Enough crop trees should remain un-cut in the thinned stands so that they (the residual trees) sufficiently respond (within increased growth rates) to the increased availability of light, water, and nutrients for the remaining 10–15 years prior to final harvest.
3. Commercial thinning will only be applied to even-aged forested stands.
4. Commercial thinning operations should remove, at a minimum, 10 MBF per hectare (about 4 MBF per acre).

Because the managers have specified a minimum residual volume level the timber volume per unit area prior to thinning should be substantially greater. The criteria for the query that the managers of the forest decide to use includes the age of the stands that could be thinned must be between 30 and 40 years old, the land allocation should include only the even-aged stands, and the timber volume prior to thinning must be above 9 MBF per acre. The criteria, placed within the structure of a query then becomes:

(age ≥ 30) and (age ≤ 40) and (MBF ≥ 9) and (land allocation = 'even-aged')

The resulting eight stands (42 hectares) on the Brown Tract that conform to this query are illustrated in Figure 5.1. These areas can be considered the potential commercial thinning opportunities for the forest in the near future.

Selecting features from a previously selected set of features

Rather than develop a long, complex query containing multiple criteria, you can design a set of less complex queries that are hierarchical in nature and that reduce the landscape features contained in the set of selected landscape features with each additional query. This process may help you stay organized and prevent the occurrence of mistakes that may be difficult to understand when using a long and complex query. To select landscape features from a previously selected set of landscape features, a number of single criterion queries are assembled.

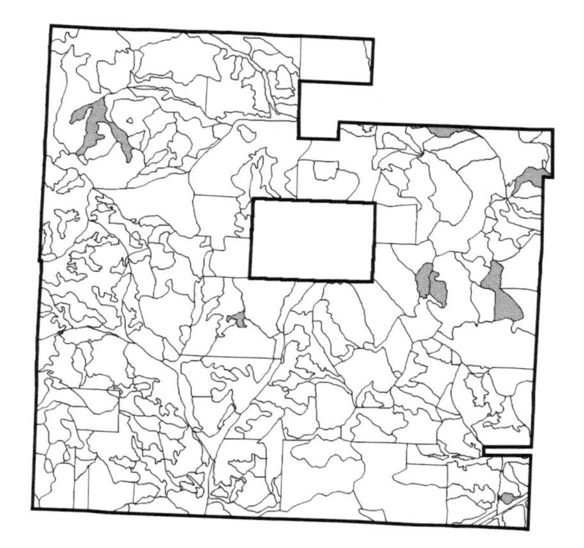

Figure 5.1 Stands on the Brown Tract that meet the following criteria: age ≥ 30 and age ≤ 40 and MBF ≥ 9 and land allocation = 'even-aged'.

Example c, presented earlier, involved the following multiple criteria query,

(age ≥ 20) and (age ≤ 30) and (TPH > 500)

which could be subdivided into three single criterion queries:

age ≥ 20
age ≤ 30
TPH > 500

Each of these can be performed in sequence; the first from the full set of stands GIS database landscape features,

age ≥ 20 (6 stands [1,2,3,6,7 and 8])

the second from the set of 6 landscape features that were selected (stands 1,2,3,6,7 and 8),

age ≤ 30 (5 stands [1,3,6,7 and 8])

and the third from the remaining 5 landscape features (stands 1,3,6,7 and 8),

TPH > 500 (1 stand [7])

resulting in the same landscape feature selected as when the multiple criteria query was used. The preference for a particular technique (selecting landscape features from a previously selected set or selecting landscape features using a multiple criteria query) will vary from user to user, depending on each user's confidence and experience.

If you were to try this hierarchical process of selecting landscape features on the Brown Tract thinning example from above, where the criteria was,

(age ≥ 30) and (age ≤ 40) and (MBF ≥ 9) and
(land allocation = 'even-aged')

you could subdivide the querying process into four steps.

Step 1: Select from the entire set of stands those stands where age ≥ 30 (result: 212 stands shown in Figure 5.2).
Step 2: Select from the 212 previously selected stands, those stands where age ≤ 40 (result: 23 stands shown in Figure 5.3).
Step 3: Select from the 23 previously selected stands, those stands where MBF ≥ 9 (result: 9 stands shown in Figure 5.4).
Step 4: Select from the 9 previously selected stands, those stands where the land allocation is even-aged (result: 8 stands shown in Figure 5.1).

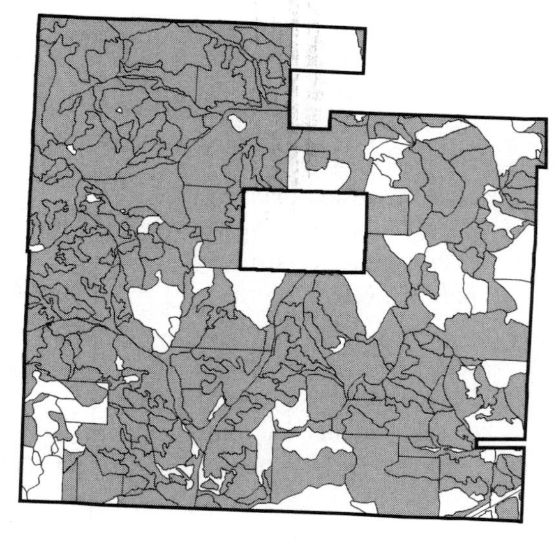

Figure 5.2 Stands on the Brown Tract that meet the following criterion: age ≥ 30.

Breaking down a complex query into smaller, single criterion queries may not work when the logical operator involved is 'or'. In the following example, the complex query cannot be broken down into three single criterion queries.

(age = 29) or (age = 30) and (TPH > 500)

The set of stands that might comprise TPH > 500 can be subdivided into those that are 29 years old. However, the resulting set cannot further be subdivided into stands that

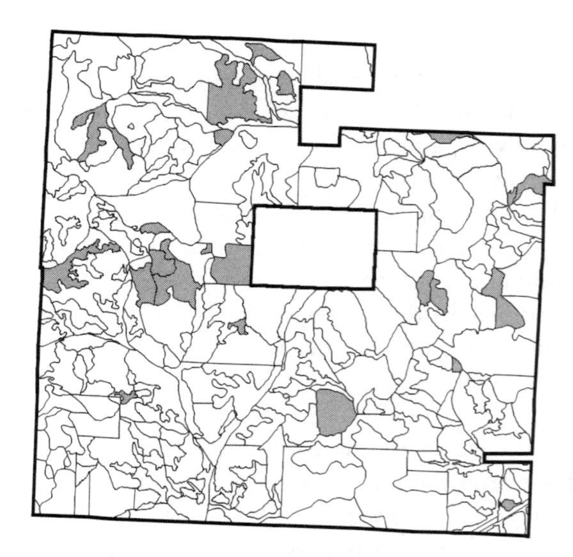

Figure 5.3 Stands from the previously selected set (age ≥ 30) on the Brown Tract that meet the following criterion: age ≤ 40.

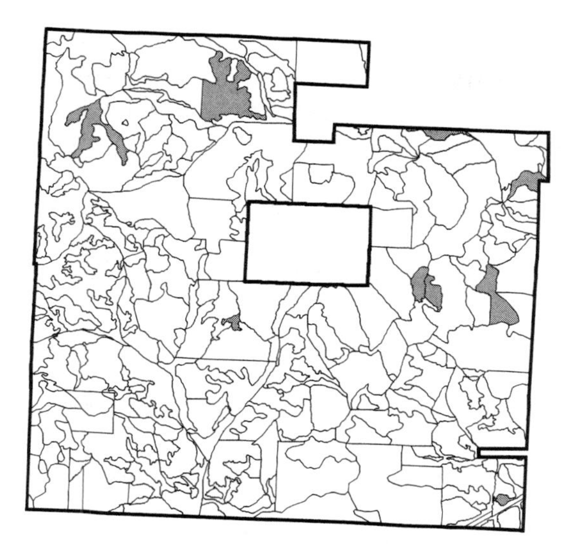

Figure 5.4 Stands from the previously selected set (age ≥ 30 and age ≤ 40) on the Brown Tract that meet the following criterion: MBF ≥ 9]

are 30 years old (they are 29 years old). Similarly, in the following example, the complex query cannot be broken down into three single criterion queries:

(age = 29) or (soil_type = 'PR') and (TPH > 500)

Again, we can locate the stands where TPH > 500, and from those we can locate the stands that have an age of 29 years. However, there may be many other stands beyond those in the resulting set that have TPH > 500 and a soil type of 'PR' (yet an age that is not 29 years).

However, the following multiple criteria query could be broken down into three single criterion queries:

(age > 28) or (age < 31) and (TPH > 500)

Here, the set of stands that might comprise TPH > 500 can be subdivided into those that are greater than 28 years old. The resulting set can further be subdivided into stands that are less than 31 years old.

Inverting a selection

Occasionally, you may find yourself in a situation where you need to understand two aspects of the spatial features contained in a GIS database: what is the condition (state or characteristic) of one set of features, and what is the

In Depth

One of the most common mistakes made when asking questions of databases is that results are often accepted as 'truth' without considering whether results are reasonable. For example, the Brown Tract timber stands GIS database contains a number of polygons that, when summed, describe a 2,123 hectare area. Within the Brown Tract a variety of ages of forests, ranging from recent clearcuts (age = 0) to older stands, are present. To describe the current structure of the Brown Tract, you could develop an age class distribution that indicates the area within, say, 10-year age classes. After performing queries of the various forest age classes, the sum of the area queried should not result in more or less than 2,123 hectares (the size of the property). You should always ask yourself whether the results obtained seem reasonable, given the resources being queried. Whenever possible, if a method of verifying results is available, it is advisable to check your work or have a colleague check your work. If multiple queries are performed that are designed to completely describe the resources of the forest, as in this case, the sum of area represented by the multiple queries should equal the sum of the resources in the original GIS database. If the sum of the area in the age classes is greater than the size of the Brown Tract, some areas were double-counted, perhaps using queries such as these,

Age class 1: (age ≥ 0) and (age ≤ 10)
Age class 2: (age ≥ 10) and (age ≤ 20)

where the area of 10-year-old stands is included in both classes. If the sum of the area in the age classes is less than the size of the Brown Tract, some areas were not counted, perhaps using queries such as these,

Age class 1: (age > 0) and (age < 10)
Age class 2: (age > 10) and (age < 20)

where the area of 0-year-old stands (clearcuts) is not included in age class 1, and the area of 10-year-old stands is not included in either age class.

condition of everything else. Two sets of queries can be developed to identify these two sets of features; however, if the second set contains 'everything else', simply inverting the selected features after the first query will produce the second set. For example, if you were interested in understanding how much land area was considered 'reserved' on the Brown Tract, and then how much land area remained 'un-reserved', you could first develop the query for the reserved areas,

(land allocation = 'meadow') or (land allocation = 'research') or (land allocation = 'oak woodland') or (land allocation = 'rock pit')

and find that it contains 42 stands covering about 229 hectares. By then inverting the selection, you will find that what remains is a set of 241 stands covering about 1,893 hectares. A second query for even-aged, uneven-aged, and shelterwood stands was not necessary.

The invert selection technique simply switches the GIS database selections, so that features previously selected are no longer selected, and vice versa. Some GIS software will make this capability available through a menu choice in the tabular database window while other programs may make this capability available through a menu or button in the spatial database viewing window. Some GIS software programs include both capabilities.

Example 1: Find the landscape features in one GIS database by using single and multiple criteria queries and by selecting features from a previously selected set of features

A combination of query processes can be used if you believe that they are necessary to accurately arrive at the desired set of GIS database features. In this example, we use a GIS database created from the set of GIS databases available for the Pheasant Hill planning area of the Qu'Appelle River Valley in central Saskatchewan. Here, we have created a GIS database that contains soils, topography, and land classification information. There are 168 polygons within the GIS database. Assume that we, as natural resource managers, are interested in understanding the land areas that contain clayey soils, have steep or undulating topography, and that have been categorized as having no significant limitation as they pertain to agricultural practices. Initially, we could develop a multiple criteria query to select from the larger set of features only those that have clayey soil types. There are a number of soil types in the Pheasant Hill planning area, thus the query would be designed something like this:

(Soil_type = 'Indian Head Clay') or (Soil_type = 'Indian Head Clay Loam') or (Soil_type = 'Indian Head Heavy Clay') or (Soil_type = 'Oxbow Clay Loam') or (Soil_type = 'Rocanville Clay Loam')

Given that a polygon is assigned only one soil type, we needed to use the relational operator 'or' in the query rather than 'and'. As a result of this multiple criteria query, we find that only 69 of the original 168 polygons have a clay component in their associated soil type. In order to locate those areas within this sub-set of the landscape features that are also located on undulating or steep slopes, we can perform a second multiple criteria query,

(Topography = 'STEEP') or
(Topography = 'UNDULATING')

and here only select features from the previously selected sub-set of landscape features (not from the larger, original set of landscape features). In this case, we find that 28 of the polygons have both the soil characteristics and topo-

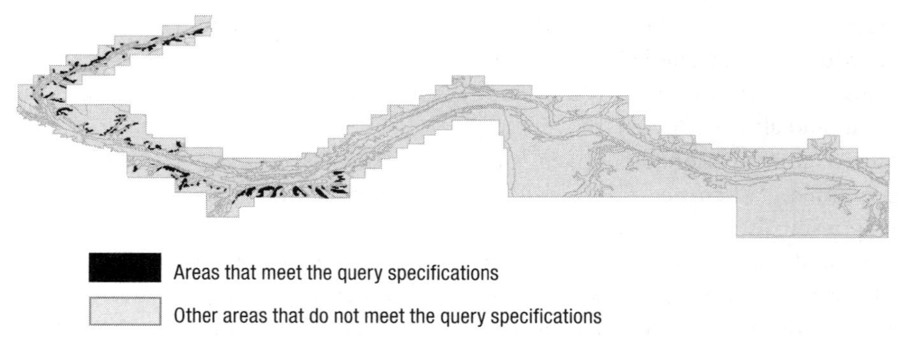

Figure 5.5 The result of a query for areas with clayey soils, located on steep or undulating topography, and with no limitations for agricultural practices using GIS databases developed for the Pheasant Hill planning area of the Qu'Appelle River Valley, Saskatchewan (1980).

graphic characteristics of interest to us. Finally, we perform a single criteria query to determine how many of the remaining 28 polygons also have no significant limitations for agricultural practices:

(Land Class = 'No Significant Limitations')

Again, this query is made by selecting from the previously selected set of 28 polygons.

As a result of this final query, we find that 18 of the original 168 polygons have soil, topographic, and land classification characteristics suitable for our original natural resource management analysis (Figure 5.5). We could have arrived at the same answer by performing one long, multiple criteria query. Alternatively, we could have arrived at the same answer by using single criteria queries to build up the selected soil types (adding to the selected set each time additional polygons that have soil type attributes of interest to us), then selecting from the previously selected set those that have the desired topographic and land classification attributes.

Selecting features within some proximity of other features

In addition to selecting landscape features based on the set of attributes available within the tabular portion of a GIS database, you can select landscape features based on their spatial relationship to other landscape features. This allows you to ask, for example, which landscape features are within a threshold distance of, adjacent to, or in close proximity of other landscape features. For example, you may want to know which research plots are in older forest stands, what forest stands are next to research areas, or which water sources are within a certain distance of a road. The ability to ask questions in spatial terms is but one indication of the power of GIS. The following three examples provide a description of three common forms of spatial queries.

Example 1: Find the landscape features in one GIS database that are inside landscape features (polygons) contained in another GIS database

In this example, a natural resource manager may be interested in examining two GIS databases: one has landscape features that he or she is interested in knowing something about; the other has landscape features that represent areas within which he or she is only concerned. Obviously the second GIS database suggests that it consists of polygons features, since line or point features do not describe areas. Alternatively, the first GIS database could contain point, line, or polygon features. In this example, assume that the first GIS database, the one containing features the manager wishes to know something about, contains points.

From a natural resource perspective, managers of the Brown Tract may be interested in understanding the habitat conditions within which certain wildlife species reside. In order to collect habitat information it may be necessary to locate and install forest inventory plots, and sample the characteristics of the forests. Permanent forest inventory plots, those that have already been installed and are periodically re-measured, can also be used for this purpose. From a review of the natural history of the wildlife species of interest, you may decide that only those research plots that are contained within older forest stands (those at least 100 years old) require measurement. Thus the problem becomes one of selecting the plots that reside within older stands. The GIS database that contains the landscape features of interest is the research plot GIS database, and the GIS database that represents the older forest areas is the forest stands GIS database.

In order to complete the spatial query, you would first select the older stands in the forest stands GIS database of the Brown Tract, using a single criterion query:

age $\geq$ 100.

The focus of the analysis is then shifted to the research plot GIS database, where the question is posed: how many plots are located within the selected landscape features of the stands GIS database (the older stands)? The entire spatial query process, in generic terms, can be described as this two-step process: (1) select the older stands from the forest stands GIS database using a single criterion query, and (2) develop a spatial query on the research plot GIS database where the selection is performed using the spatial location of selected landscape features from within the forest stands GIS database. This spatial selection ability may be described as 'selection by location' or some other similarly named menu choice or button, depending on the GIS software being used. Landscape features in the research plot GIS database are selected if they intersect the space covered by the selected landscape features in the stands GIS database.

The result of this spatial query process should yield 40 research plots that fall within the boundaries of older forest stands (Figure 5.6). Similarly, if you were interested in knowing how many research plots were located in young

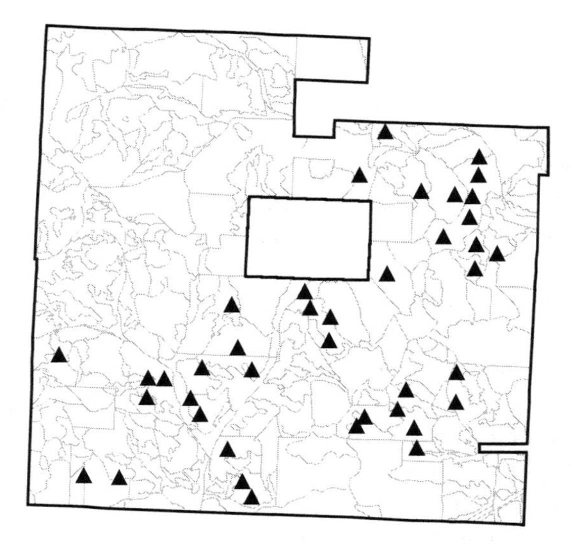

Figure 5.6 Permanent plot point locations within older stands on the Brown Tract.

stands, you would first query the stands database for young stands (perhaps age ≤ 30), then perform the spatial query similar to the process noted above. The result should yield 1 research plot.

Example 2: Find the landscape features in one GIS database that are close to the landscape features contained within another GIS database

In this example, the interest is again in examining two GIS databases: one has landscape features of interest; the other contains landscape features that represent those areas around which (not just within which) there is concern. The second GIS database suggests that it consists of polygon features, but here it could also consist of point or line features, since the area of concern is the area represented by a zone of proximity around landscape features. The first GIS database could also contain point, line, or polygon features. Assume that the GIS database of interest contains point features, and that the GIS database that will represent the area of interest contains line features.

The managers of the Brown Tract may be interested in developing a fire management plan for the forest, and thus would need to understand the types of water resources that are in close proximity to roads. Therefore, the problem becomes one of selecting the water sources that are within some distance of a road. The GIS database that contains the landscape features of interest is the water sources GIS database (because of the need to know where the appropriate water sources are located). The GIS database that represents landscape features around which one can define an area of concern is the roads GIS database.

In order to perform this spatial query, you must first determine the distance from the roads that is critical for meeting the needs of the fire management plan. Assume here that it is 30 meters, suggesting that water sources within 30 meters of a road may be of benefit to forest fire fighting efforts. This assumes that fire-fighting vehicles can draw water from these sources and transport the water to the fire area. In the development of the fire management plan, you may have also assumed that only certain types of roads can support fire-fighting vehicles, although this example will proceed under the assumption that all roads on the Brown Tract can support these vehicles. A generic description of the spatial query process might then include the following two steps: (1) select all of the landscape features in the water sources GIS database, and (2) develop a spatial query on the water sources GIS database where the selection is performed using the spatial location of landscape features contained in the roads GIS database. Landscape features in the water sources GIS database are selected if they are located within 30 meters of any road contained in the roads GIS database.

The result of this spatial query process yields 5 water sources that lie within 30 meters of a road (Figure 5.7). As you might imagine, this example, as well as the previous example, would also be helpful to those concerned with the proximity of certain resources (water sources, home sites) to potential management activities (herbicide or fertilization applications), or even to potential sites for wildlife or fisheries studies.

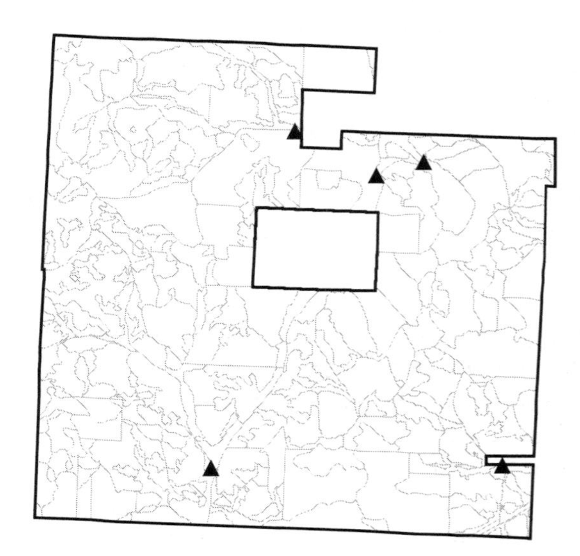

Figure 5.7 Water source point locations within 30 meters of roads on the Brown Tract.

Example 3: Find landscape features from one GIS database that are adjacent to other landscape features in the same GIS database

In this example, the interest is in performing a spatial query that uses landscape features within a single GIS database. Adjacency issues in natural resource management usually concern the placement of harvests or the location of habitat, and imply that activities may be prohibited from being implemented next to (or nearby) other recently implemented activities. In the case of habitat development, natural resource managers may desire to develop habitat next to (or nearby) other good wildlife habitat areas. Alternatively, an investment in research may need to be protected by limiting activity in nearby or surrounding areas. Since this example concerns adjacency issues, the GIS database used also suggests that it contains polygon features. This example will assume that a natural resource manager is interested in understanding the extent and number of stands that are adjacent to research areas. Since the Brown Tract is a working forest that contains some research areas, coordination of both research and harvesting activities is paramount, particularly if the harvesting activities affect a resource being studied in the research areas (for example, species of wildlife or hydrologic conditions under canopy).

A generic description of the spatial query process might then include the following steps: (1) select the stands in the stands GIS database that are designated a 'research' land allocation, and (2) develop a spatial query on the stands GIS database where the selection is performed based on how far away other stands are from the research areas. In this case, you can assume that the stands to be queried are 0 meters away from the research areas, and essentially touch a research polygon. Depending on the GIS software program used, the result of this spatial query process may yield 44 stands, including the research areas. To remove the research areas from this set of selected landscape features, you can perform a single criterion query from the previously selected set of landscape features, such as,

Land allocation <> 'Research'

where

attribute: land allocation
relational operator: <> (not equal to)
threshold value: research areas

By performing this single criterion query on the previously selected set of landscape features, you can select and identify just those stands adjacent to the research areas. Figure 5.8 illustrates the spatial location of the 37 stands that are adjacent to research areas on the Brown Tract.

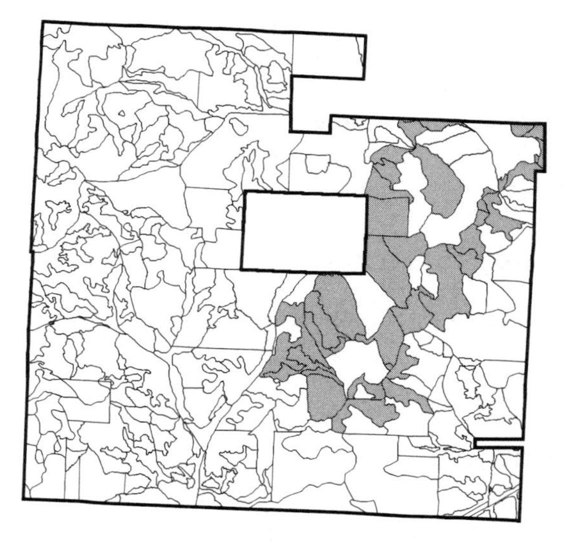

Figure 5.8 Stands adjacent to research areas on the Brown Tract.

In Depth

Structured Query Language, or SQL, is the most popular computer language for querying and manipulating data contained in relational databases. Sometimes simply called 'sequel', the language allows you to develop queries similar to those presented here to access data from large databases. Although IBM, Oracle, and Microsoft have led the recent developments of SQL, and many other organizations have tailored the SQL language for various applications, the American National Standards Institute (ANSI) and the International Organization for Standardization (ISO) have developed standard versions and offer them for sale. Some GIS software programs support the use of the SQL language, and extend it to the management and manipulation of spatial data features.

Advanced Query Applications

Advanced applications of GIS-related database queries have concentrated on limiting the focus of queries only to the features inside a spatial or temporal window defined by the user. In addition to simply providing a summary of the resources contained within a specific area, as the user-defined window (location or time frame) slides is expanded, or is contracted, the query is updated to reflect those features that have left the window and those that have entered the window. Queries can be completely re-evaluated as a window slides (or otherwise changes), or iteratively evaluated by updating the query by considering only the changes that have occurred (Edelsbrunner & Overmars, 1987; Ghanem et al., 2007). Dynamic queries can be designed that allow users to adjust questions asked of GIS databases by incorporating slider objects in a window rather than asking the user to redefine (by typing) the adjustments needed. For example, a graphical user interface can be designed to allow users to easily adjust the upper and lower bounds of a quantitative query along a scale, using a computer mouse (Domingue et al., 2003), and to provide those results quickly.

Syntax Errors

Syntax errors that occur when developing queries can discourage some users of GIS. Most of these problems occur because brackets or parentheses are missing from a query; this results in an incomplete query, such as in the two cases shown below from ArcMap queries.

[Height] >= 50) and ([Age] >= 25	(beginning and ending parentheses missing)
([Age] >= 25) and (<= 30)	(the attribute 'age' is missing from the second part of the criteria)

Most software programs require that every bracket or parentheses be balanced, so, if you begin with an opening bracket or parentheses, you must end with one as well. These parentheses and brackets can also be used to control the order of operations in queries that include mathematical operations. The dialog boxes that are generally used to help develop queries will assist with the placement of parentheses and brackets as long as the fields, operators, and values are selected in a logical manner using the computer mouse. Syntax errors usually arise when a user erases part of a query, then continues with its development (i.e., through keyboard intervention). Queries with syntax errors usually need manual keyboard intervention to correct the resulting problems. One of the easiest methods to alleviate syntax errors during queries may be to simply close the query dialog box and begin anew.

Summary

There are a variety of methods you can use to select landscape features from a GIS database, from using a mouse to manually pick the landscape features to selecting them with queries. The most common requests of natural resource managers that involve GIS queries include: (1) a map illustrating the landscape features that conform to some criteria, (2) a GIS database containing only the landscape features that conform to some criteria, and (3) summary statistics of the landscape features that conform to some criteria. Many of the applications described later in this book require you to develop a map showing some landscape features of interest, and to describe some characteristics of those landscape features. Queries will enable you to accomplish these tasks quickly and efficiently. An important habit to develop is to ask yourself whether the results of an analysis (here, a query) are reasonable. It has become very easy in modern GIS software to create spatial and attribute reference queries, and to instantly receive the results. This ease of use may lead you to believe that the results should not be questioned. Most GIS programs are designed to perform complex queries provided that the query syntax is correct. However, they simply provide you with results based on your instructions, and do not have the ability to question whether a query was correctly designed. It is therefore important to be critical of the output from a query, and to examine whether the results are within the bounds of reason. This will help you to detect errors either in the GIS databases being analyzed or in the methods being used to perform an analysis.

Applications

5.1. Daniel Pickett Forest Annual Report. For the Annual Report of the Daniel Pickett forest, you have been asked by Hugh Davenport (District Forester) to provide some information related to the forest's resources. Mr Davenport poses his request as a series of questions:

From the stands GIS database:

a) How much area of land contains forests ≤ 20 years of age?
b) How much area of land contains forests > 20 years of age and ≤ 40 years of age?
c) How much area of land contains forests > 40 years of age?
d) How much area of land contains vegetation type A?
e) How much area of land contains vegetation type B?
f) How much area of land contains vegetation type C?
g) How much area of land contains average timber volumes ≥ 49.4 MBF (thousand board feet) per hectare (20 MBF per acre)?
h) How much area of land contains average timber volumes ≥ 74.1 MBF per hectare (30 MBF per acre)?
i) How much area of land contains average timber volumes ≥ 98.8 MBF per hectare (40 MBF per acre)?

From the soils GIS database:

a) How much area of land might have a high response to fertilization?
b) How much area of land might have a medium response to fertilization?
c) How much area of land might have a low response to fertilization?

From the streams GIS database (map units are feet):

a) How many *miles or kilometers* of Class 1 streams are in the database?
b) How many *miles or kilometers* of Class 2 streams are in the database?
c) How many *miles or kilometers* of Class 3 streams are in the database?
d) How many *miles or kilometers* of Class 4 streams are in the database?
e) Why might these values be misleading, and what caveat might you provide to Mr Davenport?

Graphics for the report:

a) Develop a histogram of 10-year stand age classes, showing the amount of land in each age class.

5.2. Information for a new supervisor of Brown Tract. A new supervisor (Sharon Gillman) was recently selected to manage the Brown Tract, and she wants to get familiar with the natural resources located there. She has asked you to provide some information about the Tract and its resources:

From the streams GIS database:

a) How many *miles or kilometers* of fish-bearing, large streams are in the database?
b) How many *miles or kilometers* of fish-bearing, medium streams are in the database?
c) How many *miles or kilometers* of fish-bearing, small streams are in the database?
d) How many *miles or kilometers* of non-fish bearing, large streams are in the database?
e) How many *miles or kilometers* of non-fish bearing, medium streams are in the database?
f) How many *miles or kilometers* of non-fish bearing, small streams are in the database?

From the roads GIS database:

a) How many *miles or kilometers* of road on (or near) the Brown Tract are rock roads?
b) How many *miles or kilometers* of road on (or near) the Brown Tract are dirt roads?

From the soils GIS database:

a) How much area of land of 'PR' soil type is there on the Brown Tract?
b) How much area of land of 'DN' soil type is there on the Brown Tract?
c) How much area of land of 'WL' soil type is there on the Brown Tract?

From the water sources GIS database:

a) How many beaver pond water sources are on the Brown Tract?
b) How many hydrant sources are on (or near) the Brown Tract?
c) How many water tank sources are on (or near) the Brown Tract?

From the trails GIS database:

a) How many *miles or kilometers* of authorized trails are there on the Brown Tract?
b) How many *miles or kilometers* of unauthorized trails are there on the Brown Tract?

c) How many *miles or kilometers* of proposed trails are there on the Brown Tract?

5.3. Brown Tract Annual Report. In addition to the requests by Sharon Gillman to get her acquainted with some of the resources located within and around the Brown Tract, she asks you to supply the following information for the Annual Report:

a) How much area of land is in even-aged forests?
b) How much area of land is in uneven-aged forests?
c) How much area of land is assigned to the 'research' category?
d) How much area of land contains board foot volumes ≥ 74.1 MBF per hectare (30 MBF per acre)?
e) How much area of land has a density of trees ≥ 988 per hectare (400 per acre)?

5.4. Annual operating plan, Brown Tract. In order to develop a budget for management activities next year, the staff of the Brown Tract needs information regarding of the amount of land area that can be treated with various silvicultural treatments. Through conversations with them, the criteria have been narrowed down to the following:

Pre-commercial thinning candidate stands:

a) How much area of land contains even-aged stands that are ≤ 20 years old and have ≥ 988 trees per hectare (400 trees per acre)?

Commercial thinning candidate stands:

a) How much area of land contains even-aged stands that are ≥ 30 and ≤ 40 years old?
b) How much area of land contains even-aged stands that are ≥ 30 and ≤ 40 years old, with ≥ 494 trees per hectare (200 trees per acre)?

Final harvest candidate stands:

a) How much area of land contains even-aged stands that are ≥ 45 years, ≤ 100 years, and have a board foot volume ≥ 74.1 MBF per hectare (30 MBF per acre)?

5.5. Proposed recreation area. You received an e-mail a few days ago from Erica Douglas, forester, which read: 'We are considering developing a campground or trail system on the Brown Tract. Are there any stands or groups of stands ≥ 150 years old and ≥ 25 hectares (61.8 acres) in size?' What will your response be?

5.6. Wildlife habitat. A note was placed on your computer's keyboard, from Will Edwards, the Brown Tract wildlife biologist, which read: 'Could you perform some queries of the Brown Tract stands database for me? I am working on some wildlife habitat suitability models, and am interested in the following:

a) Sharp-shinned hawk habitat: area of land containing ≥ 25 and ≤ 50 year old stands,
b) Cooper's hawk habitat: area of land containing ≥ 30 and ≤ 70 year old stands,
c) Goshawk habitat: area of land containing ≥ 150 year old stands, and
d) Red tree vole habitat: area of land containing ≥ 195 year old stands.

Also, could you make me a map of the sharp-shinned hawk habitat?'

5.7. Fire management plan. In preparation for the development of a fire management plan the district manager of the Brown Tract is interested in knowing the following:

a) How many water sources (all types) are within 30 meters (98.4 feet) of rocked or paved roads?
b) How many pond water sources are within 30 meters (98.4 feet) of rocked or paved roads?

5.8. Research plots. One of the research scientists associated with the Brown Tract is interested in measuring a few of the research plots on the forest, to study the growth of certain stand types. They want to understand the following:

a) How many research plots are located in research plots on stands with ages ranging from 30 to 50 years?
b) If the query were expanded to include stands with ages ranging from 25 to 50 years, how many research plots would this query contain?

5.9. Potential Fertilization Project. The managers of the Brown Tract are considering fertilizing all stands that are aged between 25 and 40 years old. The managers are not only concerned about the proximity of fertilization operations to the stream system, but also to water sources. How many water sources are within 60 meters (196.9 feet) of the stands that could potentially be fertilized?

References

Domingue, J., Stutt, A., Martins, M., Tan, J., Petursson, H., & Motta, E. (2003). Supporting online shopping through a combination of ontologies and interface metaphors. *International Journal of Human-Computer Studies, 59*, 699–723.

Edelsbrunner, H., & Overmars, M.H. (1987). Zooming by repeated range detection. *Information Processing Letters, 24*, 413–17.

Ghanem, T.M., Hammad, M.A., Mokbel, M.F., Aref, W.G., & Elmagarmid, A.K. (2007). Incremental evaluation of sliding-window queries over data streams. *IEEE Transactions on Knowledge and Data Engineering, 19*, 57–72.

Plum Creek Timber Company. (2001). *Plum Creek annual report 2000*. Seattle, WA: Plum Creek Timber Company.

Chapter 6

Obtaining Information about a Specific Geographic Region

Objectives

This chapter is designed to provide readers with examples and applications from natural resource management that will allow you to analyze the resources contained within specific geographic regions. Specific geographic regions can be defined in a number of ways, for example, (a) by using query processes (the subject of chapter 5), (b) by using buffer processes (the subject of chapter 7), or (c) through GIS overlay processes (the subject of chapter 11). From a land management perspective, there is a wide range of reasons why we would want to understand what is contained within these regions, and we will illustrate a few of these. At the conclusion of this chapter, readers should have acquired knowledge of:

1. how a clipping process works, and what products should be expected when it is used;
2. how an erasing process works, and what products should be expected when it is used; and
3. how to use both clipping and erasing processes to obtain information about specific geographic regions, and obtain information that is relevant to natural resource management planning.

As natural resource managers, we are often interested in understanding the characteristics of the land resources we manage within, or perhaps outside of, specific geographic areas. For example, if you were to manage riparian areas, where limited amounts of activity can be prescribed, you might be interested in the type and quantity of resources within these areas, as well as the type and quantity of resources outside these areas—subject to a wider range of management. Understanding the soil conditions within a property is another example, since many soils GIS databases are acquired from governmental organizations, where the coverage of data extends well beyond the boundary of the land you might manage. The two main GIS processes that can be used to obtain information about a specific geographic region, and hence the focus of this chapter, are the **clipping** and **erasing** processes.

To some people, the term 'clipping' conjures up thoughts of American football athletes using an illegal blocking maneuver on their opponents; to other people it conjures up thoughts of snipping pieces of vegetation from a plant. It is also used invariably as both a noun and a verb, such as (a) a way to hold something in a tight grip, (b) a device to hold cartridges for a rifle, (c) a single instance or occasion, or (d) the act of cutting (Merriam-Webster, 2007). In a GIS context, a clipping process implies something similar to cutting cookies from a sheet of dough, when baking holiday-related cookies, although nothing is actually baked here. Imagine a landscape rolled out flat, like a map on a table. If you were to cut out a portion of the landscape with a pair of scissors, you would have essentially 'clipped' it from the landscape. There are a number of reasons why you would do such a thing, and we will explore some of the more common applications in this chapter.

The term 'erase' also has several meanings, and is mainly used a verb in the English language, in manners such as (a) to rub out or scrape away, (b) to remove written marks, (c) to remove recorded data from a magnetic medium, and (d) to nullify an effect (Merriam-Webster, 2007). When the term 'erasing' is used in a GIS context, it is much more closely aligned with the notion most people have about rubbing or scraping things away: some features are being removed (erased) from the landscape. Imagine two GIS databases, a forest stand GIS database and a stream buffer GIS database. If you wanted to visualize all of the forest stands areas outside of the stream buffers, you could use the polygons that describe the stream buffer area to erase all of those areas from the timber stand GIS database. As you can see you can use clipping and erasing tools to obtain resource information about specific geographic regions.

The Process of Clipping Landscape Features

One of the assumptions behind the use of a clipping process is that you are interested in creating a new GIS database that contains only those features within a specific geographic region. A clipping process involves the use of two GIS databases (Figure 6.1), and results in one new output GIS database. The process involves the location of the intersection of lines (in the case of line and polygon features being clipped) and the location of features wholly contained within an area (in the case of all types of GIS features). The location of line intersection points is an essential part of GIS (Clarke, 1995). Clipping processes can be manually called upon within GIS, and in some cases are automatic and transparent to GIS users, as in the case of websites that are designed to allow users to specify an area within which data will be extracted.

When using vector GIS databases, one of the input GIS databases needs to contain polygon features (the cookie cutter); the other (the GIS database to be clipped) can contain either point, line, or polygon features. The cookie-cutter GIS database is overlaid on the GIS database to be clipped, and only those features within the boundaries of the polygon(s) in the cookie-cutter GIS database are retained in the output GIS database. Thus, the output GIS database contains the same type of spatial features as the GIS database being clipped. In addition, the size of features (lines or polygons, but not points) may be spatially altered in the output database, as they may have been cut at the edges of the polygons contained in the

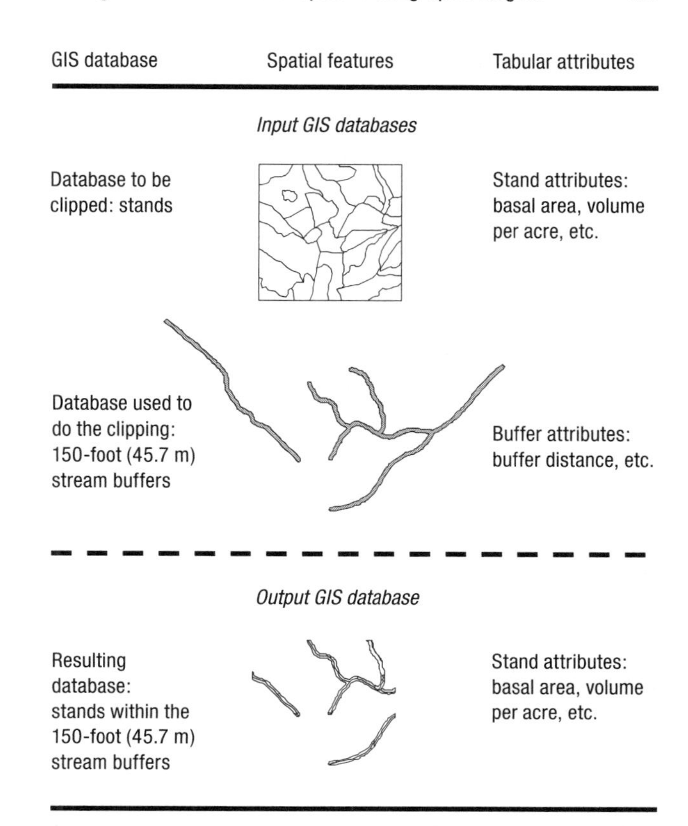

Figure 6.1 Clipping the stands within the 150-foot (45.7 m) stream buffers on the Daniel Pickett forest.

cookie-cutter GIS database. This implies that lines could be shortened and the shape of polygons altered. The output GIS database contains all of the attributes of the GIS database that was clipped, but generally none of the attributes of the GIS database that was used to do the clipping. The spatial extent of the output GIS database is limited to the boundary of the polygons contained in the cookie-cutter GIS database. For example, in Figure 6.2, a road and a fire area overlap the boundary of a land ownership polygon. The land ownership polygon can be used as a cookie-cutter to clip the portions of the road and the fire area that actually reside within the ownership boundary. In doing this, the road (initially described by line 2), is clipped at the intersection with lines 1_1 and 1_4 of the property boundary polygon (creating line 2_1), and the fire area is clipped with lines 1_3 and 1_4. Line 3_1 of the fire area is shortened, original line 3_2 discarded, and lines 3_3 and 3_4 created based on the location of lines 1_3 and 1_4 of the property ownership boundary.

Clipping a GIS database with a very large extent (such as a national soils database) to the boundary of a managed property is one common application of this type of

a. GIS databases prior to a clip process.

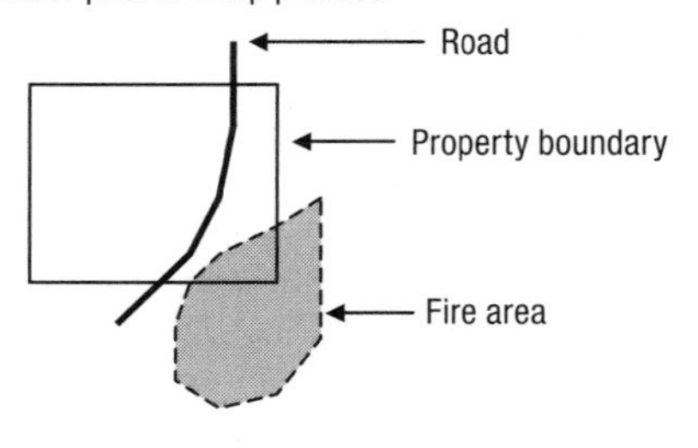

b. GIS databases during a clip process. (o = node)

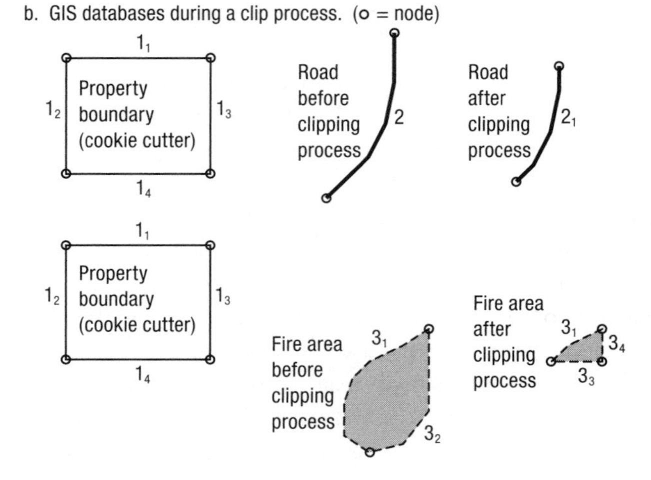

c. GIS databases after a clip process. (o = node)

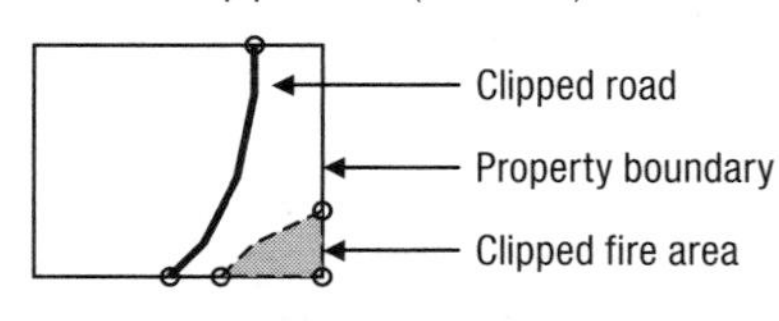

Figure 6.2 Clipping a road and a fire area to the property boundary of a land ownership.

In Depth

A clipping process is essentially the same as making cookies from a sheet of cookie dough. It is called 'erasing outside' by some GIS software programs, which is simply another way to describe how it works: a map of a landscape is laid flat on a table, a solid object is placed on top of it (perhaps a book), and everything that is visible outside of the boundary of the book (except what is under the book) is erased. The process of 'erasing outside' (or clipping) is the inverse of another process we describe shortly, called the 'erase' process. We expect some level of confusion due to the similar terminology, but are confident that with study and hands-on practice with GIS, readers will be able to grasp the differences.

In an erasing process, you seek to remove from one GIS database everything that is spatially located under the features contained in another GIS database (which contains polygons). Using the example of the map and the book that lies upon it, everything on the map that is under the book would be removed in an erasing process. Thus an erasing process is the inverse of a clipping (or erase outside) process. The example provided using the municipality boundaries and floodplain, wetland, and treed areas GIS databases illustrates how a single GIS database (the municipalities) can be divided into two completely separate, non-overlapping databases using the clipping and erasing processes (Figure 6.3).

Municipality boundaries in the Pheasant Hill planning area of the Qu'Appelle River Valley in central Saskatchewan

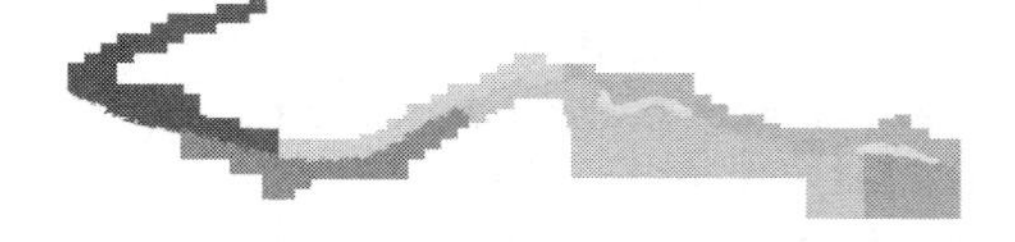

Floodplains, wetlands, and areas with trees in the Pheasant Hill planning area

Floodplains, wetlands, and areas with trees for each municipality (municipalities clipped using floodplains, wetlands, and treed areas)

Areas that are not floodplains, wetlands, nor areas with trees for each municipality (municipalities erased using floodplains, wetlands, and treed areas)

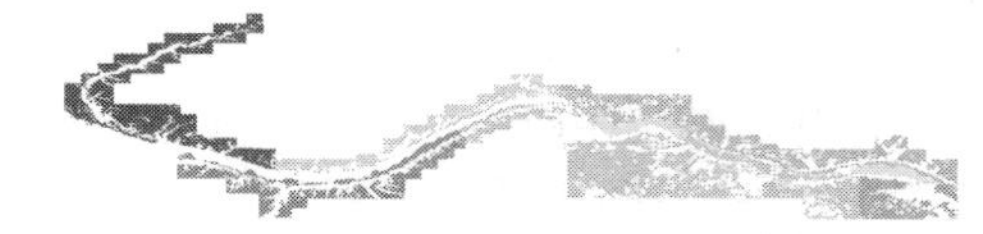

Figure 6.3 Application of clipping and erasing processes.

process. There are a number of other reasons why you would want to clip a set of spatial features to the boundary of a property. One of them relates to the accuracy and consistency of an organization's GIS databases. GIS databases can be digitized in-house or by contractors, created through other spatial operations, developed with GPS technology, obtained from organizations that sell databases, downloaded for free off of the Internet, or simply passed from one person to another. Given the wide variety of ways organizations can acquire GIS databases, it is not unreasonable to imagine that the extent of the coverage of the GIS data will likely not perfectly fit the extent of an ownership's boundary. Some organizations require that the extent of each GIS database fit perfectly with the boundaries of their land ownership, and they ensure this by clipping each GIS database to their ownership boundary GIS database. Granted, there are some GIS databases, such as roads and streams, that you may not want clipped to an ownership boundary (interest may center on where the roads and streams come from and where they go beyond a property boundary), but there are GIS databases, such as soils and land cover, where the argument may hold (interest may not center on the soils of other land owners).

Other types of polygon features could be used in a clipping process. For example, you may be interested in the resources contained within a riparian zone, or within a watershed. Further, when natural resource management organizations share GIS databases, they may decide to limit what is shared. For example, in Washington State, where watershed analysis has been an important aspect of forest management, a coordinated effort among organizations leads to the identification of the limits of appropriate management activities within watersheds. In many watershed analyses, a single organization will perform the GIS analysis tasks, and thus acquire all of the GIS databases related to a particular watershed regardless of landowner. Usually private natural resource organizations are very hesitant to share their GIS databases with their competitors—some GIS databases are considered proprietary and may contain sensitive information. In addition, organizations may be hesitant to release GIS databases that are dated or that may contain unverified information. However, since watershed analysis may benefit all landowners in the long run, some GIS databases are usually shared among land management organizations. In cases such as these, organizations may decide to share as little GIS data as possible to meet the goals of the watershed analysis, and thus avoid revealing any other information regarding the status of the natural resources that they manage. Therefore, a clipping process is used to limit the amount of information shared to that concerning specific geographic regions, such as individual watersheds.

Obtaining information about vegetation resources within riparian zones

As a first example of using a clipping process, let's assume that we are interested in the vegetation resources contained entirely within some pre-defined riparian zones within a forest. In chapter 7, we will describe the process of taking spatial features, such as streams, and creating physical zones (buffers) around them. For now, let's assume that a GIS database containing polygons that describe 50-meter riparian zones (stream buffers) for the Brown Tract already exists (Figure 6.4). The type of information we may be interested in knowing includes the amount of land and the volume of timber within the riparian zones, which are generally limited-use management areas. A clipping process would allow us to develop a GIS database containing only those timber stand areas that are within the boundaries of the polygons that describe the riparian zones (Figure 6.4). The original timber stand polygons that intersect the stream buffer polygons would be redesigned such that their boundaries now coincide with the edges of the riparian zones. Those timber stand polygon boundaries that fall entirely within the riparian zones would be left intact, those timber stand polygons that fall entirely outside of the riparian zones would be eliminated.

What you should find in the tabular database (attributes) of the clipped GIS database is the same set (number and type) of attributes that can be found in the original vegetation GIS database. However, the values of the attributes associated with 'area' (acres and hectares) may need to be adjusted to reflect only the area of the polygons within the clipped GIS database. In addition, some polygon GIS databases may also contain a perimeter measurement that describes the linear distance around each polygon border. The perimeter distance may also need to be recalculated in order to accurately represent perimeters that were affected by a clipping operation. Some GIS software programs perform these adjustments automatically depending on the type of spatial database; other GIS software programs require users to recalculate the areas with a second (albeit automated) process. Attributes that describe something other than an 'area' of land (e.g., tree

(a) 50-meter stream buffers.

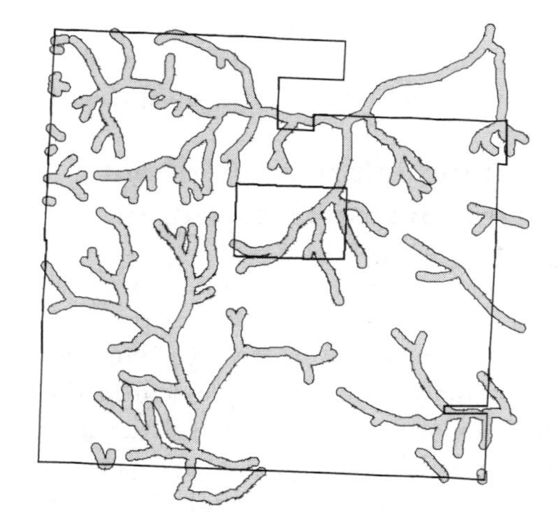

(b) Vegetation polygons.

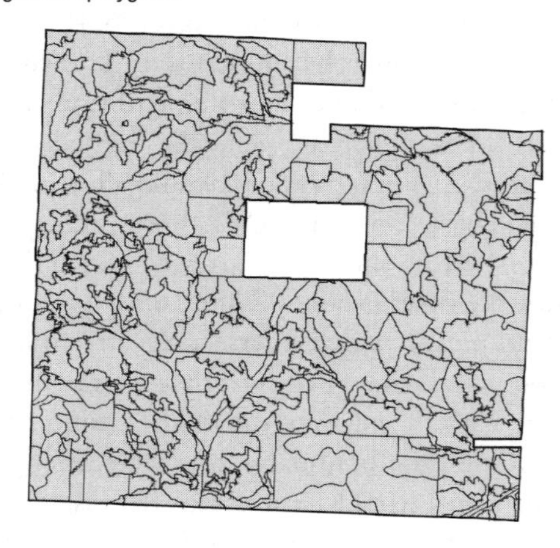

(c) Vegetation polygons within 50-meter stream buffers.

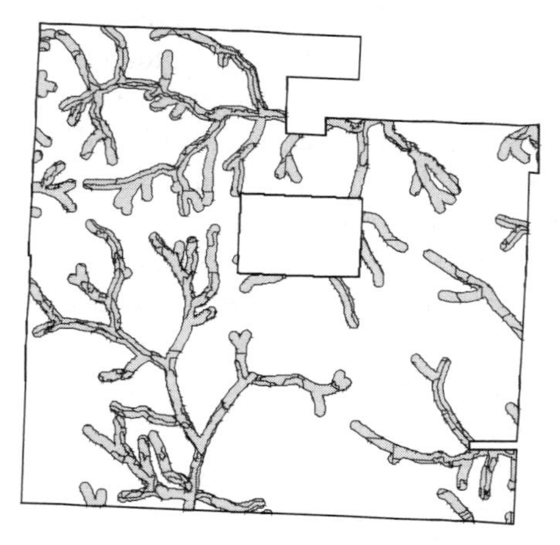

Figure 6.4 Clipping the stands within 50-meter stream buffers on the Brown Tract.

age or stand volume), are not adjusted during the clipping process. With the use of a clipping process, you should be able to understand how much area is contained within the riparian zones, 446 ha in the case of the Brown Tract (Table 6.1). The resulting clipped GIS database also provides information necessary for subsequent analyses, such as those that involve understanding the average age of the riparian zone vegetation, or the timber volumes contained within the riparian zones.

Obtaining information about soil resources within an ownership

Soil resources for North America have been mapped at various scales for each Canadian province and US state. While special soils surveys have been conducted for individual landowners, the most widely used soils databases were developed by governmental agencies. For example, in the United States, the USDA Natural Resources Conservation Service (2007) provides an online interactive process that allows you to acquire soils (and other) GIS databases for a specific portion of the country. One

TABLE 6.1 A subset of the tabular data contained in the GIS database that resulted from clipping Brown Tract stands within 50-meter stream buffers

Stand number	Acres	Hectares	Age	Volume[a]
1	0.63	0.25	52	12.7
1	3.06	1.24	52	12.7
2	12.37	5.01	46	13.3
2	2.16	0.87	46	13.3
2	0.06	0.02	46	13.3
2	1.80	0.73	46	13.3
2	0.53	0.21	46	13.3
3	4.47	1.81	51	16.6
3	3.24	1.31	51	16.6
....				
270	0.14	0.06	2	0.0
283	4.03	1.63	43	1.5
Total	1,101.39	445.52		

[a] thousand board feet per acre

product that can be obtained is the soil survey geographic database (SSURGO) for most counties in the US. The majority of SSURGO data were mapped at a 1:20,000 scale, and the minimum mapping area is 1–2 hectares. This level of detail in the mapping of soils was designed for use by farmers, landowners, and other natural resource organizations.

The STATSGO soils database is another national-level soils GIS database for the United States (USDA Natural Resources Conservation Service, 2006). However, the STATSGO data is a general soil map, and not as spatially refined as the SSURGO data. STATSGO was mapped at a 1:250,000 scale, and the minimum area mapped is about 600 hectares. This level of detail in mapping soils was designed for broad natural resource planning and management uses.

The National Soil Database of Canada (Agriculture and Agri-Food Canada, 2006) contains databases on soils, landscape features, and climatic data for each Canadian province, and is the national archive for land resource information collected by federal and provincial field surveys. The GIS databases contained in the National Soil Database range from the more general, mapped at a 1:1,000,000 scale or smaller (like STATSGO), to the more detailed, mapped at a 1:20,000 scale or larger (like SSURGO).

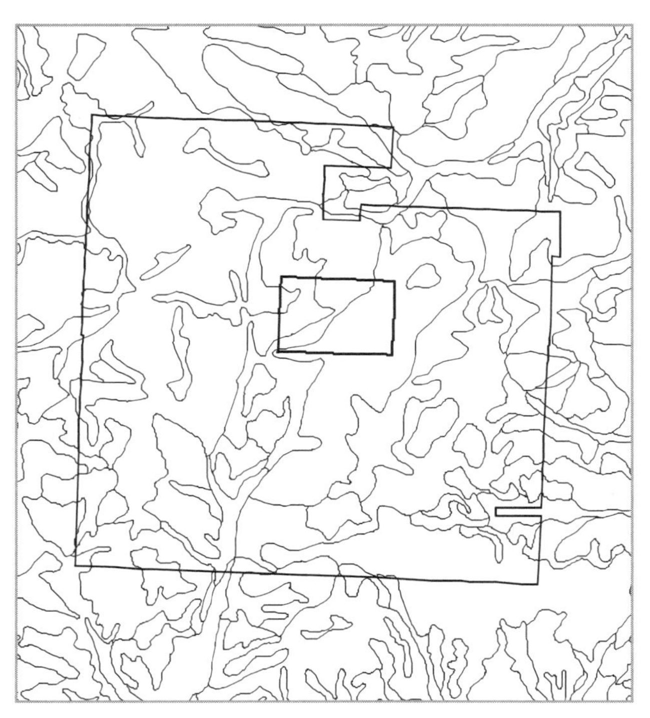

Figure 6.5 Soil polygons in the Brown Tract and surrounding area.

Assume you were to acquire the SSURGO data for the county within which the Brown Tract is located (Figure 6.5). If the managers of the Brown Tract were interested in the type of soils that the Natural Resources Conservation Service has delineated for the land that they manage, a clipping process can be used to obtain that information. In this case, you would use the boundary of the Brown Tract as the polygon theme to perform the clip, and the SSURGO soils database as the theme on which the clip would be performed. The resulting GIS database provides an indication of the major types of the SSURGO soils that are being managed on the Brown Tract (Figure 6.6).

Obtaining information about roads within a forest

In the previous examples provided in this chapter, interest was placed on obtaining information about polygon features (timber stands and soils) that were located within some geographic region. This type of information is informative for land managers, yet it is also possible to

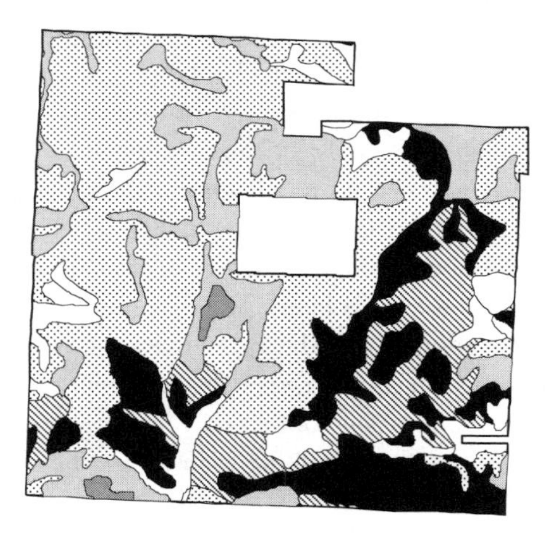

Figure 6.6 Major soil types within the boundary of the Brown Tract.

obtain information about other types of landscape features (lines or points) within certain geographic regions. Most forest road GIS databases, for example, contain a system of roads that extend well beyond the boundary of the property that is being managed by a natural resource management organization. This allows these organizations and their managers to see how the road system they manage is integrated with the road systems managed by states, counties, or other natural resource management organizations. In addition, it allows organizations to weigh their options: If you were to harvest timber stand X, which route could be used to deliver the timber to the mill? If you were to fertilize timber stands Y and Z, how would you get the fertilizer to those stands? If you were to perform an owl survey in watershed A, how could you get to (and get around) watershed A?

There are times, however, when an organization might need to understand only the characteristics of the road resources within the boundaries of the land that they manage. Each year, for example, a land manager may need to develop a budget for road maintenance expenditures, and over a longer period of time, a plan for the continued maintenance of the road system. When developing a long-term plan for maintaining rock surface roads (for example), you may first want to understand the extent of rock-surfaced roads within the land being managed.

As an example of a clipping process, the roads GIS database of the Brown Tract can be clipped to the property boundary. If you were to open the roads GIS database in a GIS software program, you would find that the database contains over 79 kilometers of paved, rock, and native surface roads (Table 6.2). These roads extend well beyond the boundary of the Brown Tract in some cases, and allow the forest managers to view the landscape that they manage in a larger context.

However, after clipping the roads GIS database to the ownership GIS database, and making sure that road lengths were updated in the output database, you would find that only about 66 kilometers of road (Table 6.3) actually reside within the boundary of the Brown Tract (Figure 6.7).

This exercise also produces an example of a potential problem related to this type of information extraction process. If you were to look closely at the resulting clipped GIS database, the eastern edge of the forest (Figure 6.8) contains a discontinuous piece of a road. Upon inspection, you might find that this road is a rocked road and that one of four situations has occurred:

1. The road was incorrectly digitized into the GIS database (which could be verified by viewing the digital orthophotograph associated with the Brown Tract).
2. The road was incorrectly laid out in the field, and

TABLE 6.2 **Length and type of road within the roads GIS database developed for the Brown Tract**

Road type	Miles	Kilometers
Paved	5.7	9.3
Rock	41.3	66.4
Native Surface	2.3	3.6
Total	49.3	79.3

TABLE 6.3 **Length and type of road within the boundary of the Brown Tract**

Road type	Miles	Kilometers
Paved	2.0	3.3
Rock	37.5	60.3
Native Surface	1.5	2.5
Total	41.0	66.1

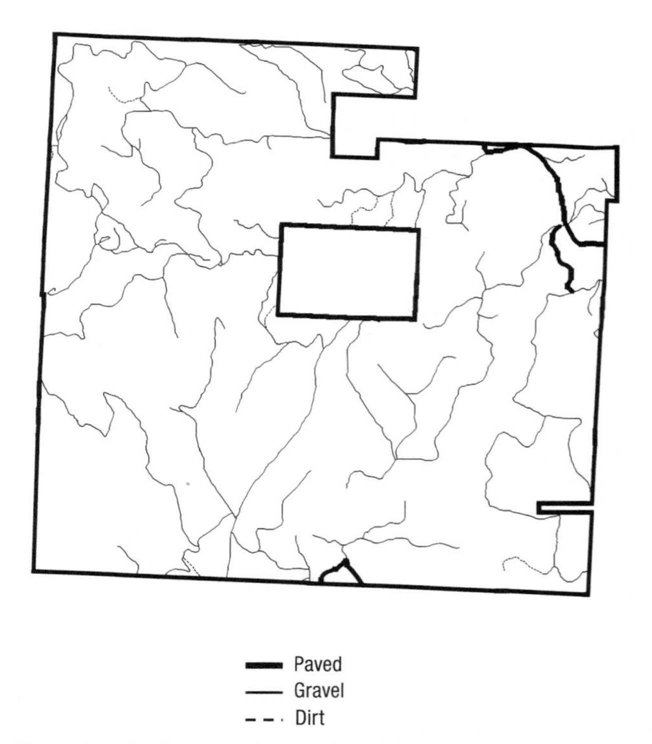

Figure 6.7 Roads within the boundary of the Brown Tract.

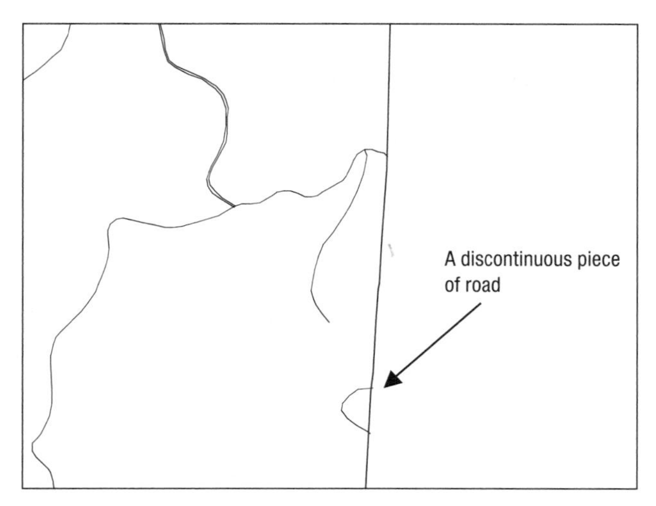

Figure 6.8 A potential error in the clipped roads GIS database.

actually does reside within the boundary of the Brown tract.

3. The boundary of the Brown Tract was incorrectly digitized into the boundary GIS database.
4. The road and boundary are correctly located and the road may be the result of an use easement, a remnant from a previous road network, or a potential ingress by an adjacent property owner.

In any event, developing a maintenance plan that includes this, and other, small pieces of road may not make sense from an operational perspective.

Obtaining information about streams within a forest

Most streams GIS databases managed by natural resource organizations are designed in a fashion similar to roads GIS databases: they contain a system of lines or links that extend well beyond the property boundary of the organization so that the managers can understand how their activities are integrated within a larger watershed system. For example, you may be interested in knowing where water flows, in the event that the need to monitor forest operations (after logging, fertilization, or herbicide operations) is important. There are times, however, when you may need to quantify the stream characteristics only within the boundary of the land that your organization manages. For example, in order to develop a budget for stream surveys, you may need to know how many and what type (along with their length) of streams are located within the boundary of land that your organization manage. This implies that you are interested in understanding

TABLE 6.4 Length and type of streams within the streams GIS database used by the Brown Tract

Stream type	Miles	Kilometers
Fish-bearing / large	0.9	1.4
Fish-bearing / medium	3.1	5.0
Fish-bearing / small	4.9	7.8
Non-fish-bearing / large	0.0	0.0
Non-fish-bearing / medium	0.0	0.0
Non-fish-bearing / small	25.7	41.4
Total	34.6	55.6

the extent of certain landscape features within a specific geographic region, and a clipping process can be used to extract those resources from other GIS databases.

Using the Brown Tract as an example, prior to clipping the streams to the boundary of the forest, approximately 56 kilometers (35 miles) of streams are represented in the streams GIS database (Table 6.4). The resulting GIS database, after clipping the streams GIS database to the forest boundary GIS database and updating the stream lengths (Figure 6.9), contains only about 45 kilometers (28 miles) of streams that are actually located within the boundary of the forest (Table 6.5). You find with this analysis that there are no large, fish-bearing streams

Figure 6.9 Streams within the boundary of the Brown Tract.

TABLE 6.5 **Length and type of streams within the boundary of the Brown Tract**

Stream type	Miles	Kilometers
Fish-bearing / large	0.0	0.0
Fish-bearing / medium	2.1	3.4
Fish-bearing / small	2.8	4.5
Non-fish-bearing / large	0.0	0.0
Non-fish-bearing / medium	0.0	0.0
Non-fish-bearing / small	23.3	37.5
Total	28.2	45.4

within the Brown Tract, although some are present in the more extensive streams GIS database. In addition, while only about one-quarter of the streams (by length) in the more extensive streams GIS database are fish-bearing, almost half of the streams (again by length) removed as a result of the clipping process are fish-bearing. Since the Brown Tract contains the headwaters of several stream systems, it is not unreasonable to assume that the fish-bearing portions of these systems are located in the lower reaches (i.e., off of the Brown Tract).

In Depth

One note of caution about clipping processes: prior to performing other types of spatial analyses, such as the buffering processes, you must consider whether or not a clipping process is appropriate. For example, if you wanted to understand the extent of riparian areas on the Brown Tract, you might not want to clip the streams to the boundary of the Brown Tract as an initial step in the analysis. By doing this, streams outside of the forest boundary are ignored, to which you probably would reply 'So what?'. Well, those streams may have a riparian area about them that extends inside the boundary of the Brown Tract. Put another way, just because a particular stream resides outside the boundary of the forest you manage does not necessarily imply it can be ignored: part of its area of influence (the riparian area), and an assessment of activities that you might be considering within this area, may need to be included in your management plan.

The Process of Erasing Landscape Features

Thus far, our interest has been centered on understanding the extent of resources that are located *within* certain geographic regions. Now, our focus will shift to obtaining information about the resources located *outside of* certain geographic regions. The erasing process is well suited to this task, and is essentially the opposite of the clipping process. When using an erasing process, you are interested in creating a new GIS database that contains landscape features located outside of a specific geographic region. Just as with a clipping process, an erasing process involves using two GIS databases as input databases (Figure 6.10), and the process results in one output GIS database. When using vector GIS databases, one of the input GIS databases (the eraser) needs to contain polygon features; the other GIS database (the database in which landscape features will be erased) can contain point, line, or polygon features. The eraser GIS database is overlaid on the GIS database containing the features of interest, and only those landscape features located outside of the

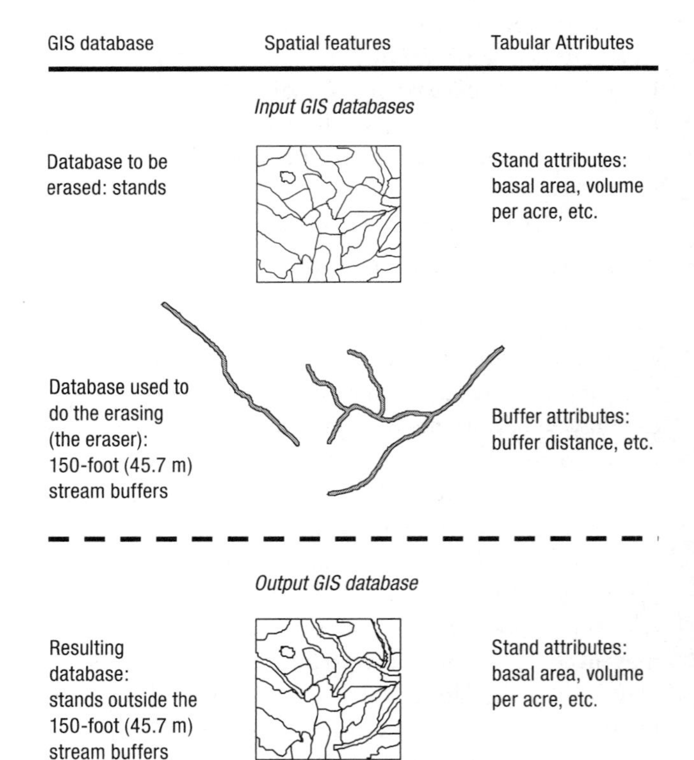

Figure 6.10 Erasing the stands within 150-foot (45.7 m) stream buffers from the Daniel Pickett stands GIS database.

boundaries of the polygons contained within the eraser GIS database are retained. Thus the output GIS database contains the same type of features as the database being erased. In addition, the size of features (lines or polygons, but not points) may be spatially altered where they overlap with the edges of the polygons contained within the eraser GIS database.

Figure 6.11 illustrates a small example and utilizes a fire (the eraser) and a timber stand (the polygon to be erased). After the erasing process has been performed, you can see that two of the original lines that defined the boundary of a timber stand (1_3 and 1_4) were shortened to the point of intersection with a fire area, and a third line was created (1_5) to describe the edge of the fire area that is common with the timber stand. The resulting erased timber stand GIS database has all of the attributes of the original timber stand GIS database, yet the spatial extent is equal to the original timber stand database minus the overlap with the fire database. As with the clipping process, feature measurements in the output database from an erasing process should be updated to reflect changes in area or perimeter of polygons, or length of links in line databases. If your GIS software does not make these updates automatically, you should ensure that you use software or other approaches to update the measurements.

a. GIS databases prior to an erase process.

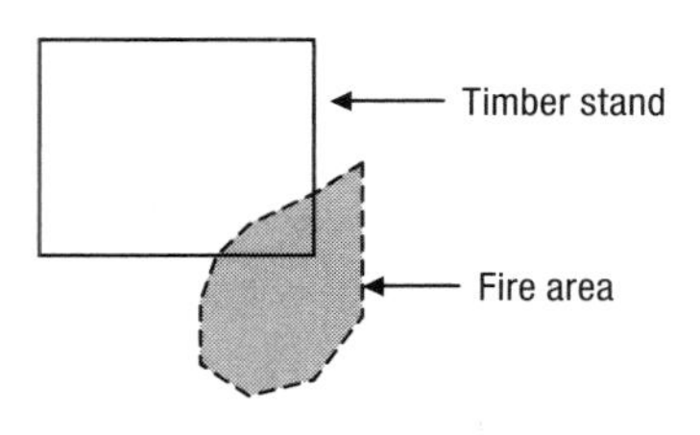

b. GIS databases during an erase process. (o = node)

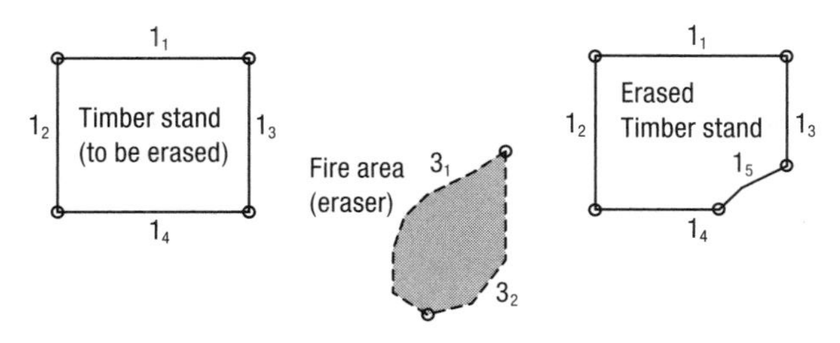

c. GIS databases after an erase process. (o = node)

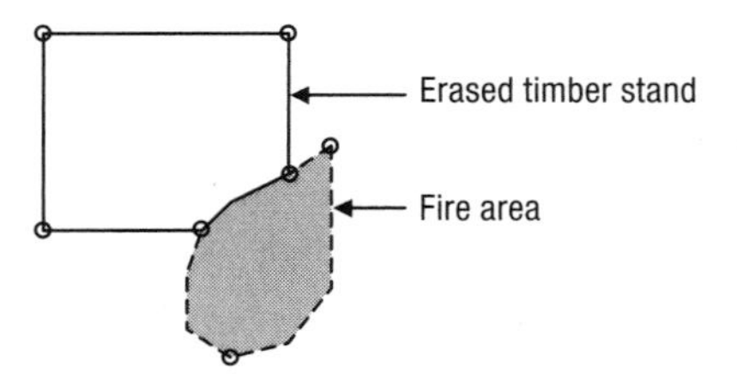

Figure 6.11 Erasing a fire area from a timber stand.

There are a number of reasons why you would consider using an erasing process. One reason involves the need to understand the characteristics of a landscape outside of areas that are considered *restricted*. Alternatively, the goal would be to define the *unrestricted* areas of a landscape with regard to management activities. For example, earlier in this chapter a clipping process was used to develop a GIS database that allowed us to summarize the resources located within riparian zones. Management within riparian zones is generally *restricted* in some form or fashion. Areas outside the riparian zones could then be considered *unrestricted*, assuming there are no other constraints on land management, such as those related to owl nest locations, research areas, and so on. Understanding the extent of the landscape where management is not restricted, may be important when considering decisions related to harvesting operations, the use of herbicides or fertilizer, or other types of management practices.

Obtaining information about vegetation resources outside of riparian zones

To build upon the examples previously provided in this chapter, let's obtain information about some landscape features (vegetation polygons) located outside of the 50-meter riparian zones that were developed for the Brown Tract. An erasing process allows us to develop a GIS database containing only those features (in this case vegetation polygons) located outside the boundaries of the polygons that describe the riparian zones. The original vegetation polygons are again redesigned such that their boundaries coincide with the edges of the polygons that describe the riparian zones. Those vegetation polygons that were entirely located outside of the riparian zones are left intact, and those vegetation polygons that were entirely contained within the boundaries of riparian zones are eliminated (Figure 6.12). The tabular database related to the resulting erased GIS database should contain the same set of attributes that were contained in the original vegetation (stands) GIS database, with only the area (and perhaps the perimeter) values adjusted to

Figure 6.12 Vegetation polygons (stands) outside of 50-meter buffers on the Brown Tract.

reflect the sizes of the redesigned polygons. When performing a check of the data, you should find that the size of the resulting erased GIS database, about 1,677 hectares (4,143 acres), is—and should be—less than the size of the original stands GIS database (which was about 2,123 hectares or 5,245 acres).

Erasing processes can be used for other purposes as well. For example, if an area of the Brown Tract were designated for sale, the managers of the Brown Tract may be interested in knowing what resources would remain after the pending sale. Alternatively, if you were to develop a land classification for the Brown Tract which involved buffering streams and classifying uplands from riparian zones, you may erase the buffer zones from a stands database (for example) as a intermediate stage of classifying the landscape and identifying upland areas. These types of analyses may prove useful in the applications described in subsequent chapters of this book.

Summary

Clipping processes are spatial operations that allow users of GIS to obtain information about landscape features within certain geographic regions. When using vector GIS databases, polygon features are used to clip the features from a second GIS database containing points, lines, or polygons. The features that are contained in the output GIS database are those contained within the boundaries of the polygons of the clipping GIS database (the cookie cutter). Erasing processes can be viewed as the inverse of clipping processes. With an erasing process, you can obtain information about spatial features located outside of certain geographic regions. In fact, you could use a single GIS database to clip landscape features (used as the cookie cutter), and subsequently to erase landscape features (used as the eraser). If both processes were applied to the same GIS database (e.g., stands, roads, or streams), the two output GIS databases, when combined, should cover the same landscape area as the original GIS databases that were clipped or erased. For example, a GIS database was created, using a clipping process, to represent those vegetation polygons on the Brown Tract that were located within 50-meter riparian zones. A second GIS database was also created, using an erasing process, to represent those vegetation polygons located outside of the 50-meter riparian zones. If combined, these two GIS databases (the clipped and erased GIS databases) should equal the land area and vegetation resources that can be found within the original vegetation (stands) GIS database—no more, no less.

Applications

6.1. Obtaining information about features within a watershed. Suppose that the hydrologist associated with the Daniel Pickett forest, Michelle Rice, has been working for some time on a watershed analysis with a few other natural resource management organizations. The watershed being analyzed is the Dogwood Creek Watershed. The watershed analysis team is now at the point where they need to obtain as much GIS data as possible to describe the current condition of the watershed. Michelle asks you to provide her with the following information:

a) a summary of the area of timber stands located within the Dogwood Creek watershed, by vegetation type;

b) a summary of the length of roads located within the Dogwood Creek watershed, by road type; and
c) a summary of the length of streams located within the Dogwood Creek watershed, by stream type. In addition, summarize these values in terms of stream miles per square mile of land, by stream type.

Develop a short memo addressed to Ms Rice that details the results of your analyses.

6.2. Summarizing resources within a management area. Jane Hayes is developing an annual report on the management of the Daniel Pickett forest and has become very interested in certain aspects of the forest resources. *Since she knows that you are becoming proficient with GIS*, she has asked you to provide some information she feels is necessary for her report:

a) amount of land in each watershed,
b) length of rock road in each watershed,
c) length of dirt road in each watershed, and
d) length of stream classes 1–3 in each watershed.

In addition, she is interested in knowing something about the resources that are within 50 meters of the streams. Using the 50-foot stream buffer provided by the GIS Department, provide the following:

a) area of land within 50 meters of the streams,
b) area of older forest (age $\geq$ 60) within the 50-meter stream buffers,
c) length of paved road within the 50-meter stream buffers, and
d) length of rock road within the 50-meter stream buffers.

Provide a memo addressed to Ms Hayes that details the results of your analyses. Keep in mind the appropriate units (perhaps kilometers for length rather than meters), and the appropriate precision (to the nearest 0.1 hectare or 0.1 kilometer) for this type of report.

6.3. Fertilization possibilities. Within the soils GIS database of the Daniel Pickett forest is an attribute, fertresp, which was meant to indicate the probability of tree response to fertilization as a function of the underlying soil type. Describe the types of vegetation found in the 'high response' areas using the following format:

	Vegetation type		
Age class	**A**	**B**	**C**
0–20	______	______	______
21–40	______	______	______
41–60	______	______	______
61–80	______	______	______
80+	______	______	______

6.4. Potential sale of watershed for conservation reserve. The managers of the Daniel Pickett forest have been approached by a non-profit organization specializing in developing and managing conservation reserves. The non-profit organization is specifically interested in acquiring a specific portion of the Daniel Pickett forest—all of the land located in the Trout Creek Watershed. It seems that the non-profit group has been active in developing a larger reserve system in the Trout Creek area, and the Daniel Pickett forest just happens to contain the headwaters of the watershed. The region manager of the Daniel Pickett forest, Becky Blaylock, is interested in understanding the effect this land sale may have on the management plan for the area. She asks you to provide her a before- and after-sale description of the resources, in the following format:

a) Before-sale conditions (area) of the Daniel Pickett forest.

	Vegetation type		
Age class	**A**	**B**	**C**
0–20	______	______	______
21–40	______	______	______
41–60	______	______	______
61–80	______	______	______
80+	______	______	______

b) After-sale conditions (area) of the Daniel Pickett forest.

	Vegetation type		
Age class	**A**	**B**	**C**
0–20	______	______	______
21–40	______	______	______
41–60	______	______	______
61–80	______	______	______
80+	______	______	______

In addition, she has also requested the following:

c) a map showing the after-sale arrangement of vegetation classes (types) on the Daniel Pickett forest, and

d) a brief summary of your opinion of the effects of the sale on the management of the forest.

References

Agriculture and Agri-Food Canada. (2006). *The national soil database (NSDB)*. Ottawa, ON: The National Land and Weather Information Service, Agriculture and Agri-Food Canada. Retrieved February 11, 2007, from http://sis.agr.gc.ca/cansis/nsdb/intro.html.

Clarke, K.C. (1995). *Analytical and computer cartography*. Upper Saddle River, NJ: Prentice-Hall.

Merriam-Webster. (2007). *Merriam-Webster online search*. Retrieved February 4, 2007, from http://www.m-w.com/cgi-bin/dictionary.

USDA Natural Resources Conservation Service. (2006). *US general soil map (STATSGO)*. Washington, DC: National Cartography and Geospatial Center, USDA Natural Resources Conservation Service. Retrieved February 11, 2007, from http://www.ncgc.nrcs.usda.gov/products/datasets/statsgo/.

USDA Natural Resources Conservation Service. (2007). *Geospatial data gateway*. Washington, DC: USDA Natural Resources Conservation Service. Retrieved February 9, 2007, from http://datagateway.nrcs.usda.gov/NextPage.aspx?HitTab=1.

Chapter 7

Buffering Landscape Features

Objectives

Chapter 7 is an introduction and examination of GIS buffering processes. A number of examples and applications are presented in this chapter to provide readers with experience in several of the common GIS-related tasks in natural resource management that require the use of a buffering process. At the conclusion of this chapter, readers should understand, and be able to discuss, the pertinent aspects of:

1. what *buffering spatial landscape features* implies;
2. how different buffering techniques can be applied to point, line, or polygon features; and
3. how buffering can be applied to assess alternative management policies and to assist in making natural resource management decisions

To someone unfamiliar with this GIS process, the term **buffering** may lead to some confusion. Technically, the noun 'buffer' refers to (a) a device for reducing shock when contacted, (b) a means for cushioning fluctuations in business activities, (c) a protective barrier, (d) a substance capable of neutralizing both acids and bases, and (e) a temporary storage unit on a computer (Merriam-Webster, 2007). In GIS applications in natural resource management, we generally refer to the buffering process as a method for creating a **buffer zone**, which is defined as a land area that delineates separate management activities or emphases. Within GIS, a buffer zone is a polygon that encloses an area within a specific distance from a point, line, or polygon, and is useful in analyses that have a proximity criteria (Association for Geographic Information, 1999).

The GIS process of buffering usually infers that a boundary is about to be drawn around some selected features. There are numerous reasons why you would want to draw boundaries around selected landscape features in natural resource management. For instance, one of the guiding management policies for an organization may suggest that some management activities may be prohibited within a certain distance of a stream, a road, a trail, or a home. Therefore, as a natural resource manager, you may be interested in the appropriate limits of allowable management activity. As another example, management activities within a certain distance of nesting, roosting, or foraging sites of a wildlife species of concern may be curtailed during certain times of the year. Therefore, delineating these 'home ranges' or 'critical habitats' may be an important aspect of natural resource management. In these cases, you might develop spatial buffers within GIS to identify resources located within certain distances of important landscape features (points, lines, or polygons). You might also be interested in examining the potential impacts of the policies that suggest the use of buffers, in order to understand how the objectives of natural resource management be affected. As you may gather, buffering has been widely used in natural resource management to identify riparian management areas (streamside management zone) and to define areas where management may be restricted for one reason or another. Yet buffers have also been used to track and assess the site impacts of logging equipment on soil resources and residual stand conditions (Bettinger et al., 1994; McDonald et al., 2002).

Fortunately, GIS software programs provide the ability to easily identify features that are within some proximity of other features (Star & Estes, 1990). Developing the boundaries of a region within a specific distance from a landscape feature, or set of selected landscape features, is often called a 'proximity analysis', or buffering process. Therefore the subject of this chapter involves the identification and delineation of natural resources within certain distances of other landscape features.

How a Buffer Process Works

Buffer processes work by using mathematical algorithms to delineate the space around selected landscape features. When using vector data, one or more features of interest are selected, the desired buffer distance is specified, and a line is drawn in all directions around the features until a solid polygon has been formed. Point, line, and polygon features can all be buffered but the buffer creation process depends on the feature type. To visualize how a buffer is created around a point feature, imagine a point on a piece of paper; with a pencil and compass set, a circle is drawn around that point (Figure 7.1a). GIS software programs can perform this type of operation on thousands of points in a few seconds. Delineating buffers around lines and polygons requires a similar process but involves some additional processing. With lines, a buffer is created around each vertex (Figure 7.1b), then tangents are created between each of these buffers, and only the outside edge is kept, forming a closed polygon. With polygons, you may have the choice of creating buffers that represent only the area outside of the polygon being buffered, the area outside of the polygon plus the entire area of the polygon, the buffered area both inside and outside of the edge that define the polygon, or the area buffered inside the polygon (Figure 7.1c). The type of proximity analysis

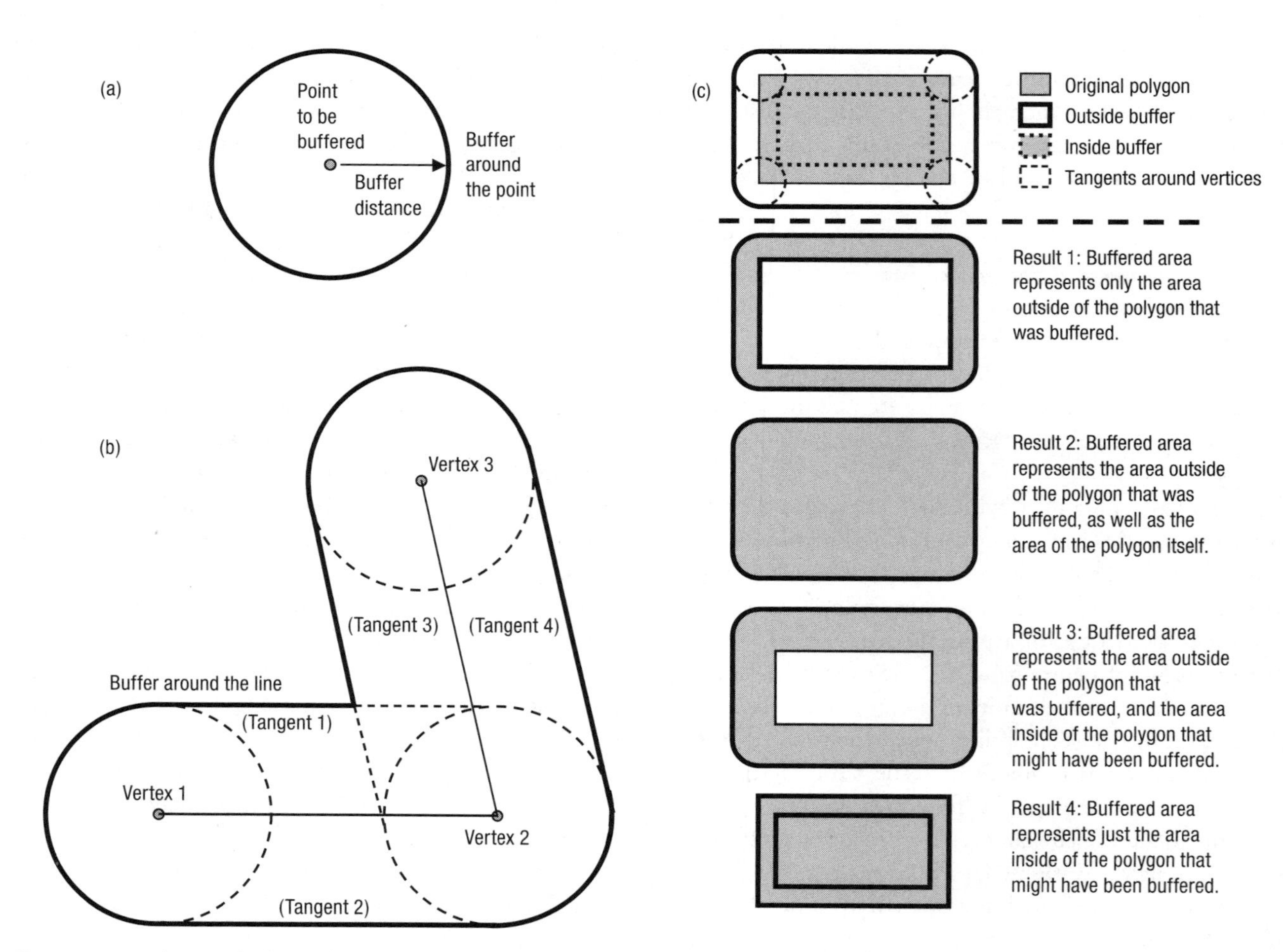

Figure 7.1 Developing a buffer around (a) a point using, (b) a line with three vertices, and (c) a polygon.

that is required will suggest the appropriate type of buffering process required when polygons are concerned. For example, if you were interested in the type of vegetation within 1,000 meters of a set of polygons that define crucial owl habitat, you would create a buffer outside the set of polygons. On the other hand, if you were interested in understanding the amount of land that is associated with a policy that prevents management activity within 100 meters of the edge of a managed property (to avoid conflict with homes and other developed areas outside the property), you would create a buffer inside the boundary of the managed property.

Buffering processes performed on vector GIS data may require some rather complex geometrical calculations, with lines and tangents to compute, and overlapping areas perhaps merged together. To remove overlapping areas, intersection and dissolving processes are used. Buffering processes performed on raster GIS data involve counting the number of pixels away from selected or specified pixels.

Buffers used in natural resource management can take on many forms, but the two most commonly employed are a constant buffer width and a variable buffer width. Constant width buffers are the most commonly used form in natural resource management (Bren, 2003), and assume a symmetrical distance around each buffered landscape feature (same distance buffered on each side of a stream, for example). Variable width buffers assume that features are buffered differently based on some inherent or assumed characteristic, such as stream size. Other types of buffers include those based on (a) environmental loading values and (b) other outside influences. Stream buffers based on loading values, for example, might take into account the amount of area contributing to an impact. For example, stream reaches that have larger water contributing areas might be buffered using a wider buffer width than stream reaches with smaller contributing areas. These are different than variable width buffers in that each section of a stream may be buffered a different distance based on the size of the watershed that contributes water to that section of the stream. Buffers based on other outside influences may include stream buffers that take into account the amount of sunlight that reaches the stream itself. In these cases, buffers on the southerly sides of streams may be wider than buffers on the northerly sides of streams (Bren, 2003).

As we mentioned, common buffer distances can be constant (fixed) distances, or they can vary for each feature in a GIS database based on an attribute of those features. Most GIS software programs can accommodate both buffering approaches. To illustrate these differences, suppose you have a GIS database that includes 10 streams, each of a different stream class (Table 7.1). With this hypothetical data, it is assumed that the lower the stream class number, the larger (wider) the stream.

One task in planning natural resource management activities may be to delineate riparian buffers around these streams. If a constant buffer distance of, for example, 30 meters is assumed, each stream would be buffered the same distance (30 m). However, many regulations pertaining to the proximity of management activities around riparian areas require wider buffers around larger streams, and narrower buffers around smaller streams. Therefore, distances specific to each landscape feature (each stream reach in this example) can be used in the buffer process, to allow the development of variable distance buffers. For example, for each of the 10 streams you could have developed an attribute to describe the appropriate buffer distance (Table 7.2), and use that attribute to guide the buffering process. Readers will examine both of these buffering assumptions (constant distance and variable width) in the forthcoming examples as well as in the 'Applications' section at the end of this chapter.

When buffering multiple landscape features, a buffer is created around each feature independent of the other features. One option available with most GIS software concerns the handling of the overlapping areas. The choices are to retain individual buffer polygons for all features to be buffered (called uncontiguous, or non-

TABLE 7.1 **Ten hypothetical streams and their stream class, length, and width**

Stream	Stream class	Stream length (m)	Stream width (m)
1	1	1000	50
2	1	750	45
3	2	500	10
4	2	450	10
5	3	375	3
6	3	450	3
7	3	400	2
8	4	300	1
9	4	250	1
10	5	275	0

TABLE 7.2 Ten hypothetical streams and their stream class, length, width, and buffer distance

Stream	Stream class	Stream length (m)	Stream width (m)	Buffer distance (m)
1	1	1000	50	30
2	1	750	45	30
3	2	500	10	20
4	2	450	10	20
5	3	375	3	10
6	3	450	3	10
7	3	400	2	10
8	4	300	1	10
9	4	250	1	10
10	5	275	0	0

contiguous polygons), or to eliminate the overlapping areas, creating contiguous polygons. The advantage of retaining individual buffer polygons is that once the buffers are created, the buffer pertaining to each individual landscape feature can be accessed, which may allow you to determine *which features* are within *what distance* of other features. This allows individual analysis for each point, line, or polygon feature that was buffered. The disadvantage is that there is a high likelihood that some of the retained individual buffers overlap; thus the overlapping area can potentially be counted more than once in any subsequent area summary calculations. The creation of a single buffer from overlapping buffered areas avoids this problem. However, the ability to understand the buffer required for each individual landscape feature is then obscured. The goals of each analysis should direct users to the choice of one buffering method or the other, but users should understand how the two methods differ in their approaches and results.

Buffering Streams and Creating Riparian Areas

Riparian areas can be defined as land areas that are in close proximity to a stream, lake, swamp, or other water body, and those that are often are occupied by plants that are dependent on their roots reaching the water table (Society of American Foresters, 1983). Alternatively, they are areas where vegetation and microclimate are influenced by seasonal or year-round water, high water tables, and soils exhibit some wetness characteristics (Oregon Department of Forestry, 1994a). The first definition includes administrative and ecological aspects, while the second is based mainly on ecological and physical aspects. When we work with riparian areas in natural resource management, they are more commonly defined administratively rather than ecologically. Generally, riparian management area widths are designated by federal, state, provincial, or organization policies, and are designed to provide adequate areas to retain the physical components, to maintain the functions necessary to meet protection objectives and goals for fish, water quality, and other wildlife (Oregon Department of Forestry, 1994b). While some policies suggest that riparian areas should be protected from logging, grazing, and other types of exploitation, other policies allow a set of limited activities within certain distances from certain types of streams. Thus it is important for land managers to know where riparian areas are on a landscape, and to understand what resources are affected by riparian area designations.

In the following examples, streams will be buffered first with fixed (constant) buffer widths, then with variable

In Depth

Generally when buffers are delineated with a GIS process, they are saved in a new GIS database that is separate from the one containing the landscape features that were buffered. To further enhance the power of buffer processes, most GIS software programs only buffer the features that are *selected* (manually or through a query). If no landscape features are selected, generally all of the landscape features are buffered but an examination of buffer output can confirm whether your GIS software follows this assumption. One processing step commonly forgotten is either to remember to select the features that need to be buffered (perhaps just all Class 1 streams), another is to clear all previously selected features (if some features were selected for a reason unrelated to the buffer process).

buffer widths (according to a set of stream buffer guidelines) to delineate the riparian management areas.

Fixed-width buffers

In this first example of a buffering process, the streams GIS database of the Brown Tract (Figure 7.2) will be used to generate **fixed-width buffers**, or buffers that do not vary based on some attribute of the landscape features being buffered. In this case, assume that an organizational policy exists that directs the managers of the Brown Tract to delineate a fixed 150-foot buffer around all of the streams. During the buffer process, a buffer polygon will be created around all stream lines in the streams GIS database. As noted earlier, most GIS software programs provide the options of either leaving the buffers as individual polygons around all stream lines or eliminating the overlapping areas. Here, we will illustrate the overlap being eliminated, so that buffer area estimates will not be overstated. At the conclusion of the buffer process, areas 150 feet on either side of the Brown Tract streams (Figure 7.3) are delineated; no other land areas are represented by the buffer polygon(s) contained in the new buffer GIS database.

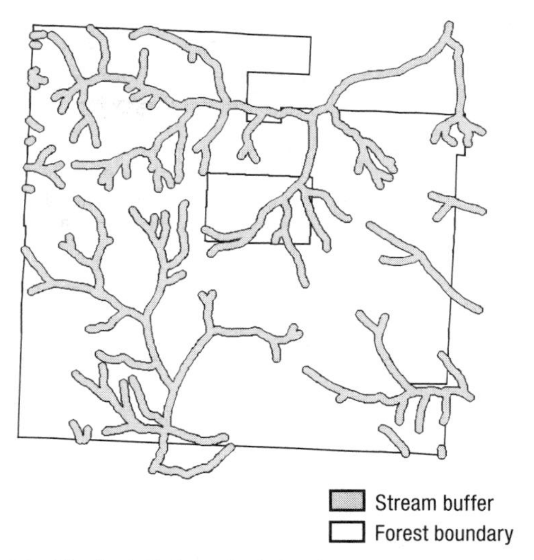

Figure 7.3 Fixed-width (150-foot) riparian management areas, generated by buffering the streams GIS database of the Brown Tract.

Variable-width buffers

Rather than a single, fixed-width buffer, some management objectives may require a buffer that varies based on an attribute of a landscape feature. For example, in some states or provinces rules exist which indicate that the size of riparian areas should be a function of the type of stream with which they correspond. In some cases, these riparian designations allow no activity within a certain distance from the stream system; in other cases, limited activity. The general notion is that, with wider streams, streams with year-round water flow, or streams with known fish populations, wider buffers are required. Once the set of buffers is based on the characteristics of the streams, they are considered **variable-width buffers** because they vary according to the characteristics of each stream. In Oregon, for example, the riparian guidelines require a 100-foot buffer around 'large' fish-bearing or domestic water use streams, and lesser buffer widths around smaller streams (Table 7.3) and streams that are not currently fish-bearing or used as sources of domestic water (Oregon State Legislature, 2005).

Since the Brown Tract is fictional, the buffer widths are assumed for the various stream classes in the streams GIS database. Stream class 1 is the largest stream, into which all of the other streams flow, so a larger buffer (100 feet) is

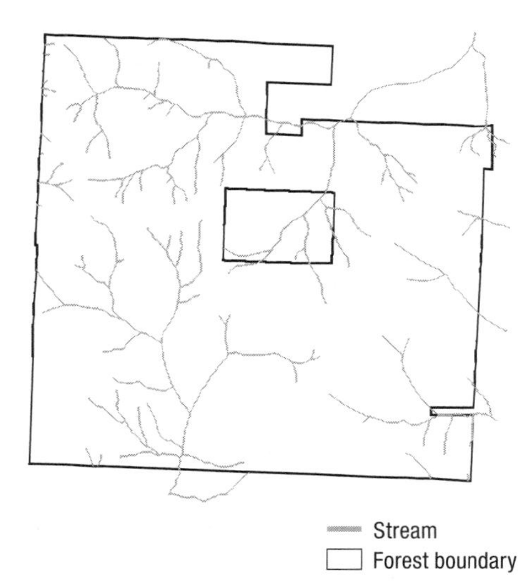

Figure 7.2 The stream system within and around the Brown Tract.

In Depth

When you want to create riparian management areas by buffering a set of streams, you usually do so by indicating how wide the buffer should be on either side of a stream, rather than by indicating the total width of the buffer (from one edge of the buffer, across the stream, to the other side of the buffer).

TABLE 7.3 State of Oregon riparian management area policy

	Riparian management area width (feet)	
	Domestic water use or fish-bearing	Non-domestic water use and non-fish-bearing
Large[a]	100	70
Medium[b]	70	50
Small[c]	50	20

[a] Average annual flow of ≥ 10 cubic feet per second.

[b] Average annual flow of ≥ 2 cubic feet per second and < 10 cubic feet per second.

[c] Average annual flow of < 2 cubic feet per second, or drainage area ≤ 200 acres.

Source: Oregon State Legislature, 2005

assumed around this stream class, and smaller buffers are assumed to be required around the other stream classes according to the direction provided in Table 7.3. Fortunately, most GIS software programs allow users to designate a field (also called column, attribute, or variable) in a GIS database to use as the reference for the desired buffer width for each landscape feature. Since each row in the tabular portion of the streams GIS database represents a stream line or reach (Table 7.4), the values located in the 'buffer width' field can be used to represent the appropriate buffer width for each stream. During the buffering process, each stream line will then be buffered according to the appropriate buffer width, based on each stream's class. The buffering process will read the buffer distance from an attribute table, one record (row) at a time, to create the buffer. Most GIS programs will do this very quickly, such that only a few seconds or less are required. Following the creation of all buffers, subsequent processing will eliminate all overlapping buffer areas, even though the size of the buffer may change in the overlapping area. A map of the variable buffer widths associated with the Brown Tract streams (Figure 7.4) shows that the amount of land area in the riparian areas will vary by stream class, and that no other land area (outside the variable width buffer) is represented in the buffers.

TABLE 7.4 Sample stream reaches represented in the Brown Tract streams GIS database, their characteristics, and resulting buffer width[a]

Stream reach	Length (feet)	Fish bearing?	Stream size	Buffer width (feet)
1	362	no	small	20
2	176	no	small	20
3	992	no	small	20
4	384	no	small	20
.....				
174	1953	no	small	20
175	2143	yes	medium	70
176	3159	yes	small	50

[a] Stream reaches are not necessarily exclusively located within the boundaries of the Brown Tract.

Buffering Owl Nest Locations

Up to this point our concentration has been placed on one of the more typical GIS buffering operations performed in support of natural resource management—buffering streams. However, any type of landscape feature (owl nest location, road, wetland, etc.) can be buffered. For example, in the western United States it is important to protect an area around spotted owl (*Strix occidentalis*) nests. The area within these buffers may either totally prohibit management activities or may limit management activities by duration and extent. Thus, it may be important to understand the amount of resources located within owl buffers. As an example, assume that an owl nest is located in the central portion of the Brown Tract, and assume that federal regulations require land

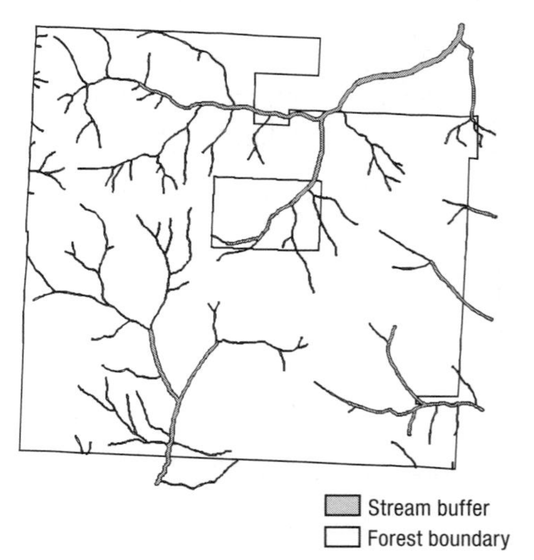

Figure 7.4 Variable-width riparian management areas, generated by buffering the streams GIS database of the Brown Tract.

In Depth

Users should bear in mind that the buffer distances and map coordinate units of GIS data layers that are being buffered must be considered. A buffer operation will typically assume that buffer distances specified either through fixed-width or through variable-width processes are in the same units as the map coordinate system of the GIS layer. If the buffer distances and map coordinate units are in different measurement systems (SI versus USCS) or are in the same system but at different scales (e.g. meters versus kilometers) the appropriate conversion should be applied to the buffer distances prior to beginning the buffer process. Some GIS software will prompt you for the buffer or GIS layer units, and will even do any conversions necessary during the buffer process. In this case, as long as the correct conversions were specified by the user, the buffer output should be correct. In other cases, however, the user must verify that all buffer-dependent measurement units are in agreement.

managers to manage the area within 1,000 feet of these nests much differently than the areas beyond 1,000 feet of an owl nest. A buffer process can be performed using the owl nest location as the selected landscape features, and a 1000-foot radius as a fixed (constant) distance around the nest locations. The result of the buffer process is a new GIS database that delineates the areas within 1,000 feet of the nest location (Figure 7.5). If more than one nest were located on the Brown Tract, and the resulting buffered areas overlapped, you could elect to eliminate the overlapping area (uncontiguous result) or allow the overlapping areas to remain (contiguous result). If your goal was to determine the amount of area and land resources associated with each individual nest then the uncontiguous result would be best. Conversely, if your goal was to determine the total area and land resources encompassed by all owl nest buffers then the contiguous result would be preferred.

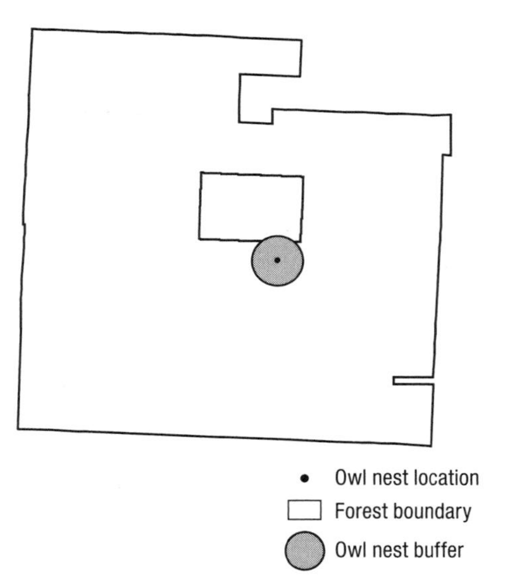

Figure 7.5 Owl nest location and associated 1,000 foot buffer on the Brown Tract.

Buffering the Inside of Landscape Features

In addition to delineating buffers outside of landscape features, you can develop buffers inside of landscape features as well. Of course, when using vector GIS databases this is only possible with polygon features. To illustrate this process, let's assume that the managers of the Brown Tract are concerned about the impact of management activities on nearby homeowners. In some cases, homes are very close to the edge of the forest. In other cases, homeowners' yards and personal belongings (sheds, etc.) are on the edge of the property. To avoid any potential instance of damage to adjacent homes or property, or any potential physical harm to nearby landowners, the managers have decided that they will allow only limited activity within 200 feet of the boundary of the property. To understand how much of the property will be allocated to a limited activity land classification, the inside of the boundary of the forest (a single polygon) can be buffered (Figure 7.6), and the resulting area (397 acres) can be compared with the total area of the forest (5,245 acres) to determine the impact of the policy (7.6 per cent of the forest shifted to limited use). As an alternative to this policy, the managers of the Brown Tract could delineate this *limited activity* zone by buffering the homes a certain distance, although locating these areas on the ground would be much more difficult than using a policy such as *200 feet from the boundary*, since the buffer around each home is circular.

Buffering Concentric Rings around Landscape Features

If you needed to generate multiple buffers around the same landscape feature(s), you could perform each buffer-

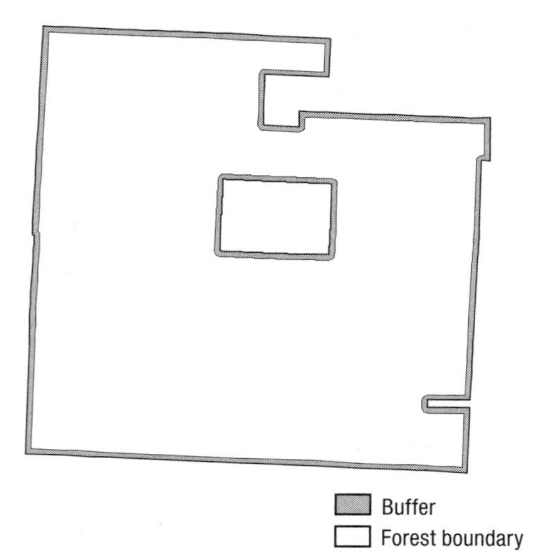

Figure 7.6 A 200-foot buffer inside the boundary of the Brown Tract.

ing operation independently using different fixed-width buffer distances. However, the resulting buffers would contain some areas that overlap. Alternatively, if you wanted to avoid the overlap that would ultimately occur with this method, most GIS software programs contain a process that enables users to easily develop concentric, non-overlapping rings. One caveat for this process is that the buffer interval for each ring needs to be constant. As an example of the use of concentric rings of buffers, the Maryland Department of Natural Resources (2005) recently established guidelines for the management of land near bald eagle (*Haliaeetus leucocephalus*) nest trees. The guidelines require managers to acknowledge three buffer zones around each nest tree.

- Zone 1 extends from a nest tree outward to a radius of 330 feet.
- Zone 2 extends from the edge of Zone 1 (330 feet) outward to 660 feet in radius.
- Zone 3 extends from the edge of Zone 2 (660 feet) outward to 1,320 feet.

Zone 1 prohibits land use changes, such as those related to development or timber harvesting. Zone 2 prohibits development (clearing, grading, building, etc.) but allows selective timber harvesting. Zone 3 prohibits any activity during the eagle nesting season. If you were interested in quickly developing buffers to represent Zones 1 and 2, the buffer interval for the concentric rings would be 330 feet. Zone 3 is not 330 feet from the edge of Zone 2 and thus would need to be developed separately. Figure 7.7

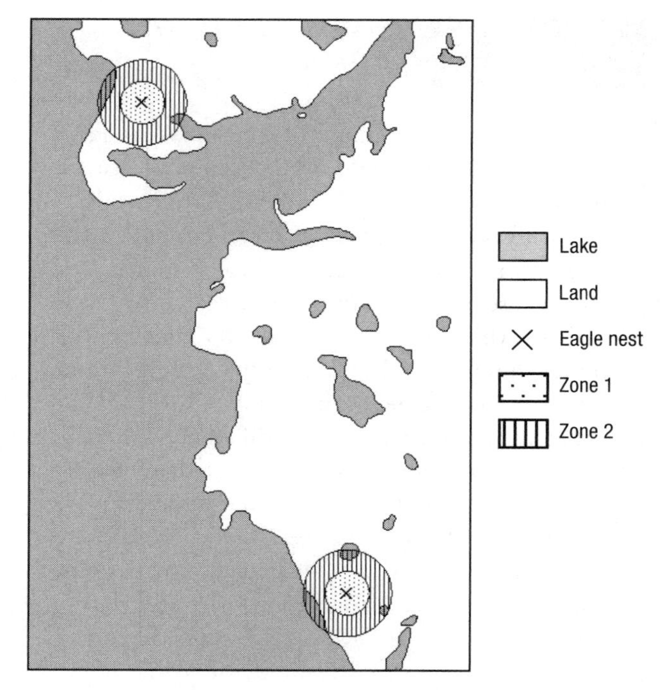

Figure 7.7 Concentric 330-foot buffers around two eagle nests.

illustrates the development of the concentric ring buffers, where Zone 1 extends 330 feet outward from the nest tree, and Zone 2 extends another 330 feet from the edge of Zone 1. Upon closer inspection, you would find that the buffers describing the two Zones do not overlap; therefore, the area underneath is not double-counted.

While these examples describe the use of concentric ring buffering around point features, the process can also be used to develop rings of buffers around line or polygon features, if management objectives suggest they are necessary. For example, riparian buffers for forested areas on Prince Edward Island need to be 20–30 meters wide, yet a 15-meter undisturbed area must be maintained (Legislative Counsel Office, 2006). If you were managing land on Prince Edward Island, and assumed that the maximum forested riparian buffer width would be 30 meters wide, you could develop 15-meter concentric ring buffers around the line features that describe the streams. This would enable the mandatory 15-meter area to be represented on a map, as well as the larger 30-meter buffer boundary.

Buffering Shorelines

The actual or planned management of areas near shorelines of lakes may be of interest to natural resource managers, local land use planners, and citizen stakeholders.

In Depth

Buffer output can take different shapes around the linear features that are being buffered depending on your GIS software. Some GIS software will allow you to specify which side (left or right) of a line to create a buffer, rather than buffering both sides. The trick in this case is to determine which side of your line is left and which is right, a condition which will often be determined by the direction in which your lines were originally digitized or created. Individual lines can also have their ends buffered through a circular shape, which is the usual default, or through a flat shape. If you chose the flat shape the output result of buffering a straight line would be a rectangle. Lines that change direction and are buffered through this process will produce output polygons with flat (non-curved) ends.

As a result, information regarding the extent of actual or planned land uses near lakes may help inform land planning processes. Just as an example, assume that the area of interest near the shorelines of lakes in the Pheasant Hill planning area of the Qu'Appelle River Valley in central Saskatchewan is 300 meters from the edge of each lake. A buffer process can be performed to create polygons that represent areas 300 meters outside of the edge of each lake (avoiding buffering the inside of each lake feature). The result is a new GIS database that contains areas within 300 meters of the shorelines. A clipping process can then be employed to clip the zoning GIS database associated with the Pheasant Hill planning area of the Qu'Appelle River Valley, and obtain the land use classes that are designated within 300 meters of the shoreline (Figure 7.8). A quantitative assessment can then be made of the amount of area of each actual or planned land use within a close proximity to the lake, as well as visual assessment of the juxtaposition of land uses near the shoreline.

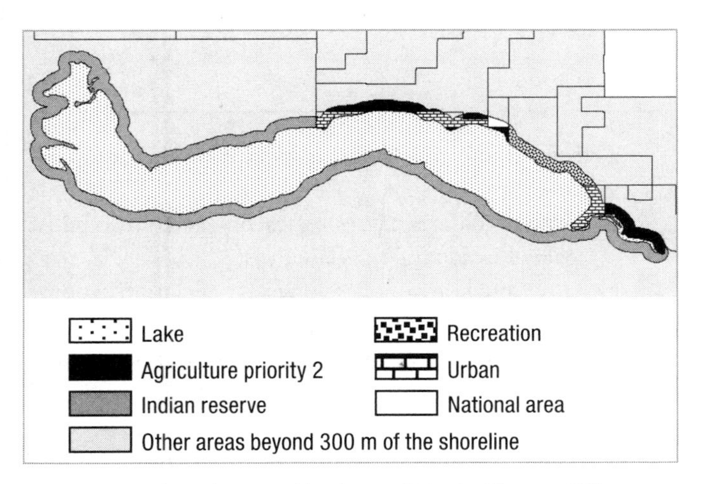

Figure 7.8 Planned or actual land uses within the Pheasant Hill planning area of the Qu'Appelle River Valley, Saskatchewan (1980).

Other Reasons for Using Buffering Processes

The examples provided in this chapter have focused on common natural resource management concerns. The delineation of riparian management areas and zones around wildlife nest locations are two typical examples of using a buffering process in natural resource management planning. There are, of course, a number of other reasons why you would use buffering operations in natural resource management, including:

1. Buffering stream systems to delineate the zone that herbicide operations must keep out of, due to the proximity to water systems. For herbicide operation planning, local buildings (particularly houses), roads, agricultural fields, and orchards may also require buffering.
2. Buffering research areas (plots or stands) to prevent the planning and implementation of logging operations within them. Generally speaking, research plots require extra protection because trees that are treated (given a researchable application) should not also be considered edge trees (trees adjacent to a cleared area), therefore the plots need to be buffered from any nearby harvest activities.
3. Buffering trail systems or roads to delineate areas of visual sensitivity within which logging operations may be limited. In many forested areas, degradation of recreation opportunities is a concern, whether the recreational activities involve humans walking (hiking) or driving. The ability to quickly delineate and visualize these areas of concern is a valuable asset in land management planning.
4. Buffering property boundaries to recognize that development codes, which may limit structures or other

landscape alternations from occurring within a threshold distance from property lines, are observed. This may also apply to easements or utility corridors, which are often subject to municipal, county, or provincial development regulations.

In addition, buffer processes may assist natural resource managers in evaluating the potential impacts of local forest regulations. Local forest regulations are generally concerned with protecting environmental quality and aesthetics, and with safeguarding local government investments in roads and other infrastructure (Martus et al., 1995). They are mainly developed as a result of the conflicts that occur with the continuing shift of the human population from urban to rural settings (Cubbage & Raney, 1987). The types of controls include requiring the development of forest plans, buffering specific landscape features, and placing restrictions on certain silvicultural practices. For example, timber harvesting ordinances developed at the local level are intended to limit site degradation and environmental quality in association with logging activities. Harvesting operations may be restricted within certain distances of public roads or other public resources, therefore requiring a buffer around these resources. Natural resource managers may need to identify these buffers in a harvest plan, and may also be concerned about the cumulative impact of local regulations on the profitability and feasibility of their management operations.

Summary

GIS buffering processes are powerful tools that allow you to investigate the nearness of landscape elements to your features of interest. These features of interest can be represented by points, lines, or polygons (as demonstrated here) or by raster grid cells. Buffering processes are spatial operations that allow users of GIS software programs to identify areas within some proximity of selected landscape features. After landscape features are selected, a zone (a polygon) is delineated around them to represent the buffered area. By default, if no landscape features are selected, all features are buffered, but you should confirm whether your GIS software uses this approach. Buffer output approaches, such as contiguous or uncontiguous results, can be customized to meet analysis objectives. In addition, some GIS software enables users to specify whether buffers created around lines have a round or flat shape at the beginning and ending of the line. The resulting buffer GIS databases allow users to visually understand the area that lies within a certain distance from some landscape feature(s) of interest, and to quantify the resources contained within the buffered area. Buffering streams and other resources of importance (e.g., owl and red-cockaded woodpecker nest locations) to create limited management zones is a common management objective included in natural resource management plans. Estimating the impact of these types of restrictions on natural resource management is important, and helps landowners investigate the effect of current and proposed policies.

Applications

7.1. Current Riparian Policy for the Brown Tract. The current organizational policy for the Brown Tract indicates that the following riparian area buffer widths should be used in conjunction with management activities:

Class 1: 100 feet
Class 2: 75 feet
Class 3: 50 feet
Class 4: 40 feet

Becky Blaylock, the Manager of the Brown Tract, wants to know the 'current situation' with regard to riparian areas:

a) How much area (acres) is located inside the riparian areas?
b) How much area (acres) of each vegetation type is located inside the riparian areas?
c) How much area (acres) of each vegetation type is located outside the riparian areas?
d) How much timber volume is located in the riparian areas?

Ms Blaylock also wants you to develop a map that illustrates the stream buffers and includes the roads, streams, and timber stands.

7.2. Proposed Organizational Policy. Becky Blaylock recently attended a meeting of policy makers, where she heard that the riparian rules might change. She now needs to understand the potential impact of a proposed 125-foot no-harvest buffer around all streams on the Brown Tract. She requests the following:

a) How much area (acres) is located inside the riparian areas?
b) How much area (acres) of each vegetation type is located inside the riparian areas?
c) How much area (acres) of each vegetation type is located outside the riparian areas?
d) How much timber volume is located in the riparian areas?

Again, Ms Blaylock also wants you to develop a map that illustrates these stream buffers and that includes the roads, streams, and timber stands. In addition, she wants you to develop a memo that describes the differences between the policy noted in Application 7.1 and the potential policy noted here.

7.3. Ballot Initiative. A proposed ballot initiative, developed by a local conservation group, suggests that the following riparian buffer widths may soon be required for the region within which the Brown Tract is situated:

Large, fish-bearing streams	150 feet
Medium, fish-bearing streams	100 feet
Small, fish-bearing streams	75 feet
Large, non-fish-bearing streams	125 feet
Medium, non-fish-bearing streams	75 feet
Small, non-fish-bearing streams	50 feet

As a result, Ms Blaylock wants to understand the following:

a) If these rules were applied to the Brown Tract, how much area (acres) would be located inside the riparian areas?
b) If these rules were applied to the Brown Tract, how much area (acres) of each vegetation type would be located inside the riparian areas?
c) If these rules were applied to the Brown Tract, how much area (acres) of each vegetation type would be located outside the riparian areas?
d) If these rules were applied to the Brown Tract, how much timber volume would be located in the riparian areas?

As before, Ms Blaylock also wants you to develop a map that illustrates these stream buffers and that includes the roads, streams, and timber stands. In addition, she wants you to develop a memo that describes the differences between the policies noted in Applications 7.1, 7.2, and the potential policy noted here.

7.4. National Forest Riparian Policy. A local National Forest uses the following riparian area guidelines in conjunction with harvesting activities:

Large streams	250 feet
Medium streams	150 feet
Small streams	100 feet

As with the other policies, Ms Blaylock wants to understand the following:

a) If these rules were applied to the Brown Tract, how much area (acres) would be located inside the riparian areas?
b) If these rules were applied to the Brown Tract, how much area (acres) of each vegetation type would be located inside the riparian areas?
c) If these rules were applied to the Brown Tract, how much area (acres) of each vegetation type would be located outside the riparian areas?
d) If these rules were applied to the Brown Tract, how much timber volume would be located in the riparian areas?

And again, Ms Blaylock also wants you to develop a map that illustrates these stream buffers and that includes the roads, streams, and timber stands. In addition, she wants you to develop a memo that describes the differences between the policies noted in Applications 7.1, 7.2, 7.3, and the potential policy noted here.

7.5. Current Owl Buffer Policy. Suppose that the current policy regarding owl nest locations is to maintain a 100-acre no-harvest buffer around owl nest site locations. Develop a memo and a map for Ms Blaylock that describes the amount of land (acres) of the Brown Tract that would be covered by the single owl buffer pertaining to this property.

7.6. Protection of Research Plots. Thousands of research plots have been established across North America to facilitate the estimation of the response of forests and wildlife to a variety of silvicultural treatments. Unfortunately, harvesting operations are usually implemented independently of research programs, and too often the research plots are harvested because the loggers were unaware of their location, and without the researchers knowing that the plots were in jeopardy of being destroyed. The resulting loss of the research plot investment (layout, tagging of trees, etc.) may be considerable, and the loss of the opportunity for one final measurement may be equally as important. In many organizations, a communication system notifying researchers of potential impacts to research plots has been developed, allowing researchers to either measure the plots one last time, or to delineate the research areas as off-limits to harvesting activities.

Assume that the permanent inventory plots on the Brown Tract were designed to evaluate the long-term growth and yield of the forest property. Assume further that these plots were designed to remain untouched, even though the surrounding forest may be harvested (clearcut or thinned). Finally, assume that the plots need to be maintained with a sufficient buffer of trees around them so that the effects (windthrow, increased sunlight, etc.) of harvesting trees outside the plot (on the plot's trees) are minimized. These plots are circular 1/5-acre plots, and require an additional 100-foot buffer from the edge of the plot boundary to be considered 'protected'. Develop a memo for Becky Blaylock that describes how much land area would be off-limits from harvesting operations as a result of protecting the research plots. Develop a map of the permanent inventory plots and their buffers. Include the roads, streams, and timber stands on the map.

In addition, for added protection of the investment in research, a second 100-foot buffer (for a total of 200 feet) could be delineated around the research plots. In this supplemental area we can also assume that the trees in the buffer are designed to remain untouched, even though the surrounding forest may be harvested (clearcut or thinned). Based on the timber volume contained in the buffers associated with the first (100-foot buffer) and second (200-foot buffer) cases, and a price of $400 per MBF, what is the cost of each case, and therefore what is the additional cost to require the added protection around each plot?

References

Association for Geographic Information. (1999). *GIS dictionary.* Retrieved February 4, 2007, from http://www.agi.org.uk/bfora/systems/xmlviewer/default.asp?arg=DS_AGI_TRAINART_70/_firsttitle.xsl/90.

Bettinger, P., Armlovich, D., & Kellogg, L.D. (1994). Evaluating area in logging trails with a geographic information system. *Transactions of the ASAE, 37*(4), 1327–30.

Bren, L.J. (2003). A review of buffer strip design algorithms. In E.G. Mason and C.J. Perley (Eds.), *Proceedings of the 2003 ANZIF (Joint Australia and New Zealand Institute of Forestry) Conference* (pp. 326–35). Retrieved September 5, 2007, from http://www.forestry.org.nz/articles/conf2003/Bren.pdf.

Cubbage, F., & Raney, K. (1987). County logging and tree protection ordinances in Georgia. *Southern Journal of Applied Forestry, 11*, 76–82.

Legislative Counsel Office. (2006). *Chapter E-9, Environmental Protection Act.* Charlottetown, PE: Government of Prince Edward Island. Retrieved February 4, 2007, from http://www.gov.pe.ca/law/statutes/pdf/e-09.pdf.

Martus, C.E., Haney, Jr., H.L., & Siegel, W.C. (1995). Local forest regulatory ordinances: trends in the eastern United States. *Journal of Forestry, 93*(6), 27–31.

Maryland Department of Natural Resources. (2005). *Sustainable forest management plan for Chesapeake forests lands, Chapter 8, Wildlife habitat protection and management.* Annapolis, MD: Maryland Department of Natural Resources, Forest Service. Retrieved February 4, 2007, from http://www.dnr.state.md.us/forests/download/sf_mgt_plan_chapters/chapter8cfl.pdf.

McDonald, T.P., Carter, E.A., & Taylor, S.E. (2002). Using the global positioning system to map disturbance patterns of forest harvesting machinery. *Canadian Journal of Forest Research, 32*, 310–19.

Merriam-Webster. (2007). *Merriam-Webster online search.* Retrieved February 4, 2007, from http://www.m-w.com/cgi-bin/dictionary.

Oregon Department of Forestry. (1994a). *Oregon forest practices rules and statutes.* Salem, OR: Oregon Department of Forestry.

Oregon Department of Forestry. (1994b.) *Forest practice water protection rules; Divisions 24 and 57*. Salem, OR: Oregon Department of Forestry.

Oregon State Legislature. (2005). *Chapter 527—insect and disease control; forest practices*. Retrieved February 5, 2007, from http://www.leg.state.or.us/ors/527 .html.

Society of American Foresters. (1983). *Terminology of forest science technology practice and products*. Bethesda, MD: Society of American Foresters.

Star, J., & Estes, J. (1990). *Geographical information systems: An introduction*. Englewood Cliffs, NJ: Prentice Hall.

Chapter 8

Combining and Splitting Landscape Features, and Merging GIS Databases

Objectives

The objectives of this chapter are to provide readers with an understanding of the opportunities related to, and potential pitfalls associated with, using a GIS to combine and split landscape features. In addition, since merging two or more GIS databases together is similar to combining landscape features, another objective is to describe the pros and cons associated with this GIS process. After completion of this chapter, readers should have the knowledge and ability to understand:

1. why, when, and how you might want to combine landscape features;
2. the reasons for splitting landscape features, and the situations where this process might be appropriate; and
3. why two or more GIS databases might be merged, and what you would expect to find within a merged database.

In the previous chapters an emphasis was placed on understanding how much of a resource (for example, the length of road or an area of land) was located within a certain geographic region, such as within a set of stream buffers. Queries were used, along with clipping, erasing, and buffering processes, to determine the size of the resources in question. In performing these analyses, interest was placed only in the end result of the set of GIS processes performed, a result that most likely included some very precise and accurate, yet perhaps unexpected, landscape features. For example, as a result of performing a clipping process, numerous spurious polygons might have been created; polygons so small that it would seem unreasonable to manage the land they represent in a cost-effective manner (let alone find them on the ground).

In this chapter GIS processes are introduced to help accomplish two goals: (1) clean up GIS databases and (2) facilitate more efficient spatial analysis processes. The GIS processes emphasized relate to combining, splitting, and merging landscape features. In introducing these GIS processes, examples ranging from facilitating wildlife habitat analysis to estimating unrestricted (from a forest management perspective) areas in a landscape are used to help readers understand the usefulness of these procedures within a natural resource management context.

Combining Landscape Features

Multiple landscape features within a single GIS database can be combined to produce a single landscape feature. The **combine process** generally begins by assessing what the landscape features of interest have in common. To make the management of GIS data more efficient, similar landscape features could be combined so that a smaller number of features are contained in a spatial database. These similarities could be associated with spatial position (features touch one another), or attribute values (they have the same characteristics—age, type, etc.). In Figure 8.1, for example, two polygons are to be combined based on their current spatial position—they share an edge.

After the combine process is completed, we find that the edge that was shared was eliminated, because the line that defined the shared edge was not needed to define the boundary of the new polygon. In a spatial database, the initial two polygons represented in Figure 8.1 would each be stored separately. In other words, two polygons and two polygon records would be represented. After combining the polygons, only one polygon record would be contained in the resulting database. Although the reduction in database complexity is modest in this example, it may be substantial when hundreds or thousands of spatial features are involved. In addition, although this example featured two adjoining polygons, the landscape features being combined may physically overlap, or may be physically separated by a gap. In the case of overlapping landscape features, the overlap is eliminated when the polygons are combined through the creation of a single polygon representing the overlap area in the output database. In the case of physically separated landscape features, the combined landscape feature is composed of two or more distinct pieces. In some GIS software programs, a combine process is known as a dissolve process.

During the management of GIS databases, or as a result of an analytical need, it might be necessary to combine landscape features. Landscape features should be of the same feature type to be combined. In general, polygons are combined with other polygons, and lines with other lines. The reasons for combining landscape features are numerous, but are generally based on the fact that it is easier to manage a smaller database (one with fewer records) than a larger one. In addition, spatial analysis and data storage considerations are typically more efficient with a smaller database. There are at least six reasons why you might want to combine landscape features:

a. Prior to combining the polygons

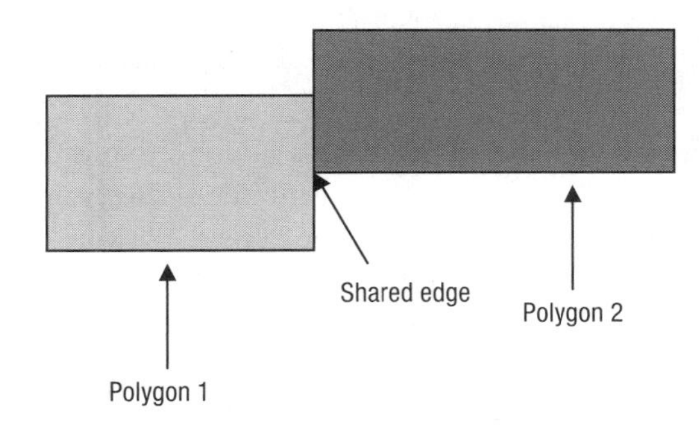

b. After combining the polygons

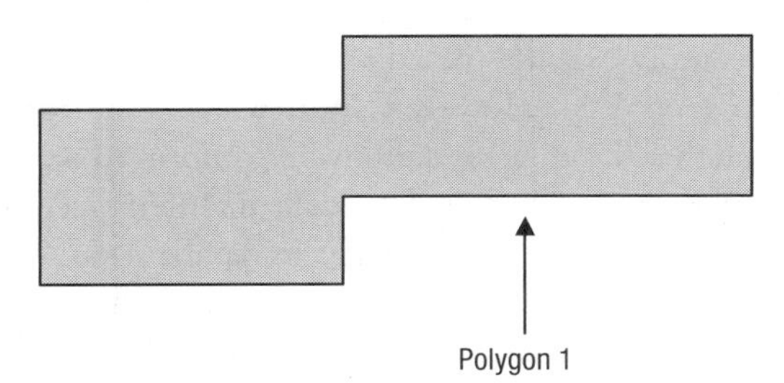

Figure 8.1 Combining two polygons, by eliminating a shared edge, to produce a single polygon.

1. You may wish to eliminate unintended small landscape features that were created through digitizing or some other GIS process (e.g., clipping, erasing) that affected the geometry of a line or polygon spatial database. Combining landscape features can then be effectively used to reduce the number of features being managed and can correct unintended features. For example, after a GIS process, spurious polygons may be present. Spurious polygons are fractions of polygons broken or created as a result of a GIS process, and you may want to combine them with other neighboring polygons to reduce the number of polygons being managed.
2. Changes in organizational policies may suggest that some spatial features need to be combined. For example, an organization may redefine the minimum mapping unit that it manages. The minimum mapping unit defines the smallest sized unit that should be present in certain types of GIS databases. A change in an organizational policy, say to increase the size of the minimum mapping unit, may require eliminating some small polygons or lines. Making the minimum mapping unit larger not only reduces the volume of data to be managed (a benefit), but also reduces the resolution of landscape features recognized (a cost). Making the minimum mapping unit smaller has the inverse effect: a higher spatial resolution is recognized, and more data needs to be managed.
3. The acquisition of GIS databases from other sources (other than developed internally within a natural resource management organization) may prompt the use of a combine process. You may find within an acquired GIS database, that the mapping unit standards are inconsistent with the standards used by their organization. In addition, existing administrative boundaries, be they socio-political (ownership) or natural (watershed areas), may not be desired in a spatial database. For example, a private company may acquire

a GIS database from the US Forest Service. The acquired database may include polygons smaller than what the private company typically manages, suggesting that some landscape features need to be combined to adhere to the minimum mapping unit standard that the company uses. A polygon layer that contains watershed boundaries at a sub-regional scale may provide unneeded detail if a regional watershed boundary will suffice. Sub-regional watershed boundaries could be aggregated through a combine process.

4. Combining landscape features may be necessary because it simply makes sense from a management perspective. For example, in a stands GIS database there may reside two polygons side-by-side that describe forested areas with trees of similar ages, similar structural conditions, similar site classes, and similar growth rates. The field personnel responsible for managing these stands might also suggest that the two stands would be treated with similar treatments, at about the same time, and with similar equipment. Thus from a management perspective, combining the two polygons into a single polygon within a GIS database might make sense.
5. Since landscape features can change in shape and characteristic over time, managing these changes may suggest using a combine process. For example, the condition of a road may change due to improvements made to it over time. A rocked road may become a paved road, a native surface road may become a rocked road, or a road of any type may become a decommissioned (obliterated) road. If conditions of landscape features change, it may seem reasonable to combine those landscape features with other adjacent landscape features of the same stature or condition. A similar example might involve stream network measurements such as what occurs during a standard watershed analysis project. Field stream crews will visit selected streams and segregate stream networks into similar hydrologic or geomorphic categories, or categories related to the surrounding land cover. These groupings are stored in a spatial database as separate line features with associated data records containing descriptive information. Some combining of streams in a GIS database might be necessary to manage the stream network more efficiently.
6. It may be appropriate to combine landscape features to facilitate a spatial analysis. For example, to delineate one category of the recreational opportunity spectrum, we may need to identify the size (area) of contiguous timber stands where the average age of trees is 50 years

In Depth

Combining processes must be used in a thoughtful manner. Once landscape features are combined, the topology that describes the resulting landscape feature is altered in the output database. Should a user decide to delete the input databases (a common thought once a new database has been created) the original description of the landscape topology is lost. Given this risk, larger natural resource organizations have generally placed the decision to edit and manage GIS databases (where combining landscape features may be necessary) to the person who has been given 'ownership' of the GIS database. All decisions regarding the development and maintenance of a given GIS database are then the responsibility of the database owner. Other individual users of the GIS database, however, can perform GIS processes, such as combining features, on copies of the original GIS database. However, once performed by someone other than the owner of the database, the altered database is no longer considered the official database of the organization. For example, John Goheen, a GIS analyst, may be the 'owner' of a roads GIS database within an organization. John may then give a copy of this database to Paul Chapman, a field forester. If Paul were to edit the database by combining or splitting roads, the copy of the GIS database that Paul uses will not be considered the official roads GIS database of the organization, even if the database helps Paul make better management decisions. Can Paul simply provide the edited roads GIS database to John? Certainly. However, John will likely need to ensure that the changes Paul made conform to organizational standards related to data maintenance, and then subsequently verify that the database does not contain any errors. If John can do these things, the data edited by Paul can be incorporated into the original (and official) roads GIS database.

or greater. Using a query, we can identify these stands, but to determine how large the contiguous area might be would require (a) either combining the queried stands, or (b) sum by hand the area of all adjacent polygons meeting the size requirement. The latter technique may lead to error, thus combining the queried polygons may be more appropriate. A combine or dissolve process (depending on the GIS software being used) would help facilitate this analysis.

Some balance must be struck between the appropriate number and size of landscape features being managed, and the amount of real-world generalization that occurs when fewer features are used to represent a landscape. The decision to combine landscape features should be made after a serious contemplation of these issues. The following three examples describe the end-result of combining landscape features.

Contiguous, similar landscape features

Suppose you queried the Brown Tract stands GIS database for even-aged stands between the ages of 40 and 45, using the following query:

(Land allocation = 'Even-Aged') and (Age $\geq$ 40) and (Age $\leq$ 45)

From this query we find that several stands touch each other, or are contiguous (Figure 8.2). From a management perspective, you may decide to combine some of these stands together (such as those illustrated in Figure 8.3), to reduce the number of management units tracked in a database. There is one important issue that must be kept in mind before combining landscape features: you should make a note of the attributes of each landscape feature before combining them. The combine GIS process, depending on the GIS software program being used, will either (a) contain the attribute data related to one of the landscape features, (b) contain the attribute data related to the other landscape feature, (c) contain an average, or some other statistical summary, of numeric data associated with all combined features, or (d) not contain any attribute data of either landscape feature. Users should take heed of the options available when running a combine process so that the intended output results. Additionally, in some GIS software programs there may be more than one process that effectively combines landscape features, and the resulting combined landscape feature may contain either no attribute data, or the attribute data of the first landscape feature selected for combining

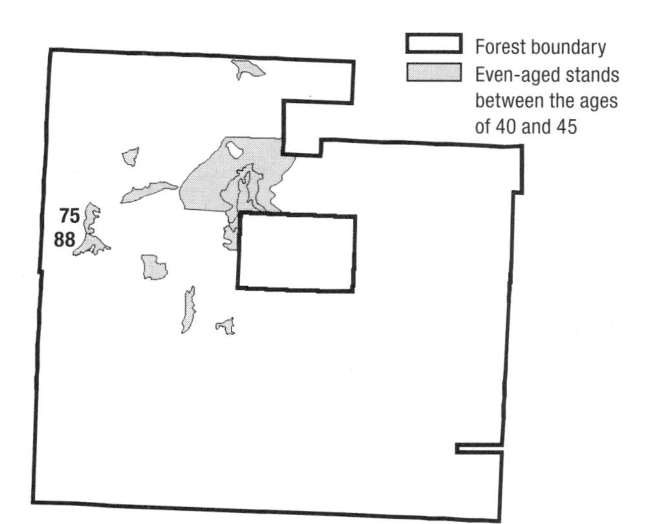

Figure 8.2 Stands on the Brown Tract that are even-aged, and between the ages of 40 and 45.

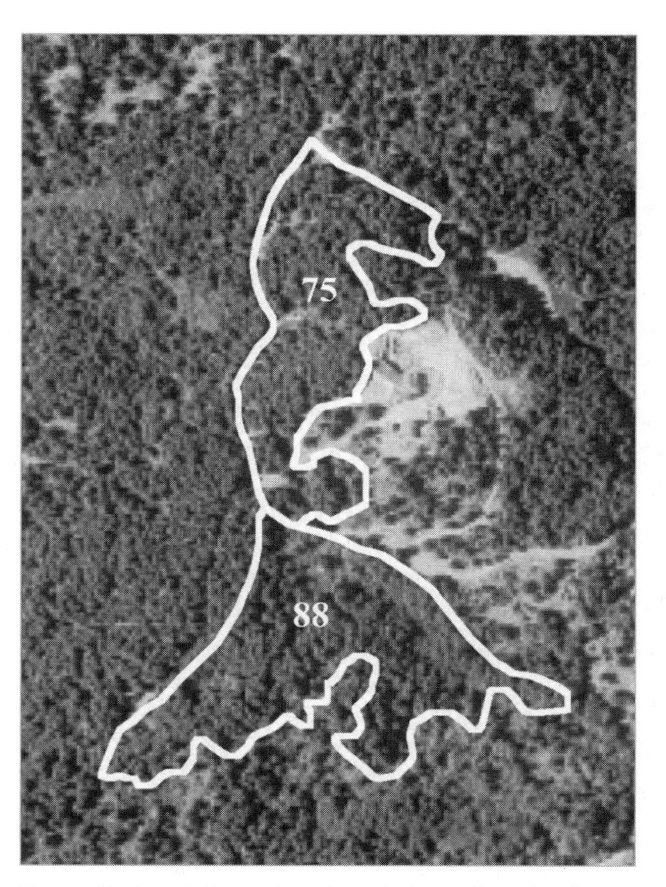

Figure 8.3 Two similar-aged stands on the Brown Tract that share a common border. Both are even-aged, and between the ages of 40 and 45.

TABLE 8.1	Results of combining two stands						
Stand	**Acres**	**Hectares**	**Age**	**Site index**	**Trees per hectare**	**Height (m)**	**Board feet per hectare**
Both stands before using a 'combine features' process							
First stand selected							
75	7.5	3.0	44	100	250	23	15,325
Second stand selected							
88	9.9	4.0	45	117	492	28	39,388
Combined stand after using a 'combine features' process							
0	0	0	0	0	0	0	0
Combined stand after using a 'union features' process							
75	7.5	3.0	44	100	250	23	15,625

(Table 8.1). In either case, the attribute data of the combined landscape feature may need to be edited if the combined landscape feature will represent a weighted average of the conditions of the original two landscape features.

Multiple spatial representations within a single landscape feature or record

At first glance, you might expect that each landscape feature, such as a forest stand, would be considered a single record in a GIS database. However, this is not necessarily true. Discontinuous landscape features can be combined to produce a single landscape feature describing a portion of the landscape (Figure 8.4). Thus, another consideration in the development and management of GIS databases is whether spatially discontinuous landscape features should be represented with one database record or with multiple records. In the case of stand 283 on the Brown Tract, for better or for worse, a single database record represents two areas (regions). Perhaps prior to the development of the rock pit that now separates these two areas, the stand was represented by a single contiguous polygon.

The Brown Tract databases are, as mentioned earlier, not perfect. However, they do allow an examination of some rather typical problems users of GIS databases must consider, such as this one. There are three options related to the GIS database management of stand 283:

1. Leave the stand as it is—represented by two spatially discontinuous regions, yet a single database record. This may be consistent with the standards used by the managers of the Brown Tract. This option would require no additional effort to manage the GIS database.
2. Split the stand into two separate parts, creating two separate stands. Separate database records would then represent each stand. The owner of the GIS database may decide that there is a sufficient amount of time and budget to comb through the database, locate inconsistencies such as this, and correct them. Splitting stand 283 into two separate stands might be considered a logical response; splitting landscape features will be discussed in more detail shortly.

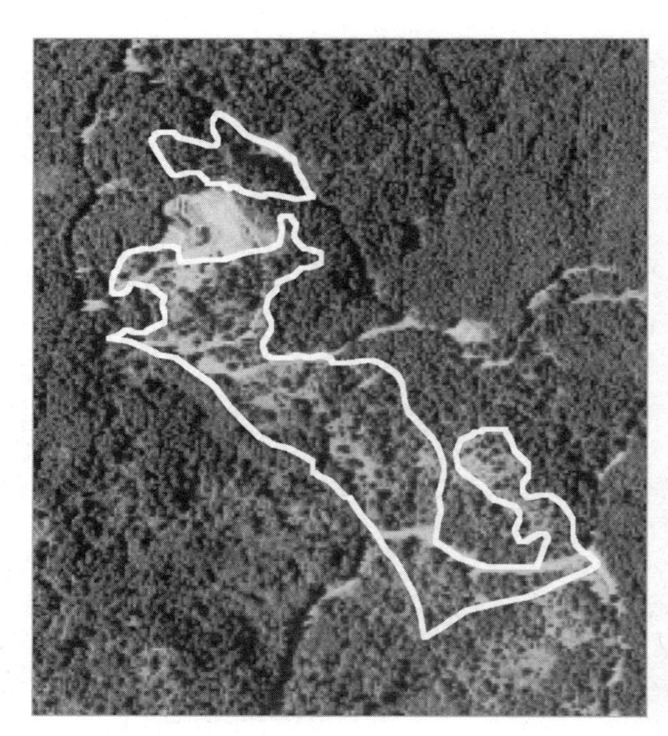

Figure 8.4 Two polygons (regions) represented by a single database record in the Brown Tract stands GIS database (stand 283).

3. Combine the small discontinuous piece of the stand into another adjacent stand. Combining the small portion of stand 283 to another stand that is adjacent to the small portion would require that the adjacent stand have similar characteristics (age, volume, density, etc.) appropriate for the management of the potential combined area.

Overlapping polygons

Although it may not be your intent to create landscape features that overlap when developing or maintaining a GIS database, overlapping features may result as the output of a GIS process. There are numerous reasons why you may find overlapping features in a GIS database; edit, buffer, and merge processes can all lead to the development of overlapping landscape features, especially with GIS software that does not enforce topological rules. When editing polygon features within GIS databases, for instance, you can easily affect the shape of polygons such that they either overlap or not touch at all (Figure 8.5). When editing GIS databases, it is wise to understand the process available within GIS software programs to 'snap' the vertices of one landscape feature to those of another. With the ability to snap vertices together, you can edit the shape or position of polygons and allow a precise match of the boundary of one to that of another. The challenge for most GIS users is to remember to activate the snapping ability. It also requires practice, once activated, to learn how to use snapping tools correctly. Practicing on a test GIS layer before editing an actual database can help reduce snapping errors.

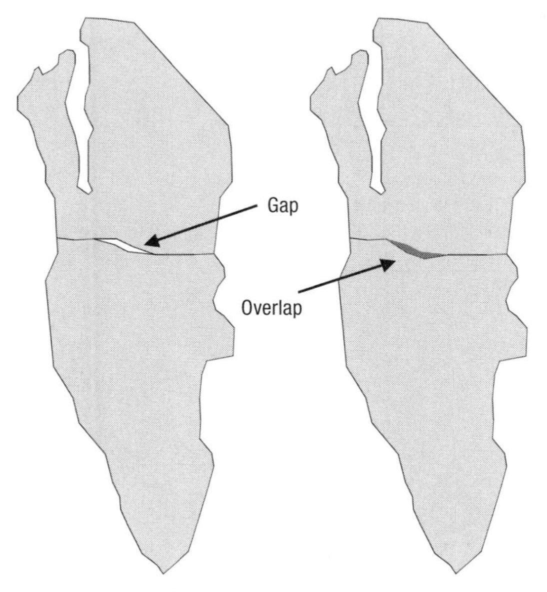

Figure 8.5 Overlap and gap remaining after editing polygon boundaries.

A discussion of buffering point, line, and polygon features was provided in chapter 7. The polygons that are created as a result of a buffer process can, perhaps, overlap. Usually you have the choice, at the time of buffering, of maintaining the overlapping areas or directing the GIS software program to remove them (Figure 8.6). In the case where the overlapping areas of the buffer remain, these polygons can subsequently be combined, removing the overlap and reducing the number of polygons and database records that describe the buffer.

Merging polygon GIS databases, as will be described later in this chapter, can also result in overlapping polygons. When a merge process is used, the overlapping areas among polygons and lines are generally not affected, meaning that no new nodes are created at the intersection of lines, and that overlapping polygon areas are not removed. Quite simply, when GIS databases are merged,

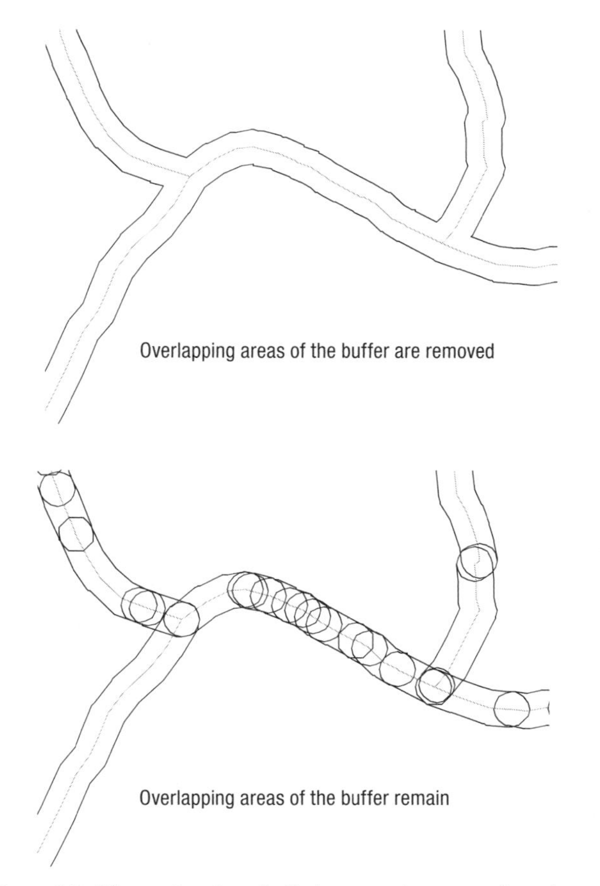

Figure 8.6 The results of two buffering operations, one where the overlapping areas of the buffer around each stream are removed, and the other where the overlapping areas remain.

In Depth

How would you know whether a polygon GIS database contains overlapping polygons? One method might be to examine very closely the boundaries of each polygon—a tedious process. Another method may be to compare the sum of the area of polygons in the suspect GIS database with another polygon GIS database of the exact same spatial extent. For example, if you were concerned about overlapping polygons in the Daniel Pickett forest stands GIS database, you could compare the sum of the area of the stands to the area of the boundary GIS database. The extent of the two databases should be exactly the same: the stands GIS database should contain polygons that are a subdivision of the ownership defined by the ownership boundary. The stands GIS database has many polygons, and the boundary GIS database has only one polygon, yet the sum of the areas should match. You could also choose to combine all polygons within a copy of the stands layer. The area of the resulting combined stands database should also match. If the sum of the areas does not match, either one or more overlapping polygons exist in the stands GIS database, or some gaps exists between polygons in the stands GIS database.

one set of landscape features is simply laid on top of another set.

Splitting Landscape Features

The decision to use a **splitting process** generally arises from some need to redefine the topology of spatial data. The word 'split' is defined as a process to divide or separate an item into parts or portions (Merriam-Webster, 2007). Within GIS, we use a splitting process to divide polygons and lines (but not points) into smaller parts or portions. Points do not describe areas; however, buffers around points can be split because they are polygons.

To illustrate several reasons for using splitting processes, consider the sub-dividing of land ownerships that regularly occurs in many rural and urban areas throughout North America. Many property owners subdivide their lands in order to earn revenue or bequeath property to land stewardship organizations or heirs. Land ownership records and GIS database are kept by county, provincial, or metropolitan organizations and must be updated to accurately represent new parcels that result from subdivisions. Splitting processes are used to separate parcels from one another and lead to the creation of additional polygons in a GIS property boundary database. As another example involving polygons, imagine a state or province where clearcut size limits are imposed. For organizations that plan clearcut activities in these regions, the forest management units represented in their stands GIS database should probably be smaller in size than the maximum clearcut area allowed. Therefore, these polygons should probably be broken down (split) into smaller management units to enable land managers to plan harvests more appropriately (Figure 8.7). In terms of linear

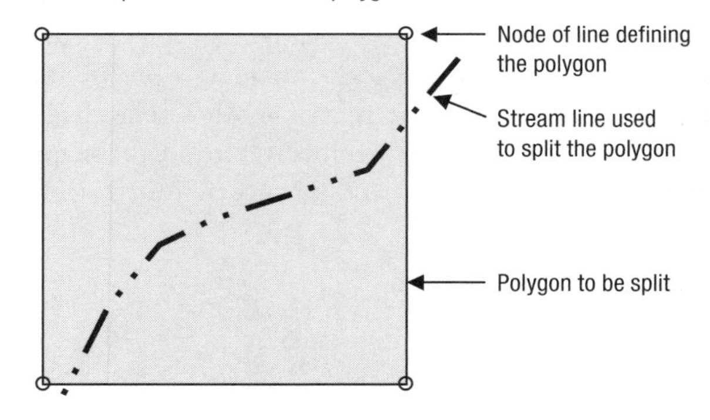

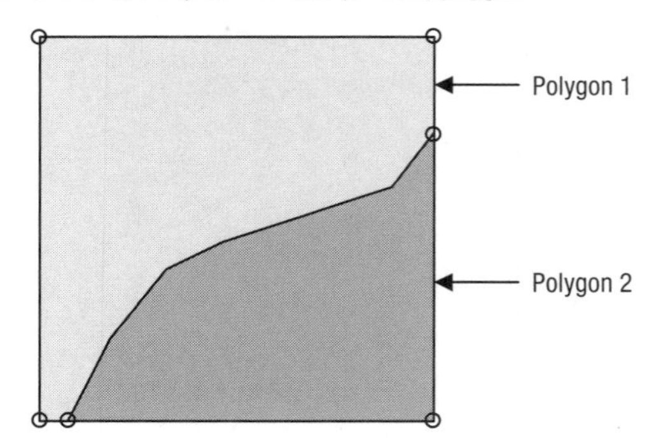

Figure 8.7 A 160-acre (64.8 ha) polygon split along a stream, forming two smaller forest management units.

features, assume that you are involved with the planning and maintenance of a road system for a natural resource management organization. Over some period of time, the status of a portion of a woods road may have changed, perhaps from a rocked surface to a paved surface (or alternatively, part of the road was obliterated in a restoration process). Within the roads GIS database, splitting the line that represents the road at the location of the status change would seem appropriate, since each resulting piece of the road should be described by different attribute data. The same argument for a splitting process can be used for stream data management. In this case, assume a recent stream survey identified some differences between the GIS data and the actual stream system. Some stream reaches, for example, may have been found to be able to support fish populations, yet the GIS data indicates otherwise. In these cases, the various lines that represent streams may need to be split to better delineate the capacity to support (or not support) fish populations.

The way you go about splitting polygons or lines (yet not points) varies according to the GIS software program being used. In some GIS software programs, splitting is as easy as drawing a line through a landscape feature (Figure 8.7). This holds true for polygons or lines that are solid or continuous. However, where a landscape feature is represented by multiple regions (or objects), and a gap separates these regions (such as that illustrated in Figure 8.4), drawing a line through the gap, without touching either of the pieces of the landscape feature, may not result in the feature being split into two separate pieces. In these cases, a more complex process using combining, clipping, editing, or pasting processes may be more applicable (Figure 8.8). Alternatively, some GIS software programs have processes for converting multipart features to single

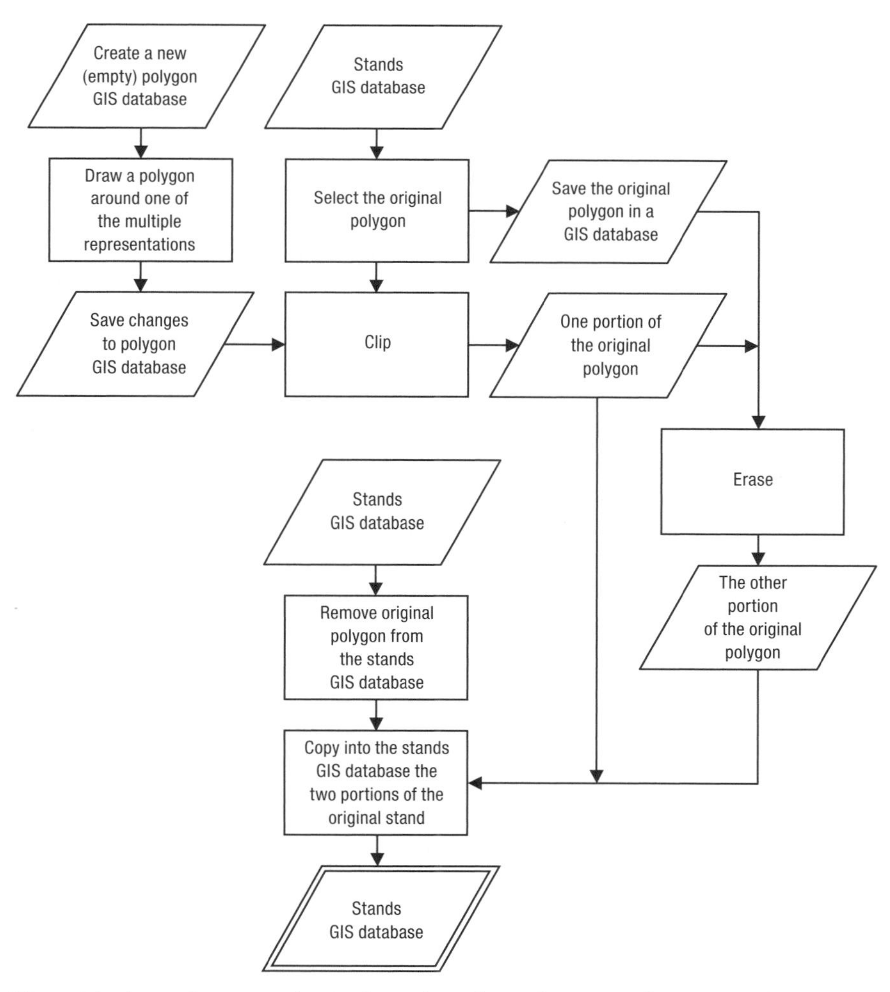

Figure 8.8 A complex process that can be used to split a polygon, initially represented by two regions, into two separate landscape features.

pieces. These types of automated processes make the task of separating non-adjacent pieces of spatial data more efficient.

Merging GIS Databases

Merging, for the purpose of managing GIS databases, is defined as the process of combining multiple GIS databases into a single database. Therefore, when you use a **merge process**, a new GIS database is created from a set of two or more previously developed GIS databases. Point, line, and polygon databases can be merged together, however, the database structure in the resulting GIS database will be mixed. Generally, merged databases contain the same GIS database structure as the input databases. For example, you might merge several polygon GIS databases together. The resulting merged GIS database would then include all of the polygons from each of the original GIS databases, and could potentially contain polygons that overlap. You must therefore be aware that a summary of the characteristics of the resulting merged GIS database may result in an overestimate of areas or distances (in the case of merged line databases).

There are two main reasons why a merge process might be used: (1) to create a template to facilitate a subsequent spatial analysis process, and (2) to facilitate mapping processes. In the first case, suppose that from a forest management perspective you were interested in defining those areas that have no restrictions placed on them either by regulatory or organizational policies. You might call these 'unrestricted areas'. In these areas the full range of silvicultural practices appropriate to the forest types and the soil and slope conditions can be considered. To define these unrestricted areas on a map, you might first attempt to define the restricted areas: those that are within a certain distance of the stream system (requiring a buffer), near some endangered species habitat (requiring a buffer), or designated as research areas (requiring a query or buffer). These are the areas within which management may be restricted to some extent. Since the intent is to delineate and summarize the characteristics of the remaining areas within the landscape being managed, the GIS databases that describe the stream buffers, endangered species habitat buffers, and the research areas (all represented by polygon GIS databases) could be merged into a single polygon GIS database. This merged GIS database can then be used in conjunction with an erasing process to delineate those areas on the landscape that are outside of the land area that they cover. When we use the merge process for this purpose, our concern is not placed on the likely overlapping polygons in the resulting merged GIS database, but on making the analytical process more efficient. With a merged GIS database that represents restricted areas, a single erasing process can be used to arrive at the unrestricted areas (full property – restricted areas = unrestricted areas). Without developing the merged GIS database, three erasing processes would have been needed: (1) full property – stream buffers = temporary database 1; (2) temporary database 1 – endangered species habitat = temporary database 2; (3) temporary database 2 – research areas = unrestricted areas.

In the second case, where a merge process facilitates a mapping process, our concern is not placed on the likely overlapping polygons in the resulting merged GIS database, but rather on the message that we communicate with the printed map. Using the previous example, the message we want to communicate to the map customers would be 'Here are the restricted and unrestricted areas.' By merging several GIS databases together and creating a single GIS database of a common theme ('restricted areas'), you could make the cartographic process more efficient. For example, rather than needing to specify the color scheme for the three GIS databases that represent the restricted areas (stream buffers, endangered species habitat buffers, and research areas), the color scheme of a single GIS database (the merged database representing the restricted areas) needs only to be specified. In addition, many desktop GIS software programs now contain automated functions to assist users in developing their maps. Without merging the restricted areas into a single GIS database, we may find several GIS databases listed in the map's legend—each representing portions of the restricted area. In this case, management of the map's legend may be required before we print it and present it to our customers.

Determining how much land area is unrestricted

To demonstrate using a merge process, the following example will expand on the discussion of developing a representation of unrestricted areas by applying the concept to the Daniel Pickett forest. Knowing what the operable 'decision space' is on the forest may be important when decisions regarding harvesting, herbicide operations, etc., must be made. As with many of the tasks presented in this text, there are several pathways you can take within GIS to complete an analysis of this type. The focus

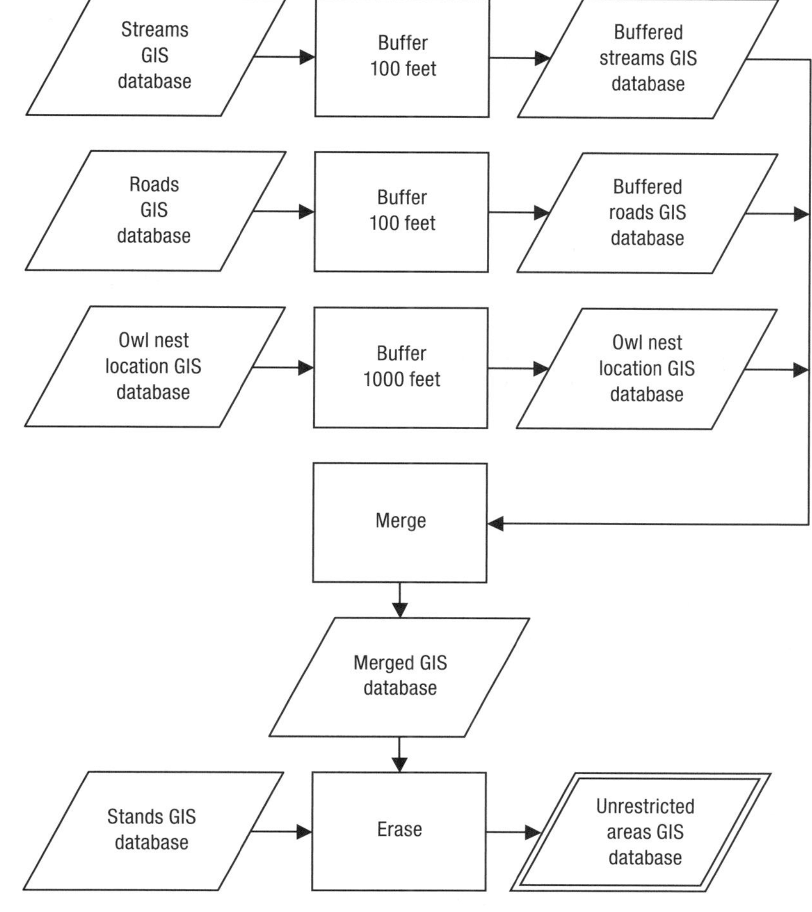

Figure 8.9 A process that can be used to delineate unrestricted areas in a forested landscape.

here will be on developing several GIS databases that describe the types of restricted areas, then on merging them together as a single GIS database. The polygons contained in the restricted area GIS database can then be erased from the boundary or stand GIS databases, leaving unrestricted areas as a result. The process can be described with a flow chart (Figure 8.9) to help visualize the steps necessary to complete the task.

In the delineation of unrestricted areas of the Daniel Pickett forest, some criteria area needed to describe the restricted areas:

- 30.48 m (100 feet) around all streams,
- 30.48 m (100 feet) around all paved roads, and
- 304.79 m (1,000 feet) around the owl nest locations.

Using the process illustrated in Figure 8.9, you will find that 140.8 hectares (348 acres) of the Daniel Pickett are restricted in some form or fashion (Figure 8.10), and 870.9 hectares (2,152 acres) are thus unrestricted. The unrestricted areas, given the criteria noted above, are open to the full suite of management activities appropriate for the forest types and landscape conditions of the Daniel Pickett forest.

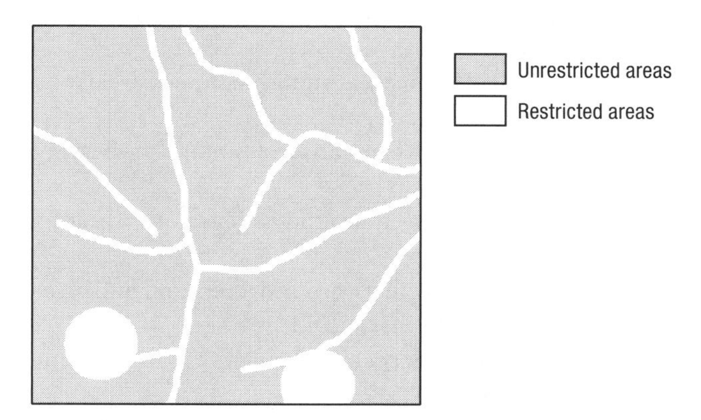

Figure 8.10 A description of the restricted and unrestricted areas on the Daniel Pickett forest.

Summary

The need to combine and split landscape features and the need to merge GIS databases together are influenced by the progression of GIS processes you choose in order to address a management issue. If there is a need to eliminate small landscape features created by some other GIS process (e.g., clipping, erasing), combining these with other landscape features would seem appropriate. If there is a need to combine multiple features within a single GIS database (regardless of their size) to facilitate further analysis, combining them would again seem prudent. If there is a need to physically separate pieces of a polygon or line, then a splitting process would be warranted. If there was a need to combine features contained in separate GIS databases, a merge process would be appropriate. The decision to combine or split landscape features, or to merge GIS databases must not be made lightly. While there may be a variety of logical reasons for using these GIS processes, there may be another reason (such as the content of the resulting attribute table) that suggests an alternative process should be used. Documenting the workflow when using these processes is therefore important.

Applications

8.1. Characterizing unrestricted areas. The Region Manager associated with the Brown Tract, Becky Blaylock, is very interested in the management opportunities associated with this land, given its proximity to an area of suburban growth. Since she knows you know something about GIS, she has again come to you for some information. Specifically, she is interested in understanding the extent of the forest resources that are outside of areas where forest management is restricted for one or more reasons (either by regulation or by an organizational policy). She defines the zones where forest management is restricted as:

- Areas within 152.4 m (500 feet) of authorized trails;
- Areas within 152.4 m (500 feet) of homes;
- Areas within 2.4 km (1.5 miles) of owl nest locations;
- Areas within the riparian zones:
  - 30.5 m (100 feet) around large fish-bearing streams,
  - 21.3 m (70 feet) around medium fish-bearing streams,
  - 15.2 m (50 feet) around small fish-bearing streams,
  - 21.3 m (70 feet) around large non-fish-bearing streams,
  - 15.2 m (50 feet) around medium non-fish-bearing streams, and
  - 6.1 m (20 feet) around small non-fish-bearing streams; and
- Stands with the following land allocations: Meadow, Oak Woodland, Research, and Rock pit.

a) How much area of land is unrestricted?

b) How much area of unrestricted land is included in the following land allocations?
   a. Even-aged
   b. Shelterwood
   c. Uneven-aged

c) Develop a map of the Brown Tract, illustrating the unrestricted areas. Include on the map the road and stream systems.

8.2. GIS processing. How would you have addressed problem 8.1 if you incorporated a process that merged the GIS databases that represented unrestricted areas? How would you have addressed problem 8.1 if you incorporated a process that combined all landscape features representing unrestricted areas into a single polygon? Provide a flow chart for each alternative process.

8.3. Combining landscape features. Your co-worker, Karl Douglas, has suggested during one of your monthly inventory meetings that all management units in the stands GIS database smaller than four hectares should be combined with another adjacent management unit. His argument is that this will make the process of managing the forest more efficient. Besides the fact that some small polygons may represent significant landscape features (rock pits, wildlife habitat, etc.), and thus should be distinctly represented in a GIS database, what argument might you provide against this potential change in GIS database management policy, particularly from the GIS processing and inventory management perspectives?

References

Merriam-Webster. (2007). *Merriam-Webster online search*. Retrieved April 28, 2007, from http://www.m-w.com/cgi-bin/dictionary.

Chapter 9

Associating Spatial and Non-spatial Databases

Objectives

This chapter provides an introduction to techniques that will allow you to associate features in spatially-referenced GIS databases with data from other sources (other types of databases) which may not have an explicit spatial reference. Once this chapter has been completed, readers should have a firm understanding of:

1. how two or more databases can be temporarily combined without creating a new database, modifying a database table, or modifying landscape features;
2. what types of GIS processes are available when there is a need to associate data from different sources;
3. how non-spatial data can be associated with spatial databases, and how data from one spatial database can be associated with data of another spatial database; and
4. what it means to relate (link) two tables, and how this process is different than joining databases.

In the last few chapters a concentration was placed on taking two (or more) GIS databases and, with a process such as erasing or clipping, creating a third (new) GIS database that featured different, or modified, landscape features. The GIS databases were, in essence, combined in a *permanent* fashion in the new database. In this chapter the databases will be combined in a way that is *temporary* but that still retains each database's original structure (both the spatial structure and attribute table structure). Since new landscape features are not being created in this process, there is no need to create a new GIS database; however, this is also possible and we will discuss this process later in the chapter. To present additional methods of combining databases, two processes of association will be introduced: the join and the relate (link) processes.

The first section of this chapter examines joining data in non-spatial databases with landscape features in spatial databases, and describes several of the common join processes you might encounter. The relationship between the non-spatial data and the landscape features is based on a common attribute value found in both databases. The second section of the chapter examines joining landscape features from two different GIS databases, and how the relationship between the landscape features in each GIS database is a function of the location and type (point, line, or polygon) of the features on the landscape. The third section of the chapter discusses how you can make the temporary joined associations among databases permanent. The final section of the chapter examines relating (linking) features or data from one database to that of another. With a relate process, you can view one of the related databases, select a landscape feature, then view the associated related data in the other database even though the database will not appear to be physically associated. Join processes do not follow this approach, but, instead, result in a single database that contains data from both joined databases.

Joining Non-spatial Databases with GIS Databases

When you want to join two databases together the objective is to associate both databases such that a single database results. Therefore, data from one database (in this case a non-spatial database) is transferred to the attribute table of the other database (the spatial database). There are several types of possible associations when joining non-spatial databases with spatial GIS databases. The two most common join associations are one-to-one and one-to-many joins. A non-spatial database is one that lacks associated landscape features and their geographic reference. Perhaps the simplest example is a text file that you can create in a word processor or text editor (Table 9.1). You might logically ask why, if a GIS database of landscape features exists, would you store other data associated with those features in a separate, non-spatial database? Perhaps there are instances where it is more efficient for analysts (or foresters, biologists, etc.) to develop data (such as wildlife habitat suitability scores) separate from GIS, knowing that a process exists to quickly associate the data developed back to the appropriate landscape features. Perhaps an analyst may want a separate tabular database that they can then import into a statistical software program. In addition, there are other software packages, such as hydrologic simulators, growth and yield models, and landscape analysis models, that can take GIS output, process the output so that additional information is added, and feed the new results back into a GIS. There are other reasons as well, including the comfort some people have with performing calculations in a spreadsheet rather than in GIS. As you will see with one-to-many joins, joining is an efficient way of associating (temporarily) non-spatial data with landscape features.

TABLE 9.1 A non-spatial database in ASCII text file format illustrating comma-delimited data

'Stand', 'HSI 2010', 'HSI 2015', 'HSI 2020'
1, 0.256, 0.312, 0.325
2, 0.458, 0.495, 0.516
3, 0.333, 0.365, 0.372
4, 0.875, 0.885, 0.889
5, 0.125, 0.215, 0.235
6, 0.468, 0.476, 0.485
7, 0.906, 0.908, 0.911
8, 0.648, 0.745, 0.753
9, 0.378, 0.425, 0.431
10, 0.096, 0.102, 0.118

In Depth

In performing join processes, three terms are important: the *source table*, the *target* or *destination table*, and the *join item* (or field). The source table contains the data that will be moved to the target table and associated with some particular landscape features stored there. The target table contains the landscape features with which the source table's data will be associated. After the join process is complete, the source table's data will be transferred to the target table so that when you view the target table, all attributes from both databases should be present. The join item is the attribute or field that is common between the source and target tables, and is the item that brings the two tables together. If no common attribute exists between the source and target tables, there is no basis for a join. Some examples of common attributes are stand numbers, road names, stream numbers, culvert numbers, and watershed names, yet any attribute specified by a GIS analyst can be used.

One-to-one join processes

A one-to-one join process assumes that there is exactly the same number of records in the source table as there is in the target table, and that each of the records in the source table is associated with exactly one record in the target table. For example, suppose you have a GIS database of permanent growth and yield measurement plots (Figure 9.1), and you want to join to this database to a file containing the installation dates of each plot. The join item in this simple example is obviously 'Plot' in the source table and the 'Plot' attribute in the target table. As you can see, there are exactly six records in both the source and target tables, and each installation date record from the source table is associated with only one unique record in the GIS database containing the permanent plots.

The original assumption behind one-to-one join processes can be relaxed, and the one-to-one join process can also be made with fewer records in the source table than in the target table. For example, assume that record

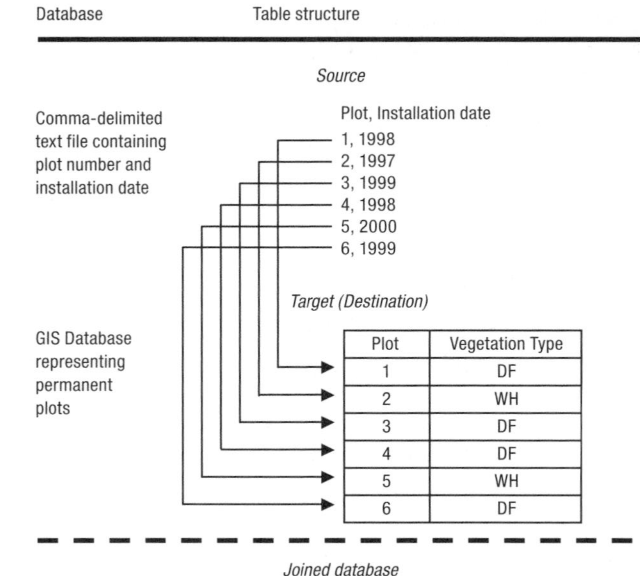

Figure 9.1 Performing a one-to-one join using a file of installation dates as the source table, and the Daniel Pickett permanent plots GIS database as the target table.

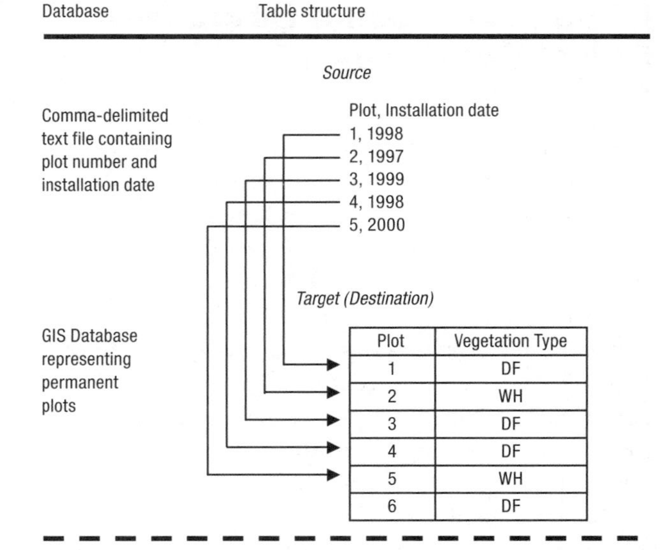

Figure 9.2 Performing a one-to-one join with one record missing from the source table.

6 (Plot 6, installation year 1999) was missing from the source table. If you were to perform a join process using the source and target tables, a one-to-one join would still occur, yet Plot 6 in the GIS database would not be joined with any data from the source table (Figure 9.2).

To work through a one-to-one join, assume you have an ASCII text file (HSI.txt) containing habitat suitability indices (HSIs) for salamanders related to every timber stand on the Daniel Pickett forest. HSIs range from 0.0 (poor habitat) to 1.0 (optimal habitat) and can be quite complex to calculate; therefore, it is not unreasonable to assume that they were generated outside of GIS, perhaps in a spreadsheet or statistical software package. The challenge, once HSI values have been calculated, is to bring them back into a GIS environment to allow the creation of a thematic map. To perform a join process of the non-spatial HSI database (HSI.txt) and a GIS database (Daniel Pickett stands) the following general steps can be taken when using ArcGIS 9.x:

1. Open the Daniel Pickett stands GIS database (this is the target table).
2. In the table of contents, right-click the target table (stands GIS database), select 'joins and relates', then select the join option.
3. Identify the join item from the target table in option 1.
4. Choose the source table (HSI.txt) for option 2.
5. Identify the join item from the source table in option 3.
6. Perform the join process (press OK).

In ArcView 3.x these general steps can be taken to perform the join:

1. Open the Daniel Pickett stands GIS database.
2. Open the Daniel Pickett stands attribute table (this is the target table).
3. Open the HSI.txt database (this is the source table).
4. In the source table, use the mouse to select the join item (stand).

5. In the target table, use the mouse to select the join item (stand).
6. Click the table join button.

Once the one-to-one join process is complete, the HSI values from the HSI.txt database should be temporarily stored inside of the stands GIS database attribute table. To confirm this, you must open and visually examine the stands GIS database attribute table. A thematic map can be created (Figure 9.3) using the HSI values, to illustrate the spatial arrangement of habitat on the Daniel Pickett forest. Areas within each habitat grouping can also be calculated to enable the development of a report concerning the amount of suitable habitat on the forest for a particular wildlife species.

One-to-many joins

In contrast to one-to-one join processes, the assumption behind one-to-many join processes is that there are more records in the target table than in the source table, and that each record in the source table may be associated with more than one record in the target table. Assume that you have a streams GIS database that contains 1,000 stream reaches. Assume also that you desire to buffer the streams to create a map of different riparian management zone policies, and that each stream needs to be buffered a distance that is based on its size. If you were interested in developing variable-width buffers (as illustrated in chapter 7) for each policy, the appropriate buffer width for each stream must be associated with each stream reach. Manually updating the attributes of each of the 1,000 streams for each riparian policy would be time consuming and would likely contain errors. With a one-to-many join process, the likelihood of errors is minimized, and the process of associating different buffer width data with each stream is fast and efficient. For example, Figure 9.4 illustrates a one-to-many join process where the source table has four records and the target table has seven records. The join item is 'Stream type' in the source table, and the 'Type' attribute in the target table. Two of the source table records ('Perennial – large' and 'Intermittent') are associated with more than one record in the target table (hence 'many' records). The source table could have included four (or more, or less) items and the target table could have included 1,000 or more items—it makes no difference to the one-to-many join process. However, similar to the one-to-one join processes, should a record be omitted from the source table, the affected records in the target table would be represented by null or missing values in the joined field.

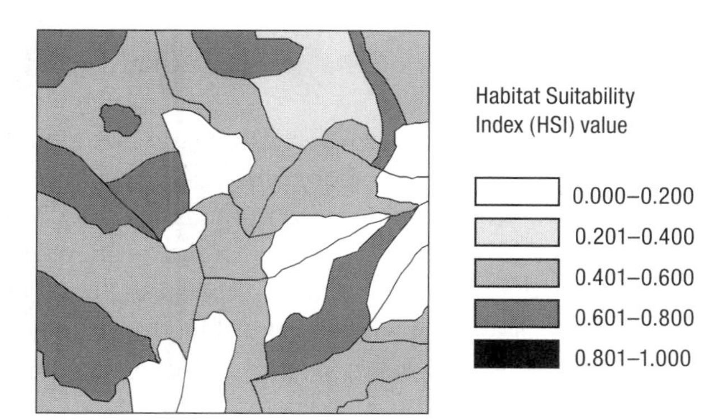

Figure 9.3 Habitat Suitability Index (HSI) values for salamanders on the Daniel Pickett forest.

Many-to-one (or many-to-many) joins

In certain cases, the source table can contain errors and lead to results that are different than what was originally

Database | Table structure

Source

Comma-delimited text file containing stream type and buffer distance

Stream Type, Buffer
"Perennial—large", 100
"Perennial—small", 75
"Intermittent", 50
"Ephemeral", 25

Target (Destination)

GIS Database representing permanent plots

Stream	Type
1	Perennial—large
2	Intermittent
3	Perennial—small
4	Perennial—large
5	Intermittent
6	Ephemeral
7	Intermittent

Joined database

Resulting database: The original streams GIS database with the temporary field "Buffer"

Stream	Type	Buffer
1	Perennial—large	100
2	Intermittent	50
3	Perennial—small	75
4	Perennial—large	100
5	Intermittent	50
6	Ephemeral	25
7	Intermittent	50

Figure 9.4 Performing a one-to-many join using a file of buffer distances as the source table, and streams GIS database as the target table.

intended. These are cases where two or more records in the source table (and one or more instances of such) are associated with a single record in the target table. These can result in a many-to-one join. For example, the many-to-one join process illustrated in Figure 9.5 shows that the fourth and fifth records of the source table have exactly the same the join item value ('Stream type' = 'Ephemeral'). When the join process is performed only one 'Buffer' value from the two source table records can be associated with the ephemeral stream in the target table (the first value in ArcGIS 9.x, the last value in ArcView 3.x). The example in Figure 9.5 shows that the first value (25) was present in the joined database, not the last value (35). Which value is present in the target table (the first instance or subsequent instances in the source table) depends on the GIS software program being used. Many-to-many join processes behave in a similar fashion (Figure 9.6).

The following two examples bring together some concepts that were introduced in this and earlier chapters. In each example, a non-spatial database is joined to a GIS

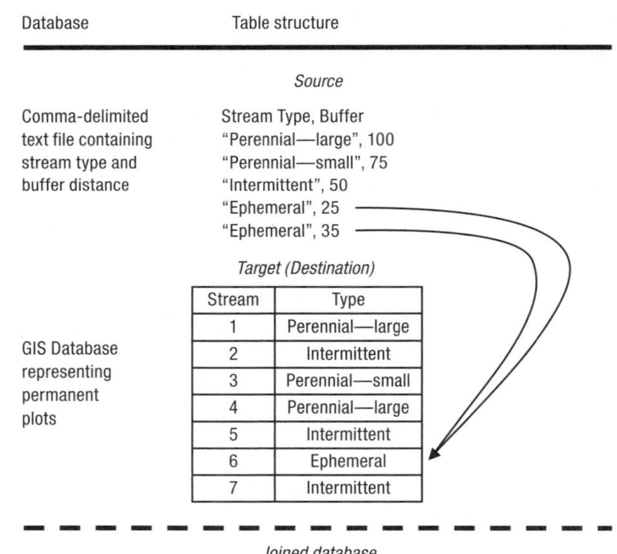

Figure 9.5 Performing a many-to-one join using a file of buffer distances as the source table, and a streams GIS database as the target table.

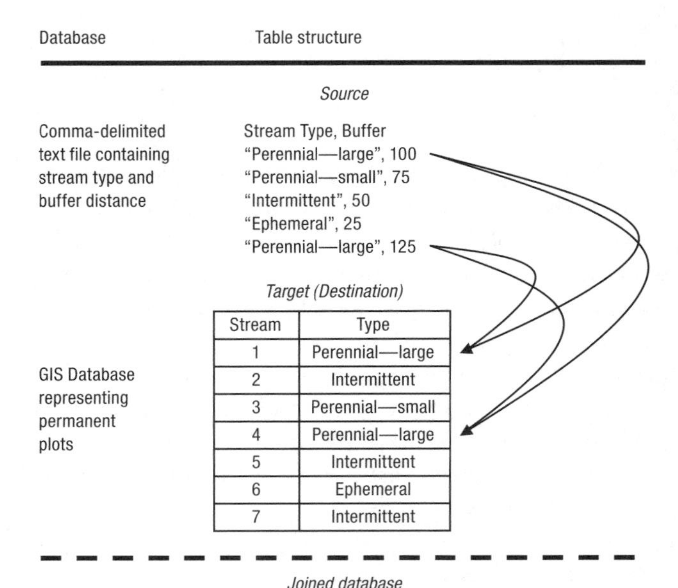

Figure 9.6 Performing a many-to-many join using a file of buffer distances as the source table and a streams GIS database as the target table.

database. Then, a spatial query is performed to determine which of the point features is contained within an area represented by a polygon feature. Ultimately, some information that was originally contained in the non-spatial database is summarized based on its associated spatial location.

Example 1: Determining the number of hardwood sawmills in a state

Assume that you currently work for Dunn and Herndon, Inc. in central Tennessee, and are considering building a new hardwood sawmill somewhere in the state. Initially, what you might find helpful is an estimate of the number of hardwood sawmills in the state of Tennessee. Another piece of information that would be valuable is an estimate of the number of people that they employ. To develop this information we will use a GIS database of the south-eastern US states, a GIS database of mill locations, and a non-spatial database of mill attributes. Mill locations and their associated non-spatial attributes were acquired from the USDA Forest Service (2006). The process includes the following steps:

1. Join the non-spatial mill attribute data to the GIS database of southeastern US mills. The non-spatial database contains a field called 'MILL_ID'. This is the join item from the source table. The mill GIS database contains a field called 'MILL-ID'. This is the join item from the target table. Once the two tables are joined, the spatial features (points) in the mills GIS database will have associated with them the attributes of mills from the non-spatial database.
2. Query the southeastern US states GIS database for the state of Tennessee. This is a query that simply uses a field (state) of the southeastern US states GIS database to locate the appropriate attribute of the states (State = Tennessee).
3. Perform a spatial query of the mills GIS database to determine (by location) which mills are completely contained within the State of Tennessee (the current selected feature in the southeastern US states GIS database).

Once these steps have been performed, we will have selected those mills that are within the State of Tennessee. The question now is whether they mainly accept hardwood tree species, and whether they are considered a sawmill. A final query using the attributes of the mills GIS database is therefore needed to select from the currently selected features (all mills in Tennessee) those that are sawmills, and mainly accept hardwood tree species (Figure 9.7). As a result, you might find that there are 329 hardwood sawmills in the State of Tennessee that employ 3,959 people, assuming that the data are current.

Example 2: Determining sawmill employment in a county

As a second, similar example, assume that you are a consultant based in Mississippi, working for Saunders Geomatics, LLC. Assume also that you are doing some contract work for the Nuxubee County Chamber of Commerce. They want to know how many sawmills are in the county, and they want an estimate of how many people these mills employ. To develop this information we will use a GIS database of the southeastern US counties, a GIS database of mill locations, and a non-spatial database of mill attributes. Once again, mill locations and their associated non-spatial attributes were acquired from the USDA Forest Service (2006). The process includes the following steps:

1. Join the non-spatial mill attribute data to the GIS database of southeastern US mills, as was described in example 1 above.
2. Query the southeastern US counties GIS database for Nuxubee County, Mississippi. This is a query that simply uses a field (county) of the southeastern US counties GIS database to locate the appropriate attribute of the counties (County = Nuxubee). You need to be careful here if there are multiple counties in the southeastern states with the same name (e.g., 'Floyd' is the name of a county in more than one state).

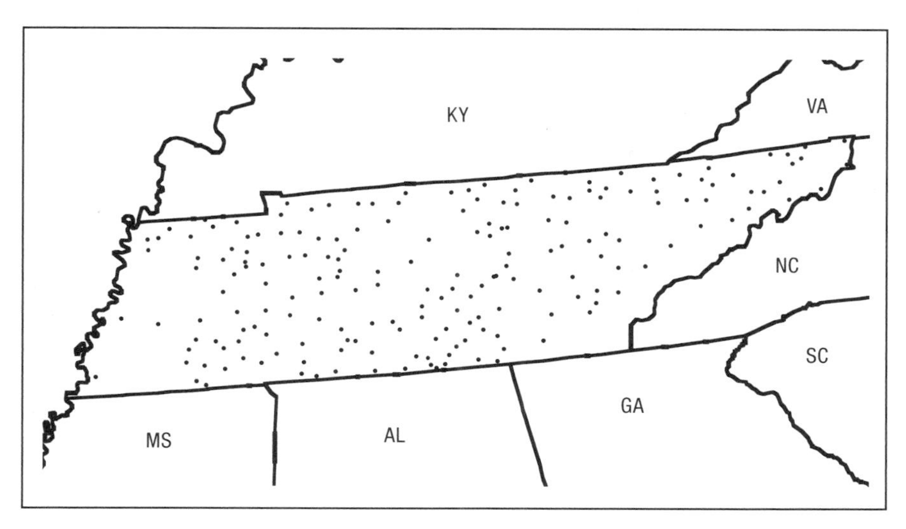

Figure 9.7 Hardwood sawmills in Tennessee.

3. Perform a spatial query of the mills GIS database to determine (by location) which mills are completely contained within Nuxubee County (the current selected feature in the southeastern US counties GIS database).

Once these steps have been performed, we will have selected those mills that are within the county. The question now is whether they are sawmills. An examination of the mills table finds that there are three sawmills in the county (although only two are seemingly visible on the map), and that they employ 231 people, assuming that the database is current.

Joining Two Spatial GIS Databases

With a spatial join process, our intent is to learn about the qualities of some landscape features that are near other landscape features of interest. In joining non-spatial databases to spatial databases, a common field (the join item) was used to associate the data in the non-spatial database to the landscape features in the GIS database. The association itself was non-spatial—the location of landscape features in the GIS database (the target table) was not used to assist in making the association. The spatial location of landscape features can, however, be used to associate landscape features from one database with those in another database. This is a powerful feature that often goes unused by GIS users. Both databases, however, must contain spatial data. For example, if you were interested in knowing the type of forest stands that sources of water (ponds, springs, etc.) fall within, you might join a GIS database containing a set of points (representing water sources) with a GIS database containing a set of polygons (timber stands) to understand the type of forest that surrounds each water source. Thus the attribute data within the timber stand database can be associated to the attribute data within the water sources database based on the spatial location of the landscape features within each GIS database.

In some software packages, such as ArcGIS 9.x, a spatial join results in the creation of a new layer that contains the joined information. Check your GIS software to determine whether new databases are created during spatial joins. You should also be aware that spatial joins will usually require that a map projection be associated with all involved layers. A key component of every map projection is the definition of a map unit that describes the coordinate division intervals, typically expressed in feet, international feet, or meters. Spatial joins are based on the comparison of feature (point, line, or polygon) locations to other feature locations. There are many GIS processes that inherently recognize the coordinate system of a layer without needing to know the real-world definition of the map unit. Without the map unit definition, however, spatial joins will usually not be possible.

As with non-spatial join processes, a source database and target database are required, yet the join item will be slightly different—it is the spatial position of each landscape feature in each GIS database. The two most common types of spatial join processes are those that are defined by which features are closest (sometimes called the nearest neighbor) and those that are evaluated by whether a feature is intersected by another feature or is completely inside another feature. A third type of spatial join involves linear features and is evaluated by determining whether one linear feature is located along or inside the extent of another linear feature. The output properties and options of the spatial join process will depend upon the feature type (point, line, or polygon) of the source and target databases and are summarized in Table 9.2.

The nearest neighbor spatial join works for almost all feature type comparisons except for line on line. Other than this exception, the nearest neighbor join process allows the characteristics of features (points, lines, or polygons) in a source table to be associated with the closest feature in a destination table. Thus you can use spatial join processes to not only identify the nearest point (e.g. tree), line (e.g. road) or polygon (e.g. watershed) feature from all point (e.g., house) features in a GIS database, but you can also determine the distance to the nearest point feature. Typically, the distance to the nearest feature and the attributes of the nearest feature are included in the output database.

A point-in-polygon or 'inside of' join process is rather straightforward: determine the polygon (from a source table) within which each point (in a destination table) is located, then join the attributes of each polygon with each associated point. For example, you may be interested in understanding the characteristics of timber stands containing the owl nest locations that exist on the Daniel Pickett forest (Figure 9.8). In the Daniel Pickett owl GIS database there are two owl nest location points. Associated with each point is information, based on owl surveys, regarding the number of adult and juvenile owls

TABLE 9.2 **Spatial join options by target and source feature type** (Italics indicate output products for each option)

Target	Source: Points	Source: Lines	Source: Polygons
Points	1. Nearest points *Attribute summary* *How many are nearest*	1. Intersecting lines *Attribute summary* *How many are nearest*	1. Falls within *Attributes*
	2. Nearest point *Attributes & distance*	2. Nearest line *Attributes & distance*	2. Nearest polygon *Attributes & distance*
Lines	1. Nearest points or points intersected *Attribute summary* *Attributes & distance*	1. Intersecting lines *Attribute summary*	1. Intersects polygons *Attribute summary* *How many are nearest*
	2. Nearest point *Attributes & distance*	2. Within other lines *Attributes*	2. Nearest polygon or intersecting polygon *Attributes & distance*
Polygons	1. Points that fall inside *Attribute summary* *How many are inside* *How many intersect*	1. Intersecting lines *Attribute summary*	1. Intersects polygons *Attribute summary* *How many intersect*
	2. Nearest point *Attributes & distance*	2. Nearest line *Attributes & distance*	2. Completely inside *Attributes*

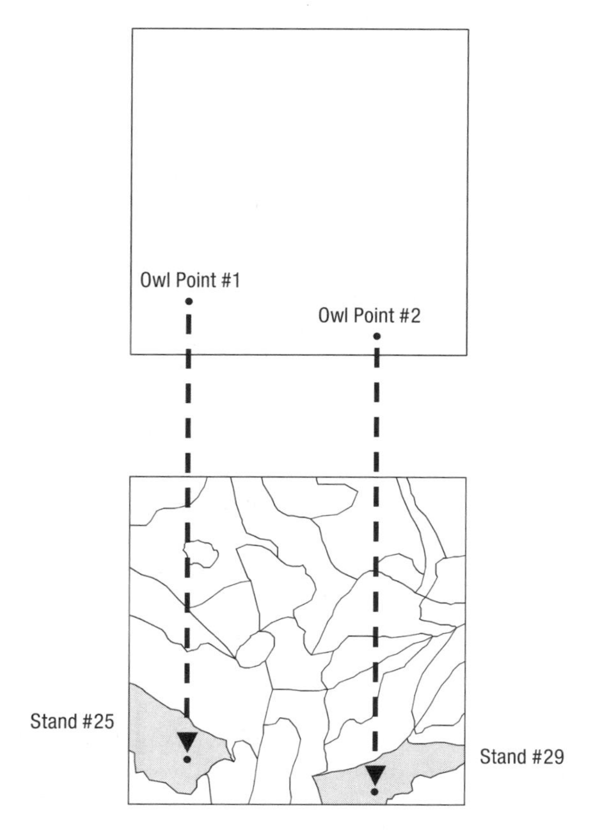

Figure 9.8 Associating owl nest locations with the timber stands within which they are located.

found there (Figure 9.9), along with the first and last sighting of the owls.

What is not known by simply viewing the owl GIS database are the characteristics of the forest surrounding each owl nest. With just two points, this can be determined by visual inspection: owl point 1 is located within stand 25, a rather densely stocked 70-year old stand; owl point 2 is located within stand 29, a 50-year old stand of trees. However, when a large number of points are present in the destination table, or the source table contains a number of relatively small features (making visual inspection difficult), an automated process may be preferred, such as the general process noted below that can be used in ArcGIS 9.x:

1. Open the Daniel Pickett owl GIS database (target table) and stands GIS database (source table).
2. In the table of contents, right-click the target table (stands GIS database), and select joins.
3. Make sure the first option in the dialog box that opens is set to 'Join data from another layer based on spatial location' (the default setting is 'Join attributes from a table').
4. Select the stands GIS database as the layer to join to this layer (use ArcCatalog to add spatial reference information if necessary). The Join Data dialog box

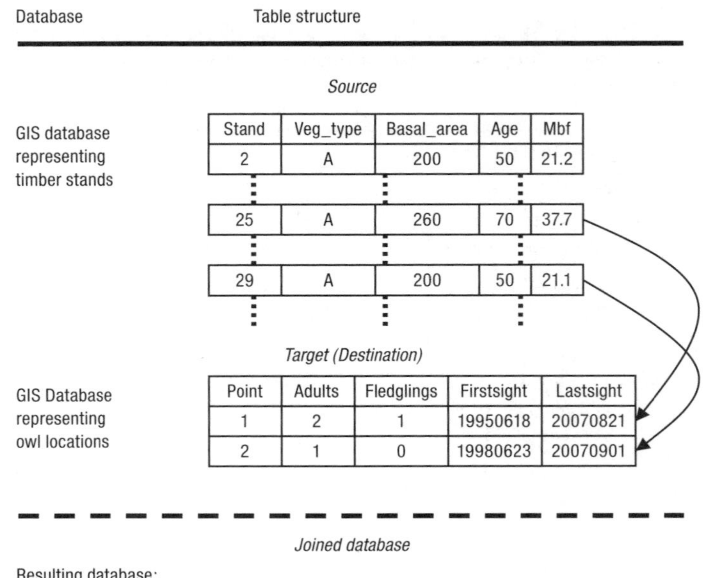

Stand	Veg_type	Basal_area	Age	Mbf
2	A	200	50	21.2
25	A	260	70	37.7
29	A	200	50	21.1

Point	Adults	Fledglings	Firstsight	Lastsight
1	2	1	19950618	20070821
2	1	0	19980623	20070901

Resulting database:
the owl points database with the appropriate stand conditions that surround each point

Point	Adults	Fledglings	Firstsight	Lastsight	Stand	Veg_type	Basal_area	Age	Mbf
1	2	1	19950618	20070821	25	A	260	70	37.7
2	1	0	19980623	20070901	29	A	200	50	21.1

Figure 9.9 Spatially joining the Daniel Pickett stands GIS database with the owl GIS database.

will update to show you the types of feature classes that you are joining, in this case Polygons to Points.

5. Use the radio or option button and choose that you want each point to have the attributes of the polygons that 'it falls inside' rather than 'is closest to it'.
6. Specify an output location and name for the resulting joined database.
7. Choose OK to initiate the spatial join.

In ArcView 3.x these general steps can be taken to perform the join:

1. Open the Daniel Pickett owl GIS database (target table) and stands GIS database (source table).
2. Open the attribute tables of both GIS databases.
3. Clear all selected landscape features in both attribute tables (to ensure no landscape features are selected).
4. With the mouse, click on the join item in the source table (e.g., the 'shape' field in ArcView).
5. With the mouse, click on the join item in the target table (e.g., the 'shape' field in ArcView).
6. Click the join table button to initiate the spatial join process.

While this example is relatively straightforward, imagine a case where you have several hundred points, such as research plots, and the goal is to quickly and accurately determine what type of stand each research plot is located within. A point-in-polygon join process would seem to be a logical and efficient option to accomplish this goal, and would likely result in fewer errors than a manually driven process.

The spatial join process can also be used to identify the number of features within another point, line, or polygon database that are closest to features within a target database. For example, suppose you had four possible trailheads within a watershed where you could park your vehicle in order to visit six rain gauges from which you need to collect precipitation measurements. Using the trailhead locations (points) as the target database, you could spatially join the six rain gauge locations (points) and determine which of the trailheads was closest to the largest number of the six gauges. This would help you at least to pick a trailhead location in which to begin your sampling. Another example might include examining a number of watersheds (polygons) and wanting to determine which was the most densely forested according to the number of trees, assuming that you had coordinates (points) of all trees throughout your watersheds. Using the watersheds as your target layer, you could spatially join the trees layer and return the number of trees, complete with an attribute summary (average tree height for example) for each watershed.

Making Joined Data a Permanent Part of the Target (Destination) Table

After a join process has been completed, you might decide that the joined data should permanently reside within the target table. One strategy to accomplish this goal would be to add the appropriate number of empty fields to the target table that will ultimately contain the joined data, and declare that the data type of the empty fields be the same as the data type of the joined data. The values of the empty fields can be calculated to equal the values of the joined data, thus filling the empty (yet permanent) fields with the joined (yet temporary) data. Saving the target table at this point results in a permanent change to the database, with the data from the source table now a permanent part of the target table. In addition, removing or changing the source table will not result in a corresponding change to data in the new fields in the target table.

Another strategy may be to perform a join process, and then save (or export) the spatial GIS database that represents the target table to a new database. In some GIS software programs the newly copied and saved GIS database will contain all previously joined data as a permanent part of the GIS database rather than as a temporary association. Finally, exporting the target table (the tabular portion of the GIS database) after a join process was performed will generally create a new file that includes all data. Unfortunately, this process does not preserve the landscape features of the target table, only the underlying attribute data. An exported target table (after join processes have been performed) may facilitate some natural resource management processes, such as report generation, metadata creation, or other subsequent non-spatial analyses.

Linking or Relating Tables

On occasion, you may want to simply link or relate two GIS databases together, allowing you to view both the source and target tables as separate entities, and to view landscape features that are associated with each other in both databases. With this process, the goal is to have the ability to display two GIS databases, and to be able to select a landscape feature from one and view the associated landscape feature(s) in the other. For example, assume you have a source table represented by a GIS database that contains multiple records related to landscape features in another GIS database (the target table). In Figure 9.10, the tables of two GIS database tables are illustrated; one represents a road system GIS database and the other represents a culvert GIS database. As you may notice, there are multiple culverts associated with each road (e.g., culverts 4 and 5 both are associated with road 602). Joining these two databases together would result in a many-to-one join process or one-to-many join process, depending on what is chosen as the source and target tables. With a many-to-one join process (culverts as source, roads as target), some of the records in the source table will not be present in the target table after the join process has been completed. With a one-to-many join process, (roads as source, culverts as target) you can identify which road is associated with each culvert, however, you will not be able to select a culvert and automatically view the associated road.

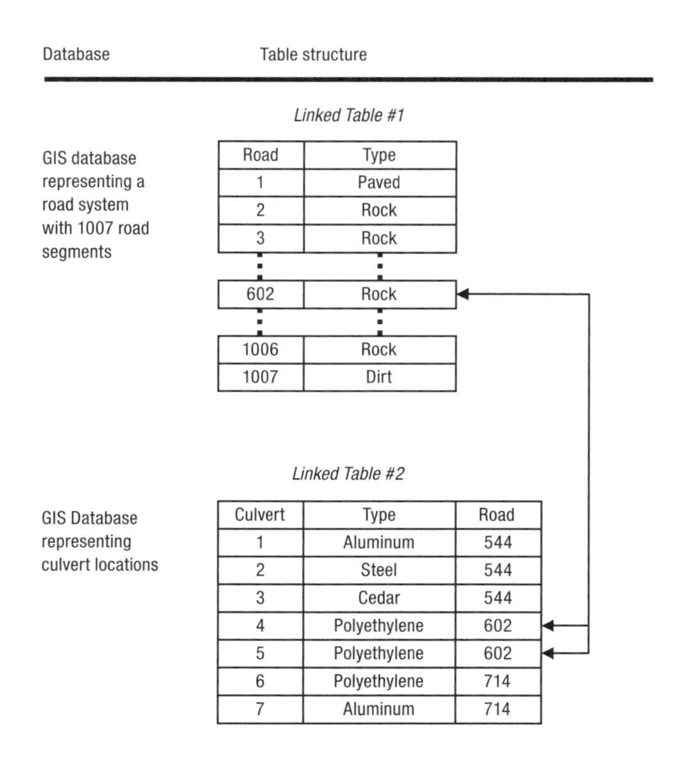

Figure 9.10 Linking a roads GIS database with a culverts GIS database.

In Depth

Which process—joining or linking—is more dynamic? Perhaps linking. With linking, you can visualize both the source and target tables, select landscape features from either table, and view the associated links from wither perspective. Further, if some data are not linked (in either the source or target table) they are still visible and available for further analysis. When a join process is used, source table data not joined with target table data are unavailable for further spatial analysis.

To further illustrate the link or relate process, a general set of steps can be followed in ArcGIS 9.x to associate two GIS databases related to the Brown Tract: a culverts database (Culverts.txt) and a roads GIS database:

1. Open the Brown Tract roads GIS database and culverts text file.
2. In the table of contents, right-click the target table (roads GIS database), and select Joins and Relates, then select Relates.
3. Select the relate field in the target table.
4. Identify the source table.
5. Select the relate field in the source table.
6. After relating the two databases, perform an identify process, selecting one of the roads that are noted in the culvert database. You should be presented with information for both the road and the associated culverts along that road.

In ArcView 3.x, the following general set of instructions may work to create a link between the roads and the culvert databases:

1. Open the Brown Tract roads GIS database.
2. Open the roads GIS database attribute table (this is the target table).
3. Open the Culverts.txt file (this is the source table).
4. Select the link item in the source table (road).
5. Select the link item in the target table (road).
6. Choose Link from the Table menu.

At this point a one-way link between the roads and culverts databases has been created. The process may need to be repeated with the roles reversed (roads GIS database as the source, Culverts.txt as the target) to create a two-way link.

By linking the two tables together, you can select records in one database (or features in either database), and the landscape features they are associated with may be selected and highlighted in the other database. For example, after linking together the culvert database with the roads GIS database, you can select one or more culverts, and the roads associated with those culverts may be simultaneously selected and highlighted in the roads GIS database. In ArcGIS you can view the related data by using the identify tool.

Summary

Join and link processes provide GIS users with a way to temporary associate two or more databases, allowing an expansion of the mapping and analysis opportunities within GIS without making permanent changes to the databases. Before utilizing one of these processes you should consider the types of databases available, and the type of association desired (e.g., one-to-one, one-to-many, and so on). In addition, the purpose of the process should be understood, which might provide guidance in the choice of a join or link process. For example, you may decide on different courses of action if the purpose is to:

A. Bring together spatial and non-spatial databases to make thematic maps.
B. Further the ability to perform other spatial operations, such as associating non-spatial stream buffer data with a streams GIS database in order to facilitate a buffer process.
C. Understand the spatial relationship between a set of points and other landscape features (through a point-in-polygon process).
D. Understand the association among multiple landscape features in one database and their counterparts in another database (through a link process).

Joining and relating processes often produce databases that are only 'virtually' combined, meaning that no new databases have been created, but that existing information appears to be associated either in one table (with a join) or in two tables (with a relate). GIS users should recognize that this temporary condition exists and that further action is necessary in order for a permanent record of the joined or linked information to be created. The exception to this virtual result is with a spatial join, in which results are usually stored in a new database that contains all spatially joined information. You should check your software package to determine whether it follows these output results in the same way. Regardless of which process (tabular-based or spatial join) is used to associate two databases, users should track their procedures and manage the output accordingly.

Applications

9.1. Salamander habitat suitability index. Bob Evans, the Brown Tract's wildlife biologist/hydrologist, developed some habitat suitability index (HSI) values for a salamander. He suggests joining the data provided in 'SAL_HSI.txt' with the Brown Tract stands GIS database. He then wants you to answer several questions about the extent and spatial distribution of these areas, which are summarized below.

a) How much land area is contained in the 0.8 to 1.0 HSI range?
b) How much land area is contained in the 0.6 to 0.79 HSI range?
c) How much land area is contained in the 0.4 to 0.59 HSI range?

9.2. Newt habitat suitability index. Bob Evans has also developed some habitat suitability index (HSI) values for a newt. He provided data in the file 'NEWT_HSI.txt'. After joining the data with the Brown Tract stands GIS database, please address his needs, which are noted below.

a) Determine how much land area of high quality newt habitat (HSI ≥ 0.65) is contained within 100 m of the road system.
b) Determine how much land area of high quality newt habitat (HSI ≥ 0.65) is contained within 1,000 m of the owl nest site.
c) Determine how much land area of high quality newt habitat (HSI ≥ 0.65) is contained in even-aged stands over 50 years of age.

9.3. Sharp-shinned hawk habitat suitability index. In addition to his previously discussed developments, Bob Evans has also developed some habitat suitability index (HSI) estimates for the sharp-shinned hawk covering the years 2010, 2020, 2030, and 2040. His tendency to provide this data in a text file continues, and you can find it in 'SSHAWK_HSI.txt'. After joining the data with the Brown Tract stands GIS database, please address his needs noted below.

a) Determine how much land area of higher quality habitat (0.6–1.0) is contained within the Brown Tract.
b) Produce a single map of the sharp-shinned hawk habitat showing HSI values for the years 2010, 2020, 2030, and 2040.

9.4. Water sources and land allocations. John Frewer, a forester associated with the Brown Tract, is interested in knowing what types of land allocations the water sources were located within. To accomplish this task, perform a point-in-polygon operation (or selection by location query), using the water sources as the source table and the stands GIS databases as the target table.

a) How many water sources are located in 'Even-aged' stands on the Brown Tract?
b) What types of water sources are located in uneven-aged stands (list the water source types)?
c) How many water sources are located in research areas?

9.5. Sawmills in a woodshed. You have recently been hired as a procurement forester for Chupp and Daughters Sawmill in Floyd County, GA. You need to understand the competition for wood in the area. To perform this task, use the southeastern counties GIS database, the southeastern mills GIS database, and the mill data DBF file. Join the mill data DBF file to the southeastern mills GIS database and determine how many sawmills are within 100 miles of Floyd County, GA. The mill locations and non-spatial mill attributes can be acquired from the book's website or the USDA Forest Service (2006).

9.6. New mill location. You work for Walker, Avery, and Houston Lumber Company, a company that is considering building a new hardwood sawmill mill in Coffee County, AL. They need to understand how much hardwood volume currently exists in the area around the county. To perform this task, use the southeastern counties GIS database and the county volume DBF file.

Join the county volume DBF file to the southeastern counties GIS database and determine how much hardwood (soft and hard) volume is contained in the counties that surround, and include, Coffee County. The county-level can be obtained from the book's website or the USDA Forest Service (2007).

References

USDA Forest Service. (2006). *US wood-using mill locations—2005*. Research Triangle Park, NC: USDA Forest Service, Southern Research Station. Retrieved April 16, 2007, from http://www.srs.fs.usda.gov/econ/data/mills/mill2005.htm.

USDA Forest Service. (2007). *FIA data mart: Download files*. St Paul, MN: USDA Forest Service, North Central Research Station. Retrieved April 16, 2007, from http://www.ncrs2.fs.fed.us/FIADatamart/fiadatamart.aspx/fiadatamart.aspx.

Chapter 10

Updating GIS Databases

Objectives

This chapter is designed to provide readers with a discussion of GIS processes that should be considered when updating GIS databases. Database updates are necessary when landscapes and associated characteristics, such as ownership, change. There are a variety of methods you can use to update a GIS database, yet only a few are presented here. The objective of this chapter, therefore, is to provide an introduction to the potential applications in this area. More specifically, at the conclusion of this chapter, readers should be able to understand:

1. why GIS databases need to be periodically updated and maintained,
2. what issues might be associated with an update process, and
3. what GIS processes could be used to physically update a database.

To accomplish these objectives, a discussion of the reasons for updating GIS databases is first presented. Two types of update processes are then examined, one where new landscape features are added to an existing GIS database, and another where the landscape features and attributes in an existing GIS database are modified. These two examples likely address the two most common forms of GIS database updates. This chapter relies heavily on the GIS processes associated with editing GIS databases. For a review of these editing processes, please refer to chapter 3.

GIS databases are rarely considered static entities: vegetation conditions change due to human manipulation and natural disturbances, roads are constructed and obliterated, and some stream characteristics (pools, sediment, fish abundance) may change as woody debris moves through the system. Although you may have developed or acquired GIS databases at one point in time, as management needs or direction change, or as the resources you manage change, GIS databases used to describe landscapes must be updated. Table 10.1 illustrates a number of events that could occur and affect the landscape being managed, suggesting that the GIS databases used to describe the landscape being managed must be updated to reflect changes. Most natural resource management organizations (as well as data development organizations) have created a set of processes and protocols to guide the updating of GIS databases. Finding and illustrating a standard protocol is, therefore, difficult because each organization generally will develop the steps they feel are necessary to integrate new data within their system of natural resource information. For example, assume a tract of land was recently purchased by a land management organization. Integrating the forest stand component of this new tract into a forest stand GIS database can be accomplished in a number of ways, such as the three processes described in Figure 10.1. As database protocols and organization strategies vary from one organization to another, there is no one update approach that will work for every organization.

The users of GIS databases are the ultimate customers of groups (GIS departments, consultants, agencies) that produce the spatial data. As GIS databases become available, and users begin to explore the usefulness of the data for assisting in natural resource management processes, the limitations of the databases will become evident. The period of time from initial GIS database availability to serious consideration of updates to the databases may last

TABLE 10.1 A sampling of reasons for updating GIS databases

Events	Examples	Update spatial data	Update tabular data
Stochastic disturbances	hurricanes, fires, insect outbreaks	✓	✓
Transitions of forests	growth and yield		✓
Management activities	harvesting, road construction, installation / removal of culverts, creation of trails, thinnings, etc.	✓	✓
Transactions	land acquisitions, donations, sales	✓	✓
Regulations	riparian management areas, owl habitat areas, woodpecker habitat areas	✓	✓
Organizational policies	special areas, personal reservations	✓	✓
Improvements in technology	digital orthophotographs, GPS capture of road data, ownership boundaries, etc.	✓	✓
Organizational initiatives	periodic / annual cruises, photo interpretation of harvested areas not normally recorded via normal processes	✓	✓
New data availability	databases developed by other organizations	✓	✓
Changing map projections	conforming to new organizational standards	✓	
Collaborative projects	watershed analyses, landscape planning efforts	✓	✓
Periodic maintenance	cleaning up databases after spatial operations, digitizing, or attributing processes	✓	✓

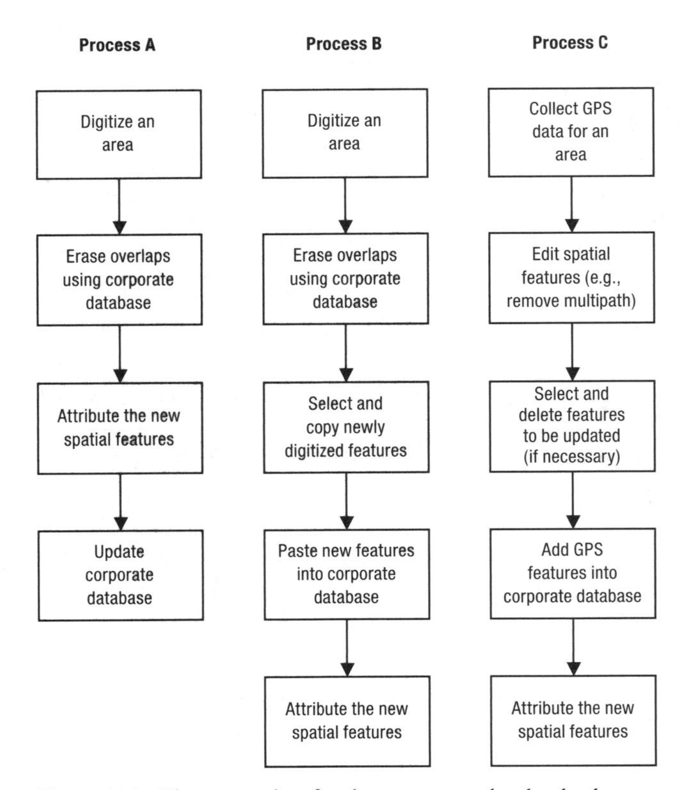

Figure 10.1 Three examples of update processes related to land acquisition.

from a few hours to a few months. Users of GIS databases will ultimately suggest a variety of enhancements to the databases that would facilitate further analyses. For example, the Brown Tract vegetation GIS database could be modified to show more explicitly the riparian areas, or could include more attributes that describe forest stand structure. A roads GIS database might also be enhanced to show the type of road surfacing or all of the trails (unauthorized roads) that weave through the property. Updating these GIS databases to include all of the information that is necessary to make natural resource management decisions may be, however, limited by the time and budget available to make the changes, the quality of the information available to make the changes, and other organizational data standards. The needs of natural resource managers, with regard to GIS databases, must eventually be considered along with the costs of data development.

The Need for Keeping GIS Databases Updated

Natural resource managers generally base management decisions on the best available data. The quality of data can range from very precise and accurate (collected with a high quality GPS receiver) to somewhat imprecise and inaccurate (drawn by hand from memory). Keeping the data used for making decisions accurate and updated is therefore important, and thus the interval between updates becomes important. For example, the **update interval** that a resource management organization uses to refresh the spatial extent (history) of their management activities and the growth of their forest inventory is important, since subsequent management decisions might be affected by previously implemented management decisions. The interval chosen can range from six months, to a year, or even two years between updates, depending on the GIS database considered. The interval chosen depends on the organization's perception of the usefulness and cost-effectiveness of such an update on a GIS database. For example, if the goal of an organization (e.g., a southern US forest management organization) were to generate revenue for its stockholders, the need for updating the data related to its primary resource (pine forest stands) may be more important, and updated more frequently than data related to secondary resources (hiking trails). Other resources, such as roads, streams, culverts, water resources, and wildlife may be more or less important, depending on the goals of the organization, thus the frequency with which these GIS databases are updated may vary according to the organization's perceived need to do so. At the extreme end of the spectrum, every GIS database could be updated continuously; however, the cost of doing so may be quite high and the task would require employees (or consultants) dedicated to the task.

Two of the more important questions an organization must address, beyond determining *when* a GIS database should be updated, are *how* the update process will be accomplished, and *who* will do the work. As mentioned earlier, the methods by which a GIS database could be updated vary considerably; the form of input could range from hand-drawn maps to LiDAR-derived measurements, and the GIS processes could involve scanning, digitizing, attributing, and other methods (Table 10.2). As you may have gathered from chapter 3, when a GIS database is being updated, the database is being edited. In some form or fashion, the intent is to change something about a GIS database—either the landscape features or their underlying attributes, or both. Two examples of GIS update processes are now presented, one related to a forest stands GIS database maintained by a forest industry organization in Florida, and the other related to a streams GIS database maintained and distributed by the Washington State Department of Natural Resources.

TABLE 10.2 Inputs and processes that can be used to assist a GIS database update

Input	
Hand-drawn maps	GPS features
Tabular databases	Field notes
A person's memory	GIS Features developed by field personnel
Digital orthophotographs and subsequent interpretation	
GIS processes	
Digitizing	Scanning
Joining	Updating
Linking	Copying / pasting
Importing	Attributing
Querying and verification	

Example 1: Updating a forest stand GIS database managed by a forest management company

A typical forest management company in Florida might update their forest stand GIS database on an annual basis. Their field personnel collect information related to changes in its forest land ownership throughout a calendar year, and the forest stand GIS database is updated near the end of the calendar year. Why would they update the forest stand GIS database once a year? The forest stand GIS database is arguably the most important GIS database for assisting industrial forest management activities, and field-level managers require high quality data (maps and inventory data) to make management decisions. In addition, most corporations require an annual estimate of the value and volume of resources, for planning and tax reporting purposes. A less frequent updating interval may not be appropriate given the short rotations typical of southern industrial forestry operations. For example, waiting two years between updates of a timber stand GIS database may represent 8–10 per cent of the length of a forest harvest rotation. A more frequent interval, say six months, may provide field personnel with higher quality information with which to make management decisions, particularly in cases where a large amount of activity takes place over a six-month period. Some have argued that continuously updating GIS databases may be appropriate, but the time and cost required to update a GIS database may make a nearly continuous update process impractical. Further, field personnel could easily become confused when faced with a continuously changing set of GIS databases, thus updating databases and leaving a window of time (a year, perhaps) between changes may be perceived as more desirable.

The changes to the forest stand GIS database that are recommended by field foresters and other natural resource managers may be indicated on hard-copy maps and timber cruise forms, or they may be contained in digital databases created in GIS or with GPS. Field foresters, timber procurement managers, or other professionals responsible for managing land will typically indicate (draw) on maps the changes (e.g., harvest and regeneration activities) that have occurred on the forest land base as these activities have been completed. This information is usually sent to a central office (Figure 10.2), which takes ownership of the timber stand GIS database. The central office checks the new data for mistakes and omissions according to a set of organizational standards, and may ask for clarification from the field staff. The information is then digitized, either in-house or by an external contractor. The resulting digitized GIS database is checked again for mistakes and omissions, and then integrated into the official (sometimes called 'corporate') GIS database, and again checked for mistakes and omissions. Finally, the forest stand GIS database is distributed back to the field office, either as a GIS database, or as hard-copy maps and tables. The field office may then have its own verification procedures for checking the updated database (or maps) for mistakes and omissions that may have arisen during the update process. Processes such as these, with a systematic method for data collection, entry, and verification, are designed to ensure that high-quality data will be developed and available for use in natural resource decision-making contexts.

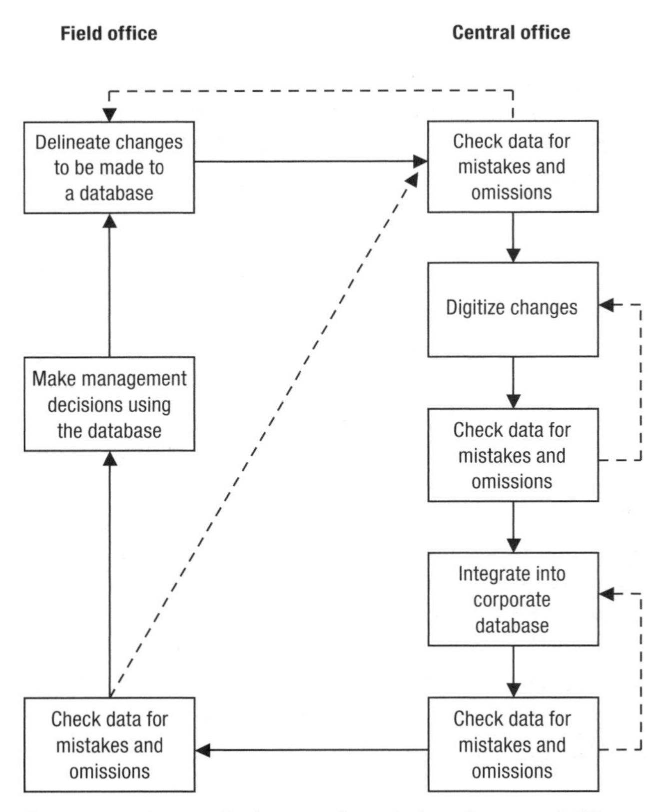

Figure 10.2 A generalized process for updating a forest stand GIS database.

Example 2: Updating a streams GIS database managed by a state agency

In the State of Washington, all forest harvest plans must be submitted to the Department of Natural Resources (DNR) for review and approval. A map must accompany each plan, and illustrate the juxtaposition of proposed activities in relation to, among other landscape features, the stream system. To ensure a consistent definition of the 'stream system', the DNR provides (at a minimal cost, as was illustrated in chapter 3) a streams GIS database for the entire state. This database is continuously updated by the DNR as new information is collected. However, processes and protocols exist that are related to each potential change to the GIS database. For example, assume a private landowner surveyed a stream reach and noted that the type of stream (and perhaps location of the stream) on the landscape is different than the type of stream illustrated in the DNR streams GIS database. The landowner has the option to submit certain documentation to the DNR in support of a request to change the DNR streams GIS database. The DNR directs each request through a review process, and based on the outcome of the reviews, decides to either accept or reject the proposed changes suggested by the landowner. The amount of time required to make a change in the streams GIS database, from initial submission by the landowner to official incorporation in the streams GIS database, may require several months. The process is considered a continuous one since approved changes to the streams GIS database can be made at any time during a calendar year. Therefore, landowners may need to continuously review the status of the DNR streams GIS database in the areas where they own or manage land, and acquire updated data as they deem necessary to reflect the latest stream information.

Updating an Existing GIS Database by Adding New Landscape Features

GIS databases can be updated with new landscape features (points, lines, or polygons) by either adding the new landscape features to an existing GIS database, or by editing the existing landscape features, or both. Two examples are provided below to illustrate updating a GIS database by adding new landscape features. The first example involves a land purchase and subsequent addition of two forest stands to a stands GIS database. The second example involves the addition of new trails to a trails GIS database. In each case, assume that the new landscape features were either digitized or collected with a GPS system and are available in a GIS format. Prior to the initiation of the update process, you should assume that the new data are contained in GIS databases that are separate from the GIS databases that need updating. Refer back to chapter 3 for a review of methods and tools for development of a new GIS database.

Updating a stands GIS database

Assume that the owners of the Daniel Pickett forest have purchased 80 acres (32.38 hectares) of land adjacent to the southwest corner of the original forest boundary (Figure 10.3). Following process B illustrated in Figure 10.1, the stand boundaries of this area have been digitized into a new GIS database that is separate from the original stands GIS database, and these features have been attributed with data fields similar to that in the original stands GIS database (Table 10.3). The edge between the newly digitized stands and original stands is seamless, implying that there is no gap between the polygons of the two GIS databases, and no overlap if the two sets of polygons were placed together. By simply copying the landscape features from the land purchase GIS database into the stands GIS database, it is possible to bring the newly digitized land purchase polygons into the stands GIS database, however, the attributes of the new stands may not be present, depending on the GIS software program being used (Figure 10.4). The new forest stand polygons would then need to be attributed a second time, after they have been pasted into the original stands GIS database.

TABLE 10.3 Attributes of stands in a 32.38 hectare (80 acre) land purchase adjacent to the Daniel Pickett forest

Stand	Hectares	Acres	Vegetation type	Basal area[a]	Age	MBF[b]
1	17.24	42.6	A	190	55	21.3
2	15.14	37.4	B	15	7	0.8

[a] square feet per acre

[b] thousand board feet per acre

To avoid duplication of effort in update processes, three options are clear: (1) digitize the new landscape features directly into the original stands GIS database, (2) use a merge process to combine the newly digitized stands with the original stands GIS database, or (3) if available in

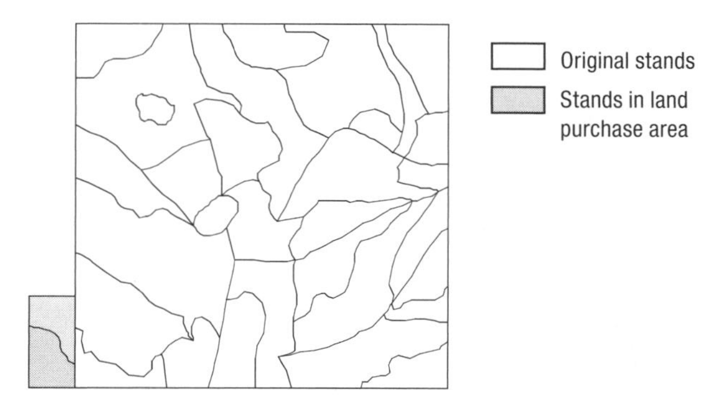

Figure 10.3 Daniel Pickett forest stands and land purchase area.

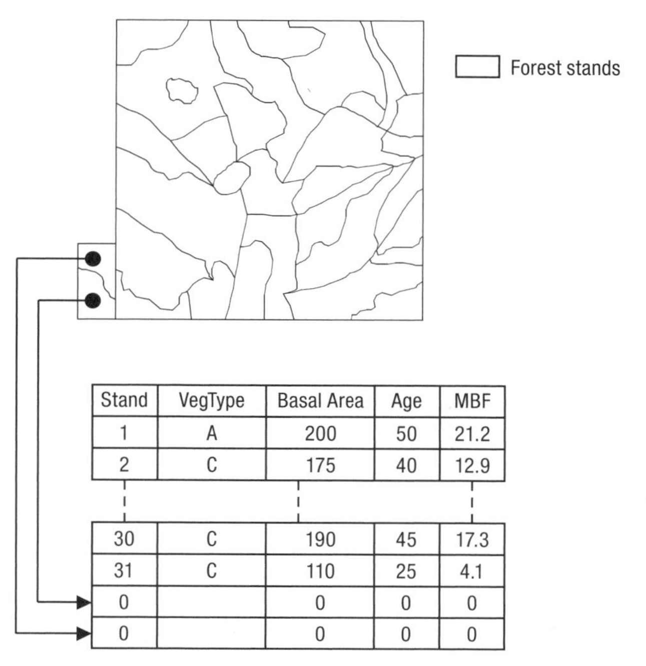

Stand	VegType	Basal Area	Age	MBF
1	A	200	50	21.2
2	C	175	40	12.9
30	C	190	45	17.3
31	C	110	25	4.1
0		0	0	0
0		0	0	0

Figure 10.4 Daniel Pickett forest stands and land purchase area after copying and pasting landscape features from the land purchase GIS database to the stands GIS database.

the GIS software program being used, use an 'update' function. ArcMap and ArcView 3.x, for example, both have the ability to use an update function made available through the XTools extension (Data East, 2007; Oregon Department of Forestry, 2003). When using a merge process or an update function, the stands GIS database will be updated with the new polygon data contained in the land purchase GIS database. If the land purchase GIS database includes fields named and formatted exactly as those in the forest stands GIS database, the attribute data within the land purchase GIS database will be moved (along with the associated purchased polygons) to the updated stands GIS database (Figure 10.5).

The assumption was made that the polygons in the land purchase GIS database seamlessly matched the edges of polygons in the Daniel Pickett forest stands GIS database. How is this possible? Matching the spatial juxtaposition of the new landscape features to the landscape features in the GIS database being updated (the stands GIS database) can be accomplished using one of at least two methods, depending on what type of process is available within the GIS software program being used: (1) copy the new polygons into the original stands GIS database and use snapping tools to properly match the new polygons with the original stands polygons, or (2) use a process as described in the first two steps of processes A and B of Figure 10.1. Here, you might first digitize the new land purchase polygons beyond the extent needed in the new land purchase GIS database, creating an area of overlap with the polygons in the stands GIS database (Figure 10.6). Then, erase from the land purchase GIS database the area of overlap with the stands GIS database, creating a second (new) land purchase GIS database. In this new land purchase GIS database, the edges of the new polygons seamlessly match the edges of the associated polygons in the stands GIS database (Figure 10.7).

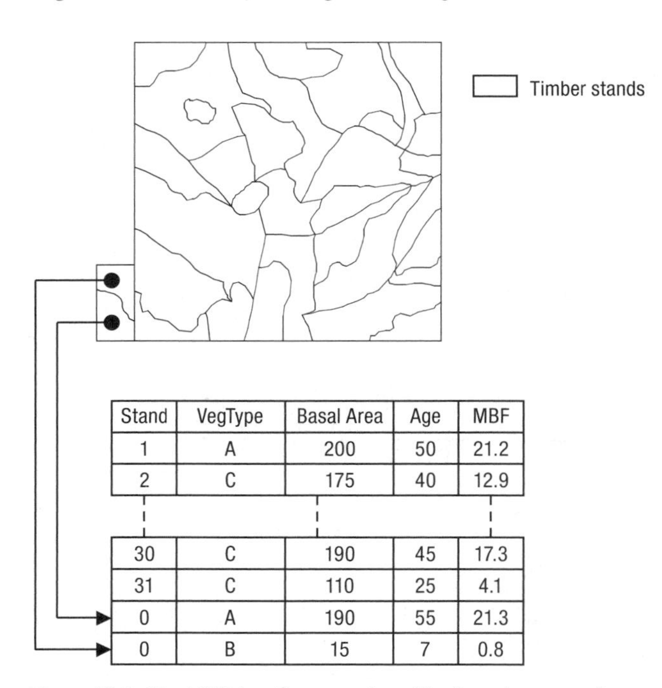

Stand	VegType	Basal Area	Age	MBF
1	A	200	50	21.2
2	C	175	40	12.9
30	C	190	45	17.3
31	C	110	25	4.1
0	A	190	55	21.3
0	B	15	7	0.8

Figure 10.5 Daniel Pickett forest stands and land purchase area after updating the stands GIS database using the land purchase GIS database.

Updating a trails GIS database

The existing trails system for the Brown Tract was digitized several years ago using hard-copy maps provided by the forest recreation planner (Figure 10.9). While suitable for recreation planning and the development of recreation maps to guide visitors around the area, the trail system described in the trails GIS database could very well be considered out-of-date. The trail system, like other features of a landscape, evolves as the managers of the forest develop new trails, or as people find different hiking or mountain biking routes through the landscape. The latter

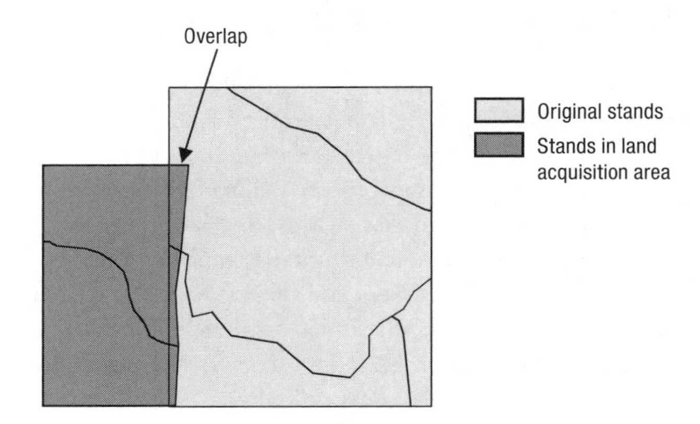

Figure 10.6 Overlap of new land purchase polygons with a GIS database that will be updated.

Figure 10.7 Land purchase GIS database after erasing the overlap with the stands GIS database.

In Depth

The term **digitizing**, as described in chapter 1, means to convert a hand-drawn (or other type of) map to a digital image of a map. Normally, digitizing is performed using a digitizing table and a digitizing puck. A map is laid on the table, taped down to ensure that it doesn't move during the digitizing process, and at least four control points on the map, for which on-the-ground coordinates are known, are entered into the computer system using the digitizing puck (similar to a computer mouse). The puck is then used to trace all of the polygons or lines to be digitized, or to note the points that need to be digitized. These landscape features are then saved as a GIS database. The person using the digitizer can control the number of vertices that describe lines or polygons. Digitizing points along 'salient features' (i.e., placing more vertices at very distinct changes in stand boundaries, or road curves) is a common method of digitizing. Other types of digitizing include allowing the creation of vertices along equal distances moved by the digitizing puck (e.g., a vertex created every millimeter that the puck moves) or at equal time intervals during the digitizing process (e.g., a vertex created every second during the digitizing process).

The term **heads-up digitizing** probably arose because a person's head is up, facing a computer screen, when landscape features are being digitized directly on a computer monitor rather than on a digitizing tablet. When performing traditional digitizing using a digitizing table and puck, a person generally has their head down, since they need to look down upon their map. With heads-up digitizing, a reference GIS database (perhaps a digital orthophotograph) is generally used as a guide for the creation of new points, lines, or polygons. A computer mouse is used to draw the new landscape features. Depending on the skill of the person performing the heads-up digitizing function and factors associated with the supporting spatial databases (scale and resolution), the accuracy of this method may be just as good as when digitizing using a digitizing table and puck. However, heads-up digitizing is much easier and faster. In addition, the chance for error through heads-up digitizing is greatly reduced because registration coordinates need not be entered and verified. Of course, this assumes that the reference GIS database (e.g., the digital orthophotograph) contains a limited amount of error. Given these trade-offs, a decision must be made with each digitizing project regarding the method of developing new landscape features. The decision is generally based on the standard protocols within an organization, the risks related to potential errors in the resulting GIS database (e.g., errors that may lead to making incorrect management decisions in the future), the quality of both hardcopy and digital products that would be used to support manual or heads-up digitizing, and the time and cost of the data development effort. Thus a balance must be struck between the organizational policies, the level of effort required to delineate landscape features, and the need to adequately and accurately describe those landscape features. For example, Figure 10.8 illustrates two attempts to digitize a young forest, with more effort (and thus time) being applied to one over the other. Which one of these is more accurate or useful to those making management decisions? Unfortunately the answer is uncertain.

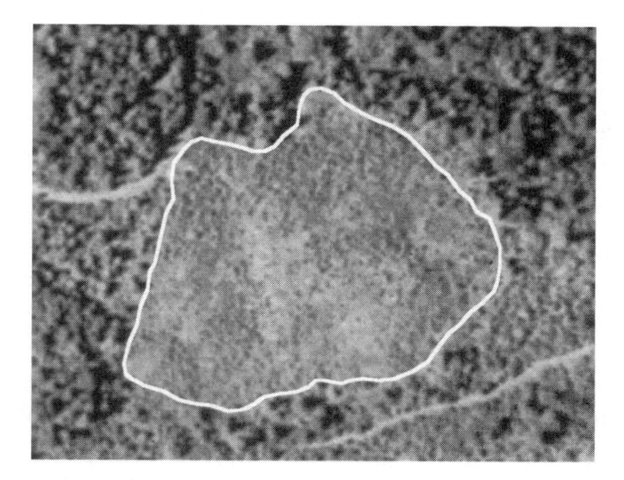

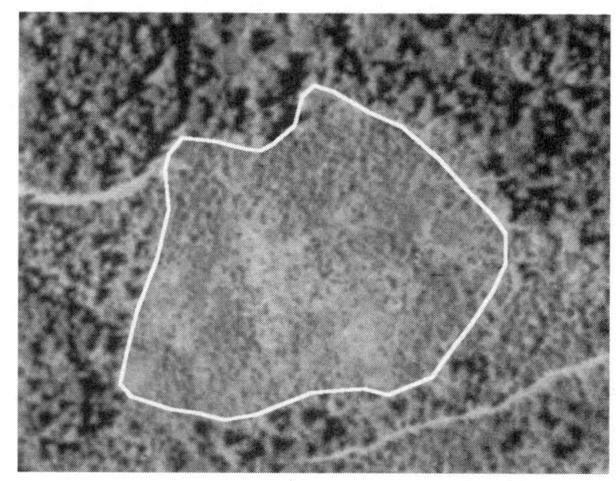

Figure 10.8 Two delineations of a young forest, one using twice as many vertices when digitizing (above) than the other (below).

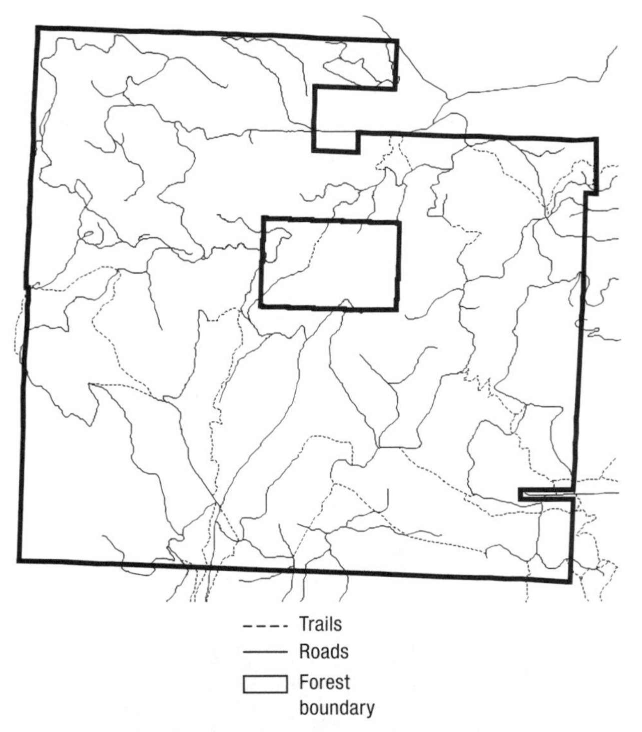

Figure 10.9 Trail system of the Brown Tract.

case usually involves the development of unauthorized trails, which the forest managers may or may not decide to more fully develop and maintain (or they may decide to develop measures to hinder the use of those trails). In addition, as resources become more popular and visitations increase (as is the case with many urban-proximate recreation destinations), there may be a need to identify additional resources and trails with which to decrease the density of use. In some cases, closing trails either seasonally or permanently is necessary to prevent resource degradation such as soil compaction and erosion.

The forest recreation planner decided that a new authorized trail would be of value to the recreation program (Figure 10.10). After the new trail was developed, the spatial coordinates of the trail's location were collected using GPS, brought into a GIS software program as a new GIS database containing line features, attributed, and saved as the 'proposed trail' GIS database. Since both the original trails GIS database and the proposed trails GIS database are composed of line features, and not polygons, they can be brought together without the worry of creating landscape features that overlap (and hence, in the case of polygons, leading to a double-counting of some areas in area calculations). However, careful attention must be paid to the connectivity of the network of lines that describe the trails.

Figure 10.10 Proposed new trail on the Brown Tract.

For example, the new trail may not end with a node that allows direct connection to a vertex of a trail in the original trail system (Figure 10.11). This creates a gap in the linear extent of the trail at the intersection point and will require some post-merging editing to correct.

To update the original trails GIS database, the original trails GIS database could be merged with the proposed trails GIS database. When doing so, only the fields (attributes) within the proposed trail GIS database that match exactly with the attributes within the original trails GIS database (in both attribute name and type) will be moved into the new merged GIS database. The spatial position of the proposed trails, as mentioned above and described in Figure 10.11, may then need to be edited. In addition, a verification of the attribute data (Figure 10.12) may suggest some alterations as well (e.g., the trail number of the proposed new trail is the same as another existing trail in the original trails GIS database). An understanding of the update process (Figure 10.13) will be of value in the planning of projects that involve alterations or updates to GIS databases.

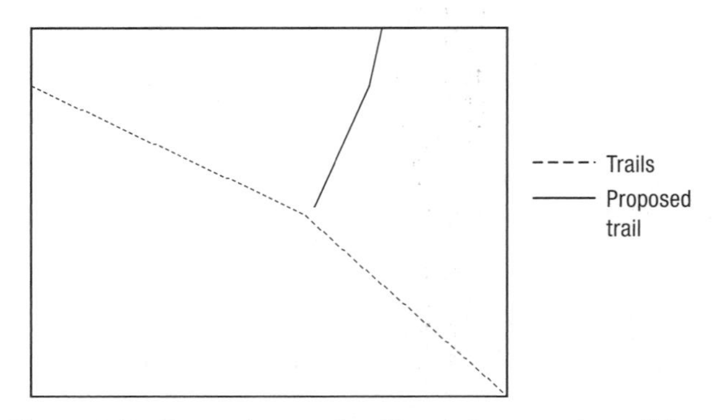

Figure 10.11 Proposed new trail and its relation to another trail from the original trails GIS database.

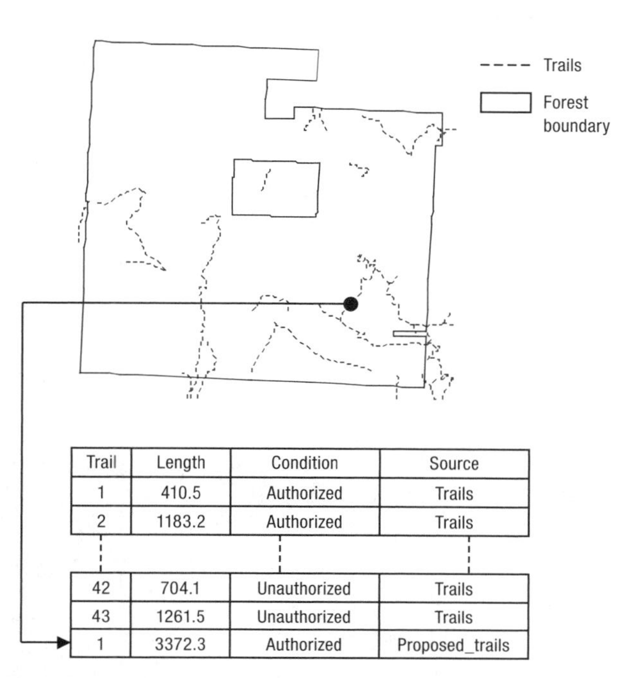

Trail	Length	Condition	Source
1	410.5	Authorized	Trails
2	1183.2	Authorized	Trails
⋮		⋮	⋮
42	704.1	Unauthorized	Trails
43	1261.5	Unauthorized	Trails
1	3372.3	Authorized	Proposed_trails

Figure 10.12 Updated trail system of the Brown Tract.

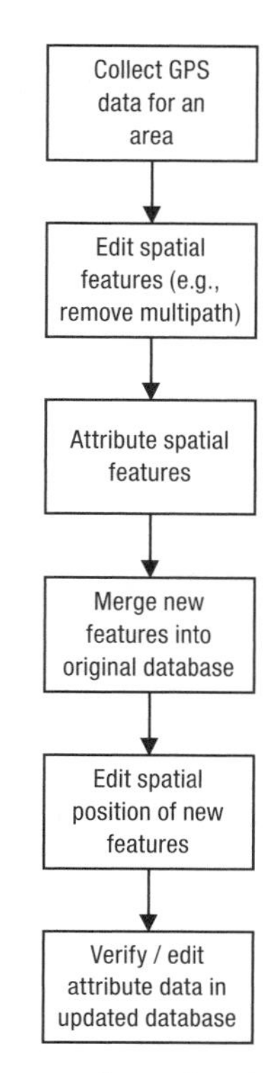

Figure 10.13 The process used to update the trails GIS database.

In Depth

Large GIS database updating projects require careful consideration of the time and cost required to successfully complete the project. For example, a roads GIS database for a 100,000 hectare forest may have been originally digitized from hard-copy maps (drawn by hand) and, therefore, might contain some spatial position errors. If you were to consider updating many of the roads in the GIS database by collecting new data with a GPS, the following should probably be considered:

1. The development of a standard protocol for GPS data collection (e.g., maximum PDOP), to ensure an acceptable and consistent level of accuracy in the data collected.
2. The need to drive (or walk) all roads that need updating.
3. The need to differentially correct and manage the GPS-collected databases.
4. The identification and elimination of error in the GPS data (such as multi-path error).
5. The removal of the old roads from the updated roads GIS database.
6. The connectivity of new roads to old roads that were not updated.
7. The development of a verification process to ensure that the attributes of the new roads are correct.

Determining the number of person-days required to accomplish each step will depend on the people, equipment, and technology available. The alternative to a large, single process for updating a GIS database is to perform it in small phases. However, while the cost of using multiple smaller phases may be lower than a single large project (assuming it requires several years to complete all of the phases), the total cost of the update process will likely be lower if the entire project were completed as a single project due to the economies of scale (fewer start-up and clean-up operations). In addition, fewer errors might arise since the same people will be working with a protocol that is clearly stated and understood.

Updating an Existing GIS Database by Modifying Existing Landscape Features and Attributes

An alternative to updating GIS databases with new landscape features contained in other GIS databases is to modify existing landscape features using the editing functions described earlier in this book. While this alternative may seem more logical than what was previously described in this chapter, the risk of damaging the GIS database being updated is greater. For example, the processes described earlier have a relatively low risk of damaging the original GIS features (the ones not requiring updating). These processes involved creating and modifying landscape features in a GIS database separate from the original GIS database, then moving the new features into the original GIS database only when it was appropriate to do so. Here, editing the original GIS database may pose a higher risk of damaging landscape features that did not require updating due to human error. And it is possible that these errors can occur without realizing a mistake had been made. In addition, unless the steps taken in editing a GIS database were carefully documented, when errors are located it may prove difficult to understand which landscape features had been verified as correct, and which may require further editing.

Two processes are next briefly described to illustrate the update of GIS databases by modifying existing landscape features. The first process requires editing the location of landscape features with the assistance of digital orthophotographs. The second process illustrates updating attribute data in a GIS database with the assistance of a join process.

Editing the spatial position of landscape features using digital orthophotographs

As described earlier, heads-up digitizing may be used to assist in the GIS database update process. Digital orthophotographs may be of benefit in updating the position of landscape features if the orthophotographs are registered appropriately to the correct landscape position, and if they have been stored in the coordinate and projection system consistent with the GIS databases to be updated. Point, line, and polygon GIS databases can be displayed in a GIS software program on top of a digital orthophotograph to examine how well the landscape features are being represented. Using the boundary GIS database of the Brown Tract as an example, you will find that a particular line may either incorrectly identify the forest boundary, or that a management operation on an adjacent landowner's property may have been incorrectly located (Figure 10.14). If this is, in fact, an incorrect boundary line specification within the boundary GIS database, then editing the appropriate vertices that define the boundary while using the digital orthophotograph as a guide can easily modify the spatial position of the boundary. If, however, you needed to be very precise regarding the location of the forest boundary, informa-

Figure 10.14 Boundary line issue on the Brown Tract.

tion from a land survey or survey-grade GPS measurements of property corners would be more appropriate in updating the spatial position of the boundary.

When updating the spatial position of landscape features in a GIS database, you must also be aware of the potential issues that may arise in other associated GIS databases. Here, for example, the intent may be to simply update the position of the boundary of the Brown Tract. However, by doing so, the spatial extent of the land ownership is no longer consistent with other polygon databases used to represent the Brown Tract; the stands and soils GIS databases being two good examples. Thus, after updating the boundary of the Brown Tract, a corresponding update of the affected polygons in the stands and soils GIS databases may also be required.

Updating the tabular attributes using a join process

In some circumstances only the attribute data of a GIS database may require periodic maintenance. For example, if over a given year no activities have been implemented on part of an ownership then perhaps only the attributes that describe the structural condition of the forest(s) need to be updated. In this process (Figure 10.15), the update might be accomplished by passing the stand-level forest inventory data through a growth and yield model, summarizing the resulting forest structural conditions, saving this data in a non-spatial database, then joining the non-spatial database to the stands GIS database. A unique stand identifier, such as the stand number, could be used to connect records in the non-spatial database to the forest GIS layer.

Using the Daniel Pickett forest as an example, a non-spatial database that represents the updated growth of stands (Update.txt) can be joined to the stands GIS database attribute table, using the stand identification number as the join item. Then, the attributes within the original stands GIS database can be replaced with the joined attributes that represent the updated basal area, age, and volume (MBF). The joined non-spatial table can subsequently be removed from the original stands GIS database, and the updated stands GIS database can be saved.

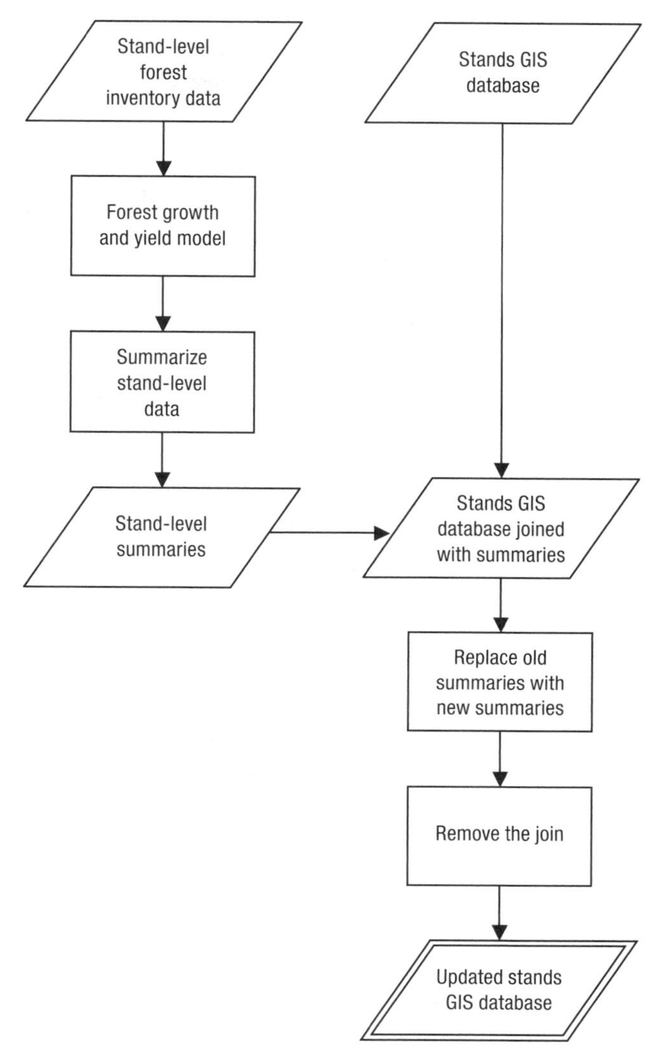

Figure 10.15 A process to update attribute data in the Daniel Pickett stands GIS database.

Summary

There are a variety of methods you can employ to update a GIS database. Several of the approaches were described with the examples provided in this chapter. GIS databases are rarely static, and in only in a few situations, such as in the delineation of national, provincial, state, or county boundaries, can you be confident that changes to a GIS database will rarely occur. Landscape features can change quite often: watershed boundaries change as the processes used to define watershed change, public land survey (PLS) section lines may change as corners are reestablished, stream locations change as they are better mapped using GPS or digital orthophotographs, and of course, vegetation stand boundaries change with management of natural resources or with natural disturbances. Attribute data associated with GIS databases can change just as often, and may need periodic updating. In addition, given continued improvements in measurement technology, it is a certainty that the effort and cost required for collecting update information for natural resources will continue to decrease in the future. The reduced resources required for collecting update information will likely lead to an increase in update frequencies for many natural resource organizations. At the very least, update costs will play less of a role in determining how often to update a spatial or non-spatial database.

The processes used to perform an update of a GIS database will likely be different from one organization to the next, and from one type of GIS database to the next. Careful consideration of the components of a system (the people involved, the databases considered, the data acquisition options, the software and hardware technology available, the budgetary constraints), and the needs of the ultimate customers (field personnel) will help determine the appropriate update process for each GIS database. In any update process, copies should be made of the databases that are subject to updates. These copies should be kept until such time that the updates are verified, and that no further need exists for the updates.

Applications

10.1. Land purchase. In the middle of the Brown Tract you may have noticed a piece of land that is managed or owned by someone else. Let's assume that the owners of the Brown Tract purchased this piece of land, in an effort to consolidate the ownership area. You have been asked to develop stand boundaries for this area, and incorporate them into the stands GIS database. Specifically, you need to delineate the stands into logical age or structural categories, although the only attribute you are asked to add to the GIS database is one for 'land allocations' (even-aged or uneven-aged). The forest staff will eventually develop an inventory for these stands and produce age, timber volume, and other statistics. To accomplish this updating task, you decide to use the digital orthophotograph associated with the Brown Tract GIS databases as a backdrop, and to use heads-up digitizing techniques to delineate the new stands.

The forest planning staff needs the following information:

a) How much area of land in even-aged stands will be added to the Brown Tract?
b) How much area of land in uneven-aged stands will be added to the Brown Tract?
c) A map of the new stands that illustrates the land allocation of each new stand, and includes the Brown Tract roads, streams, and digital orthophotograph.

10.2. Pondering the update process. In one example in this chapter an update of the Daniel Pickett forest stands GIS database was described, where a 32.4 hectare (80 acre) parcel of land was purchased. The edge between the polygons representing the land purchase and the polygons in the original stands GIS database was assumed to be seamless.

a) What GIS processes could be used to ensure that the edge would be seamless?
b) Draw a flow chart of a process that could be used to ensure a seamless edge.

10.3. Adding new roads to the Brown Tract roads GIS database. Assume locations of a new set of roads have been captured with GPS equipment. The GPS-captured features are stored in the GIS database called 'New_roads'. Update the Brown Tract roads GIS database by incorporating these new road features. Use your best judgment to attribute them in a consistent manner with the attributes in the roads GIS database.

a) How many kilometers (or miles) of dirt or native surface roads were in the original roads GIS database?
b) How many kilometers (or miles) of dirt or native surface roads are contained in the updated roads GIS database?
c) How many kilometers (or miles) of rocked roads were in the original roads GIS database?
d) How many kilometers (or miles) of rocked roads are contained in the updated roads GIS database?

10.4. Update process for a streams GIS database. A field forester working on the Brown Tract has compared mapped locations of streams to actual stream locations, and has discovered some differences. She has proposed that the mapped streams be updated and has asked for your guidance in how this process might be accomplished. Describe three options for gathering the data necessary to facilitate an update of the streams GIS database, and the merits of each approach.

10.5. Update intervals and approaches. A recreation manager, who has learned of your geospatial data skills, has asked your advice about updating the 10-year-old GIS trails system. Two trails have been added to the system during the past five years. The recreation manager is new to the area and does not know much about GIS and related technologies. The trails system is near an urban center and indications are that *use density* is increasing. Evidence of increasing use density includes a growing number of conflicts being reported by horseback riders, dog owners, and hikers throughout the forest. Additional evidence includes an unauthorized trail that is now clearly visible on the landscape, and which shows signs of significant use at one of the primary trailheads.

a) What is an appropriate update interval for trail information? Defend your choice.
b) How would you recommend that any new spatial data be collected?
c) How does any new spatial and attribute information collected be integrated into the existing GIS trails database?
d) What attributes of trails would you encourage the recreation manager to collect during field visits to the trails? Defend you recommendations.

10.6. Update tools and approaches. You've been asked to create a spatial data layer representing structures (buildings), streams, roads, and watershed boundaries within a relatively small 259-hectare (640 acres) experimental forest that is home to studies on hydrological observation and testing. At present, a polygon identifies the administrative boundary of the experimental forest and only primary roads and perennial streams are included in existing spatial databases. The existing data was created from hard copy maps at a scale of 1:24,000. You have at your disposal a 1:24,000 digital topographic quadrangle, a 30 m digital elevation model, a 1 m^2 resolution digital orthophotoquad, a consumer grade GPS, and a total station. To use the total station, two reference benchmarks are located within 2.0 km (1.2 miles) of the forest boundaries.

a) What data source and/or instrument would you use to capture the new spatial information related to:
   i. structures?
   ii. secondary roads?
   iii. smaller streams, including ephemeral streams?
   iv. watershed boundaries?
b) Which of the new spatial information sources described in part (a) would you incorporate into existing GIS databases to update them?
c) For the data you identified to be updated into existing GIS database (described in part (b)), how would you incorporate the new information into existing GIS databases?

References

Data East. (2007). *XTools pro.* Retrieved April 21, 2007, from http://www.xtoolspro.com.

Oregon Department of Forestry. (2003). *Guide to Xtools extension.* Salem, OR: Oregon Department of Forestry.

Chapter 11

Overlay Processes

Objectives

One of the great strengths of GIS is the capability to integrate landscape features and attribute information from more than one GIS database into a single GIS database. This capability is useful because it allows you to actively investigate and determine relationships between landscape features. Upon completion of this chapter, readers should have a sufficient amount of tools in their GIS toolbox to perform many of the common vector GIS overlay analyses required of field personnel working in natural resource organizations, whether they are employed in forestry, wildlife, soils, fisheries, or hydrology fields. The objectives of this chapter to provide readers with an introduction to three primary types of overlay analysis: the intersect, identity, and union processes. In addition, the ability of these overlay processes to accommodate different vector feature types (point, line, and polygon) is considered. When this chapter is completed, readers should have obtained knowledge and understanding of:

1. the outcomes from using an overlay process to accomplish one or more analytical tasks within GIS;
2. the circumstances that help you decide when each of the three overlay processes might be used to support an analysis or research objective; and
3. the differences among the three overlay processes, and between them and other similar GIS processes.

The topics discussed in this chapter relate to spatial overlay processes. Overlay processes are spatial analysis techniques that involve two or more GIS databases. More specifically, overlay processes can be used to investigate and identify spatial relationships between two or more databases. Overlay processes are powerful GIS tools and they represent what many consider to be the essence of GIS: the ability to integrate and organize information from multiple spatial data layers into a single database. Overlay processes can be thought of as modelling techniques that allow us to consider what might occur from the integration of information from multiple sources. Perhaps the most widely recognized example of pre-GIS (manual) overlay analysis is that demonstrated in Ian McHarg's book *Design with Nature* (1969). As mentioned earlier in chapter 1, this book provided examples of manual map overlays that were used to identify areas that were suitable for certain activities. *Design with Nature* inspired many people to apply manual overlay techniques for their own analysis needs and to also consider developing digital tools (GIS) that might be used to make overlay analysis more precise and efficient.

The topics discussed in this chapter relate to spatial analysis processes that are similar to merging (chapter 8), and to obtaining information about specific geographic regions (chapter 7). There are, however, some very distinct differences between the merge process and the overlay processes presented in this chapter. When two (or more) polygon GIS databases are merged, for example, the resulting GIS database may consist of overlapping polygons. The processes presented in this chapter—the intersect, identity, and union processes—all result in new output GIS databases where the landscape features do not overlap. Overlapping areas and attributes are combined, in some form or fashion, thus the potential exists for polygons to be split and combined and their topology reassessed in the resulting GIS database. Calculating the area covered by polygons in a merged GIS database may therefore be misleading, due to the presence of overlapping polygons. Further, you may be interested in the characteristics of the actual areas that overlap, such as the overlapping soils and forest stands, and how this information can help you make better (or more informed) management decisions. To accomplish this with a merged GIS database is difficult, as the overlapping polygons are independent and share no information other than that which the user can obtain visually on a map or computer monitor.

Intersect Processes

When performing an **intersect process**, you hope to acquire information about the overlapping areas of two GIS databases. The term 'intersect' implies that features meet or cross at certain points, and (in the case of polygons) share common areas (Merriam-Webster, 2007). In using an intersect process, a third, new GIS database will be created that consists of only those areas where the two original GIS databases overlap, no more, no less. For example, Figure 11.1 shows two GIS databases, one that represents vegetation (a stands GIS database), and the other representing a burned area from a fire. If you used an intersect process on the two GIS databases, the resulting GIS database would contain only the geographically overlapping landscape features. Spatially, the extent of the output will be defined by the boundaries of the polygons in the original two GIS databases: stands outside the burned area are excluded and the burned area outside of the vegetation stands is excluded. The extent of the attribute data that will be contained in the resulting GIS database can include attribute data from one database or all attribute data from both of the original GIS databases.

By comparison, a similar result can be obtained using a clipping process. Here, you might clip the vegetation GIS database using the fire GIS database. This alternative, while providing the same geographic representation of the vegetation burned by the fire, provides no attributes of the fire in the resulting output GIS database. Although the example includes only one fire, suppose you had a fire GIS database that contained the geographic boundaries of several fires that may have occurred over a summer. If each fire polygon was attributed with the day, month, and year of its inception, this information would not be contained in the output GIS database if a clipping process was used. To generate a GIS database that not only included the stands within the burned areas, but also the

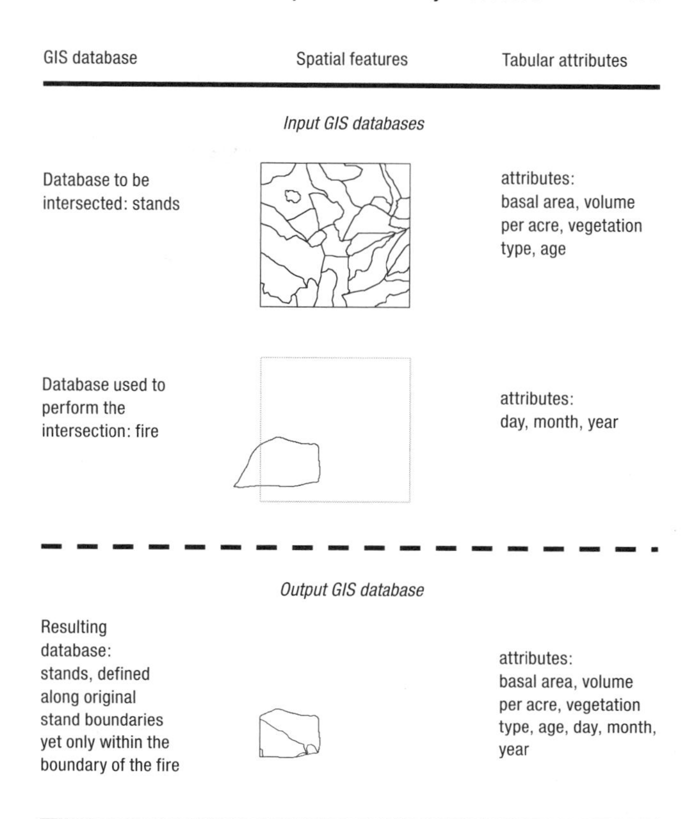

Figure 11.1 Intersecting the stands GIS database with the fire GIS database on the Daniel Pickett forest.

stand and fire attributes, you should consider using an intersect process.

The intersect process works by locating the arcs (links) that are present in one GIS database that overlap some area in the complementary database. For example, GIS database #1 in Figure 11.2 contains a single polygon defined by two arcs. GIS database #2 contains two polygons and 7 arcs. The polygon contained in GIS database #1 does overlap the polygons in GIS database #2, as arc 1_1 runs from the middle of arc 2_6 to the middle of arc 2_7, crossing over arc 2_4 in the process. Therefore, arc 1_1 overlaps an area of represented by the polygons in database #2, while arc 1_2 does not. Further, portions of arcs 2_4, 2_6, and 2_7 overlap some area represented by the polygon in database #1, yet arcs 2_1, 2_2, 2_3, and 2_5 do not. The resulting output GIS database will contain two polygons, with one side bounded by portions of arc 1_1 (split at the intersection with arc 2_4), and the other sides by portions of arcs 2_4, 2_6, and 2_7.

To further illustrate the power of the intersect process, the next example illustrates how you can use the results to assist in developing information relevant to natural

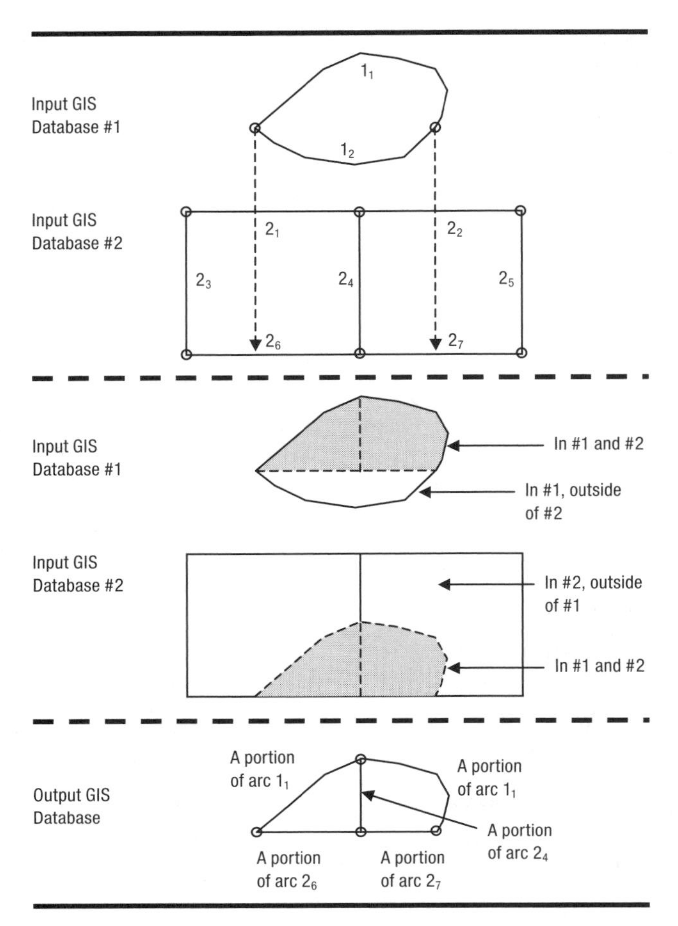

Figure 11.2 An example of the processing of landscape features during an intersect process.

resource management planning. In this example, assume you are interested in developing information for a potential forest fertilization project, and an examination of the intersection of the Daniel Pickett stands and soils GIS databases would be helpful, since the decision to fertilize may be based on both forest structural conditions and soil conditions. Separately, each of these two GIS databases contains a theme: the stands GIS database describes the forest structural conditions of the Daniel Pickett forest and the soils GIS database describes the underlying soil types and their characteristics. The stand polygon boundaries are defined by transitions in forest structural conditions (e.g., a change from young forest to older forest defines a stand boundary), roads, and perhaps streams. The soil polygon boundaries are defined by changes in soil characteristics, although some would argue that these boundaries should be considered 'fuzzy' because soils, generally speaking, do not change as abruptly as stand characteristics might, but change gradually over a larger area. In addition, the ability to accurately and precisely measure and map changes in soil types is limited by the tremendous effort that is required to sample and delineate soils. Nonetheless, by overlaying the two GIS databases using an intersect process (Figure 11.3), a GIS database is created where polygon boundaries are now defined by those lines that were present in both the stands and soils GIS databases. The outside boundary of both the soils GIS database and the stands GIS database was exactly the same, therefore no areas covered in either database are excluded in the resulting stands/soils GIS database. The resulting stands/soils GIS database contains more polygons (47 polygons) than what was found in the original stands GIS database (31 polygons) and soils GIS database (7 polygons). Upon inspection, you may find that may of these polygons are actually spurious (small and irrelevant) polygons created simply due to the happenstance location of the boundaries of polygons from the original two GIS databases (Figure 11.4).

The tabular data contained in the resulting intersected stands/soils GIS database contains all of the attributes of

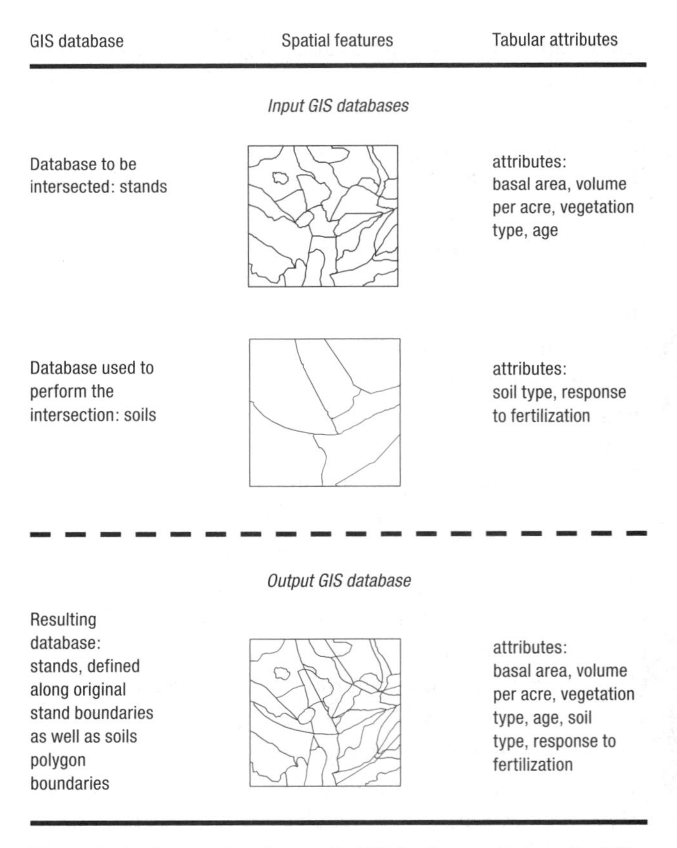

Figure 11.3 Intersecting the stands GIS database with the soils GIS database on the Daniel Pickett forest.

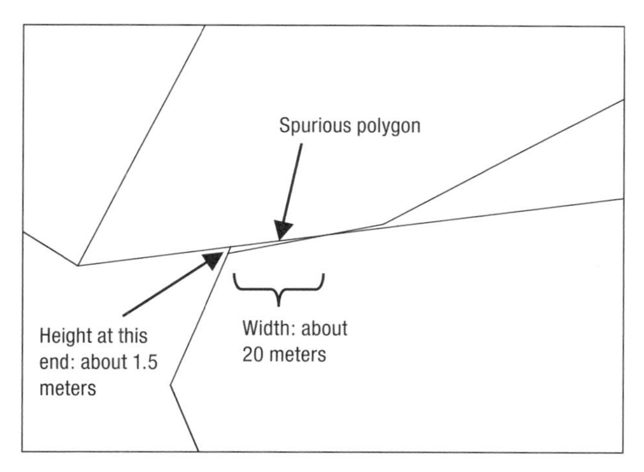

Figure 11.4 A spurious polygon that was created during the intersect process.

each of the original stand polygons, and all of the attributes of each of the original soil polygons. Thus for each of the 47 polygons in the stands/soils GIS database, you now know the stand conditions above ground as well as the soil conditions below ground level. With this GIS database you can perform queries that relate to both stand and soil conditions. For example, based on a discussion with the foresters associated with the Daniel Pickett forest, it may be decided that 20–30 year old stands on soils that were amenable to a high forest growth response to a fertilization treatment would be the most preferable areas to fertilize. To find these areas in the stands/soils GIS database, you might develop a query such as the following (for a review of queries, please refer back to chapter 5):

[Age ≥ 20] and [Age ≤ 30] and [Fertresp = 'high']

where:

Age = the stand age attribute in the Daniel Pickett stands GIS database, and
Fertresp = the fertilization response attribute in the Daniel Pickett soils GIS database

With this type of query GIS will provide the locations of areas where the intersection of stand age and soil type meeting the criteria. As you can see in the case of the Daniel Pickett forest (Figure 11.5), the potential fertilization areas may not correspond directly with the original stand boundaries. Natural resource managers using an analysis such as this will subsequently need to decide whether to fertilize whole stands (not just the part of a stand where the stand and soil conditions are appropriate, since stand boundaries are usually easy to locate) or parts of stands. The advantage of fertilizing a whole stand, when only a portion seems appropriate, is that you do not need to waste time trying to identify a vague transition in soil types. On the other hand, fertilizer, and the

In Depth

What is a **spurious polygon**? When something is labeled as 'spurious', it is meant to indicate that it is counterfeit, false, fictitious, or not legitimate (Merriam-Webster, 2007). Within GIS, spurious polygons are certainly genuine, and can be somewhat troubling to eliminate. These polygons arise simply because the computer and GIS software are taking two GIS databases and combining them according to the instructions provided by the user. They are only doing what they were told to do: break polygons along intersecting lines and create new polygons using the newly formed intersections. Spurious polygons might not be considered legitimate, however, given the management needs of an organization. Most organizations, in fact, have what they term 'minimum mapping units', and polygons below this size are eliminated by being included with other larger, adjoining polygons. In some cases, the spurious polygon could become a part of an adjoining polygon with which it shares the longest edge. The shared edge is essentially removed, and the difference between the spurious polygon and its adjacent neighbor is lost. When using GIS processes such as the intersect, clipping, and buffering processes, spurious polygons will undoubtedly be created. How you manage them (e.g., ignoring them, eliminating them, etc.) once they have been created is a matter of personal preference, or perhaps a reaction to organization standards. Most natural resource management decisions will not be affected by the presence of spurious polygons, however the presence of spurious polygons may become a database management problem and may detract from a message presented in a map.

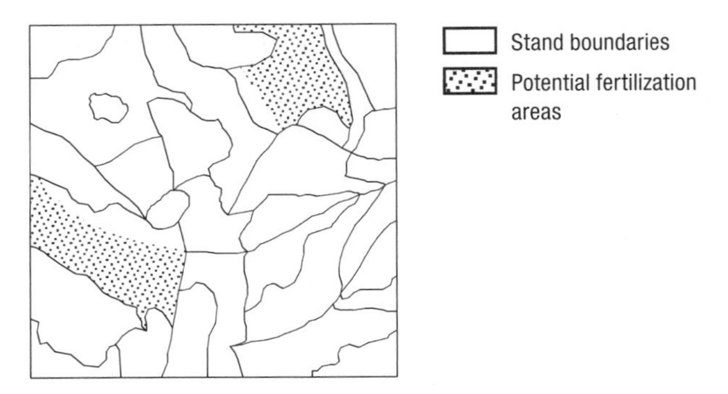

Figure 11.5 Potential fertilization areas on the Daniel Pickett forest that consist of forest stands 20–30 years of age located on soils that provide a high response to a fertilization application.

application of fertilizer, costs money, and organizations may want to operate as efficiently as possible. Therefore, the goal would be to apply treatments only where necessary. It is becoming increasingly common to provide a fertilization contractor with the exact geographic coordinates that define the fertilization area, either with maps or actual GIS databases. If a fertilization contractor uses a helicopter system to distribute the fertilizer, which is a common practice when fertilizing forested stands, it is possible that GPS technology may be employed to assist in locating the desired fertilization areas. The GPS data may facilitate the development of a digital map, which can help guide the pilot or can be coupled with an internal guidance system to automatically apply fertilizer.

Identity Processes

When performing an **identity process**, you hope to incorporate information about the overlapping area of one GIS database into a second GIS database. While the term 'identity' refers to a distinguishing character of a personality (Merriam-Webster, 2007), this process is more likely similar to the term 'identity element', where parts of the original data are left unchanged when combined with other data. This is partially true—some portion of the original GIS data may be left unchanged (and present in the output)—where a second set of GIS data does not coincide spatially.

Similar to the intersect process, one GIS database is physically laid onto another, yet there is usually a distinct difference in the resulting output when compared with the intersect process. The resulting GIS database is defined by the boundary of one of the input GIS databases, not by the boundary of the overlap between the two GIS databases. Figure 11.6 illustrates that the result-

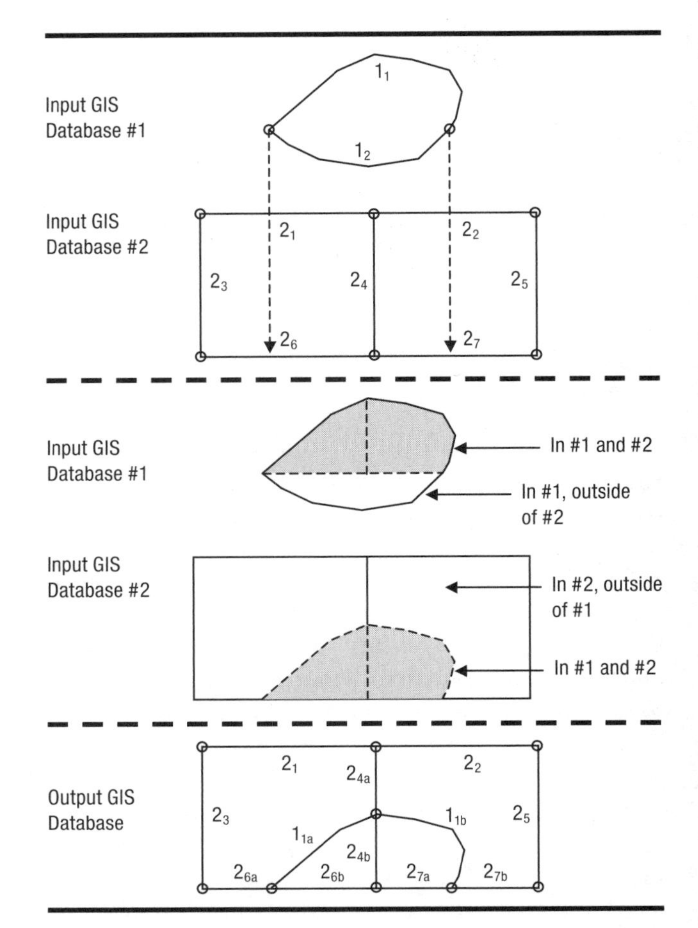

Figure 11.6 An example of the processing of landscape features during an identity process.

ing GIS database has a geographic extent representing the same area as database #2, yet the arcs that defined the polygon in database #1 are present where the polygon from database #1 overlapped the polygons in database #2. The resulting GIS database now includes 4 polygons, 12 arcs, and 9 nodes, as arcs 2_4, 2_6, and 2_7 were split into two pieces (*a* and *b*) based on their intersection with arc 1_1, and arc 1_1 was split into two pieces based on its intersection with arc 2_4.

To extend the identity process example to the Daniel Pickett forest, again examine the case of the stands GIS database and the fire GIS database. Suppose the intent was to develop a GIS database that contained the entire stands data (geographic and tabular), but with the fire boundaries being integrated into the stands database. Using an identity process you can see that some stand polygons have been split along the fire polygon boundary (Figure 11.7), thus the fire GIS database has an influence on the structure of the resulting polygons. In addition, attribute

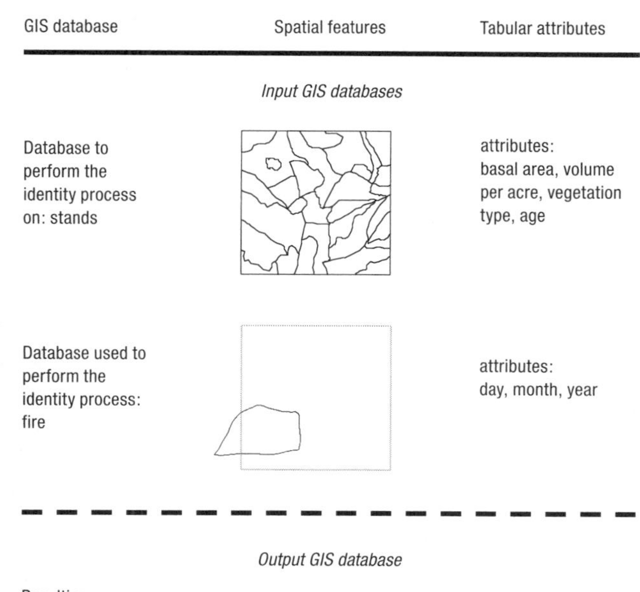

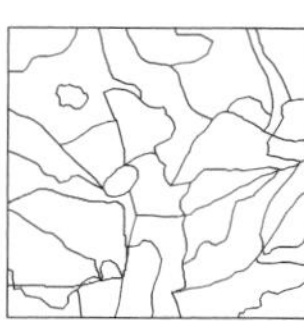

Figure 11.7 Performing an identity process on the stands GIS database using the fire GIS database.

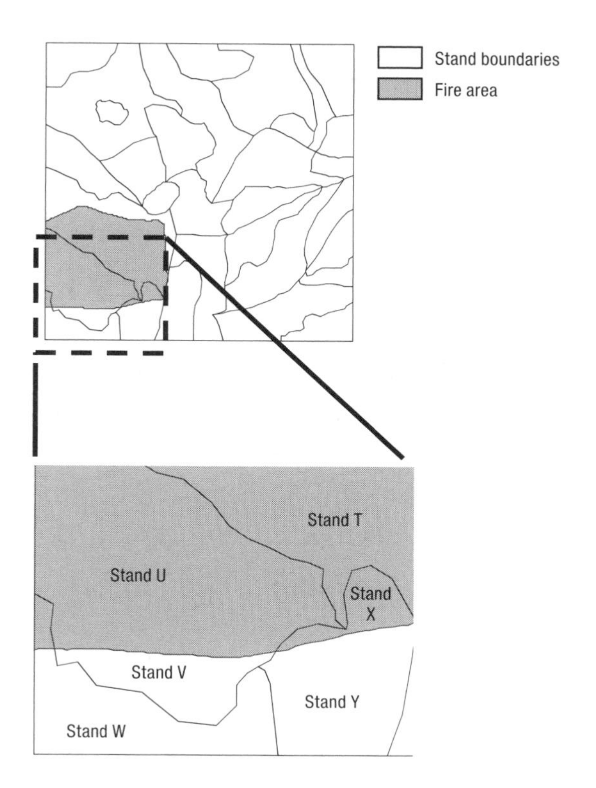

Stand	VegType	Basal Area	Age	MBF	Month	Day	Year
T	C	120	55	19.5	7	2	2002
U	A	260	70	37.7	7	2	2002
V	A	260	70	37.7	0	0	0
W	C	190	45	17.3	0	0	0
X	B	20	10	1.8	7	2	2002
Y	B	20	10	1.8	0	0	0

Figure 11.8 A more detailed examination of the results of the identity process of the fire GIS database overlaid on the stands GIS database.

fields are present in the resulting GIS database to represent those that were present in the fire GIS database. However, only the polygons within the fire boundary actually contain data related to the fire. The attribute fields from the fire GIS database related to polygons outside the fire area are empty and contain '0' values (Figure 11.8).

A key concept when performing the identity process is obviously determining the spatial extent of the two GIS databases that is to be retained. In the above example, the fire GIS database was overlaid on the stands GIS database, and the intent was to retain the spatial extent of the stands GIS database. If you were to reverse the order and overlay the stands GIS database onto the fire GIS database, the resulting GIS database would have quite a different look to it (Figure 11.9), as the spatial extent of the resulting GIS database is defined by the spatial extent of the fire GIS database. Here, only the stand boundaries within the fire remain. While the stand-level data attributes associated with the stands in the fire area are present in the resulting GIS database, no stand-level data attributes are available for the polygons outside of the area represented by the original stands GIS database (Figure 11.10).

Union Processes

In a **union process**, the intent is to overlay one GIS database on top of another GIS database, and retain all of the spatial boundaries of the landscape features contained within both GIS databases. A 'union' is the act of joining two or more features into one (Merriam-Webster, 2007). Figure 11.11 illustrates that when using a union process, the resulting GIS database has a geographic extent representing the same area as both database #1 and database #2, yet the arcs that defined the polygon in database #1 are present where the polygon from database #1 overlapped the polygons in database #2. The resulting GIS database now includes 5 polygons, 13 arcs, and 9 nodes, as arcs 2_4, 2_6, and 2_7 were split into two pieces (*a* and *b*)

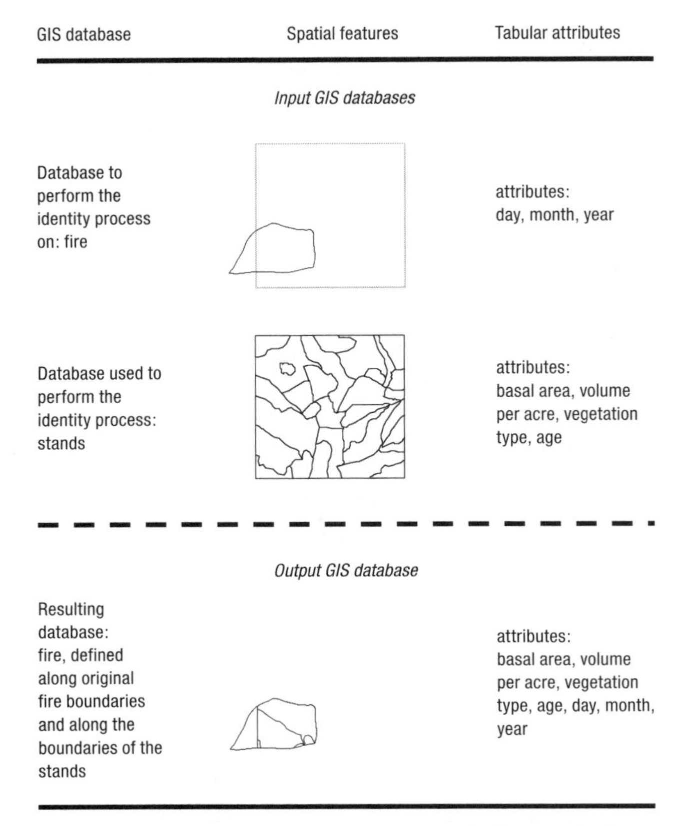

Figure 11.9 Performing an identity process on the fire GIS database using the stands GIS database.

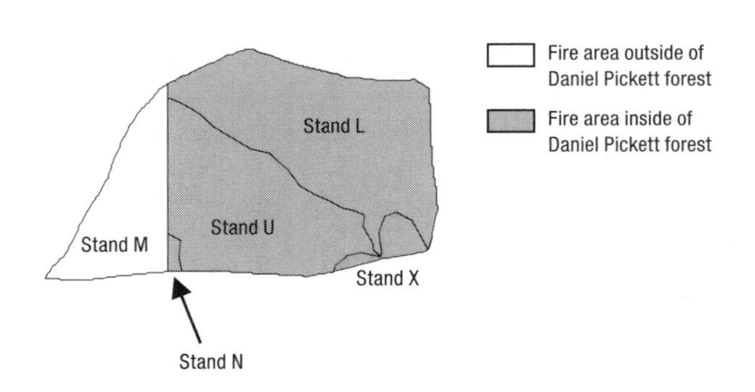

Stand	VegType	Basal Area	Age	MBF	Month	Day	Year
L	C	120	30	5.6	7	2	2002
M		0	0	0	7	2	2002
N	C	190	45	17.3	7	2	2002
U	A	260	70	37.7	7	2	2002
X	B	20	10	1.8	7	2	2002

Figure 11.10 A more detailed examination of the results of the identity process of the stands GIS database overlaid on the fire GIS database.

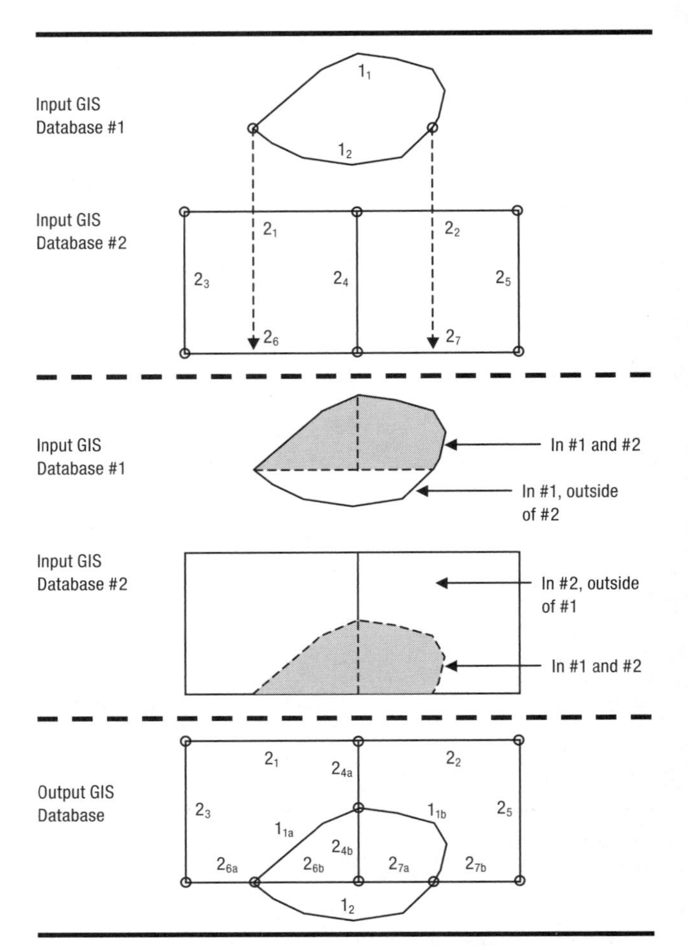

Figure 11.11 An example of the processing of landscape features during an union process.

based on their intersection with arc 1_1, and arc 1_1 was split into two pieces based on its intersection with arc 2_4.

To illustrate a union process with a more realistic natural resource management problem, suppose a union process were to be performed on the Daniel Pickett forest fire and stands GIS databases. The resulting GIS database has the combined geographic extent of the two GIS databases, and contains similar landscape features as were found in the original fire and stands GIS databases (Figure 11.12). This illustrates one advantage of using a union process: the spatial delineation of the polygons in the resulting GIS database is a function of both of the original GIS databases, thus polygon boundaries from both original GIS databases are retained. However, it also suggests a disadvantage of the process: the resulting GIS database may contain landscape features outside of the boundary

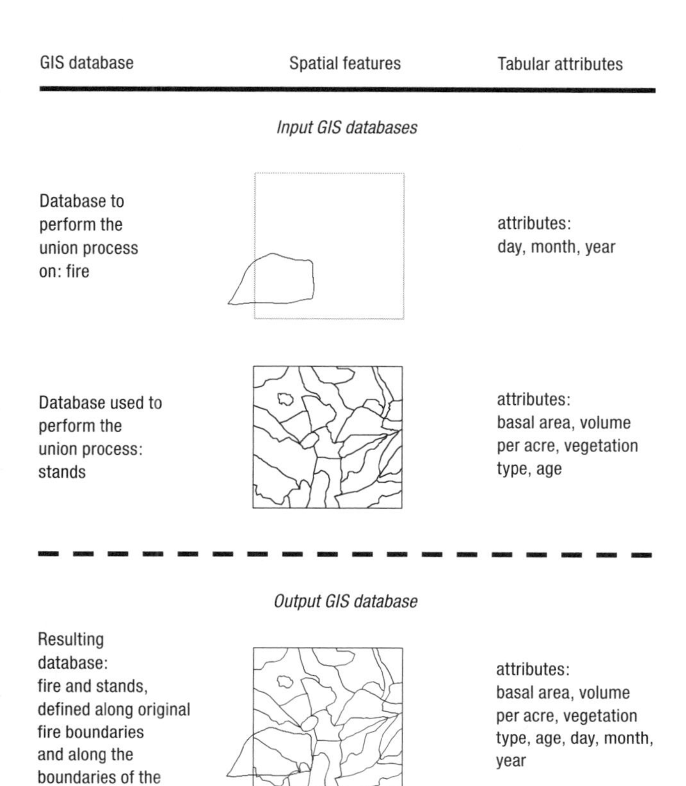

Figure 11.12 Performing a union process using the fire GIS database and the stands GIS database.

of interest, and thus perhaps may include unnecessary landscape features. Although some of the fire polygon lies outside the Daniel Pickett forest, this area will be represented in the output GIS database. The attribute fields contained in both of the original GIS databases may also be present in the resulting GIS database, however some data cells will likely be empty in the attribute table (Figure 11.13). The union process is useful for those situations in which you want to preserve all of the spatial and non-spatial data that is present in two input GIS databases.

More complex analyses can be performed using the intersect, identity, or union processes than simply bringing together the characteristics of two GIS databases. For example, suppose you were interested in locating areas suitable for a certain type of agricultural practice in the Pheasant Hill planning area of the Qu'Appelle River Valley in central Saskatchewan. While the databases we will use in this example are dated (1980), they are rich with information and they allow us to examine the usefulness of the union process for a natural resource manage-

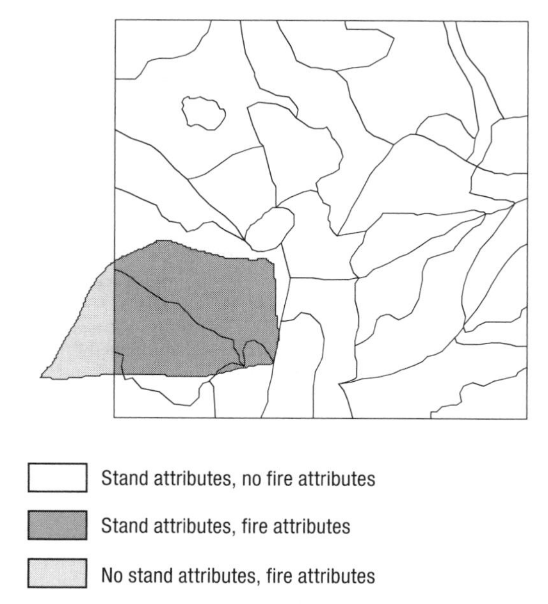

Figure 11.13 Illustration of completeness of a tabular database after a union process of the fire and stands GIS database.

ment analysis. In locating the suitable areas, you decide that the criteria should include locating a certain type of soil (loamy soils), on a certain slope condition (flat slopes), where the area is currently zoned for agricultural use, and where there are few limitations for using the land to grow agricultural products. In this assessment, there are four distinct attributes about the land: soil type, slope condition, zoning code, and land classification. These four attributes are contained within four different GIS databases associated with the Pheasant Hill planning area, and are delineated using polygons that do not necessary coincide (spatially) from one GIS database to the next. One way to accomplish this overlay analysis is to use an iterative union process to bring all four databases together so that all of the attributes are available in a single, combined GIS database. Initially, two of the GIS databases would be unioned, then a third would be unioned to the union result of the first two. Finally, the fourth database would be unioned to the union result of the first three GIS databases. Using a query that involved the criteria listed below, the areas suitable for the practice you had in mind could be identified (Figure 11.14), since the union of the four GIS databases would contain the attributes of all of the original GIS databases, and since the polygons would be split along the boundaries of the original polygons.

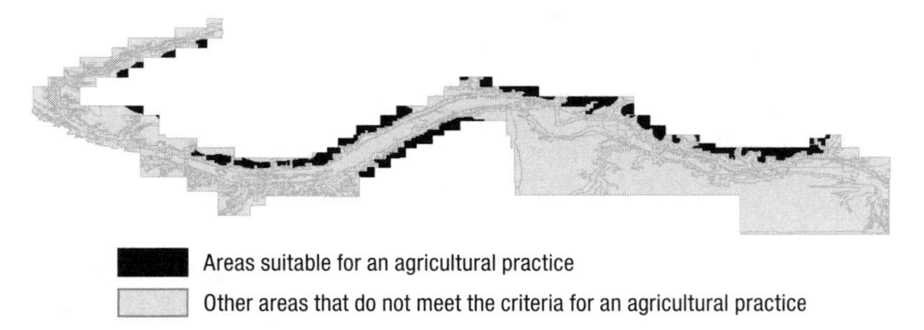

Figure 11.14 The result of a query on the union of soils, topography, land classification, and zoning GIS databases developed for the Pheasant Hill planning area of the Qu'Appelle River Valley, Saskatchewan (1980).

Query criteria	Original database
Soil_type = Canora Loam	soils
or Soil_type = Indian Head Clay Loam	soils
or Soil_type = Oxbow Clay Loam	soils
or Soil_type = Oxbow Loam	soils
or Soil_type = Rocanville Clay Loam	soils
or Soil_type = Whitesand Gravelly Loam	soils
Topography = FLAT	topography
Class = No Significant Limitations	land classification (CLI)
or Class = Moderate Limitations	land classification (CLI)
Zoning = Agriculture Priority 1	zoning

Incorporating Point and Line GIS Databases into an Overlay Analysis

Although the overlay examples thus far have focused on the analysis and manipulation of polygon GIS databases, it is also possible to incorporate other types of features (points and lines) into overlay processes. For example, point and line GIS databases can be used in association with polygon databases when performing the intersect and identity processes. The union process, however, requires that all GIS databases of interest be composed of polygon features. When using the intersect and identity overlays, the input GIS database can be composed of points, lines, or polygons but the overlay GIS database must be composed of polygons with one exception. Within some GIS software, the intersection of two line databases is possible, with the result being a new point database that has captured all intersection locations. When point or line databases are involved in an overlay process with a polygon, the resulting output GIS database will be of the same feature type as the first input GIS database (point or line).

As an example, if you use a GIS database containing lines as an input GIS database, and then intersect it with a GIS database containing polygons (Figure 11.15), the resulting GIS database will be composed of line features. The line features will be split at all intersections with the boundaries of the polygons in the polygon GIS database,

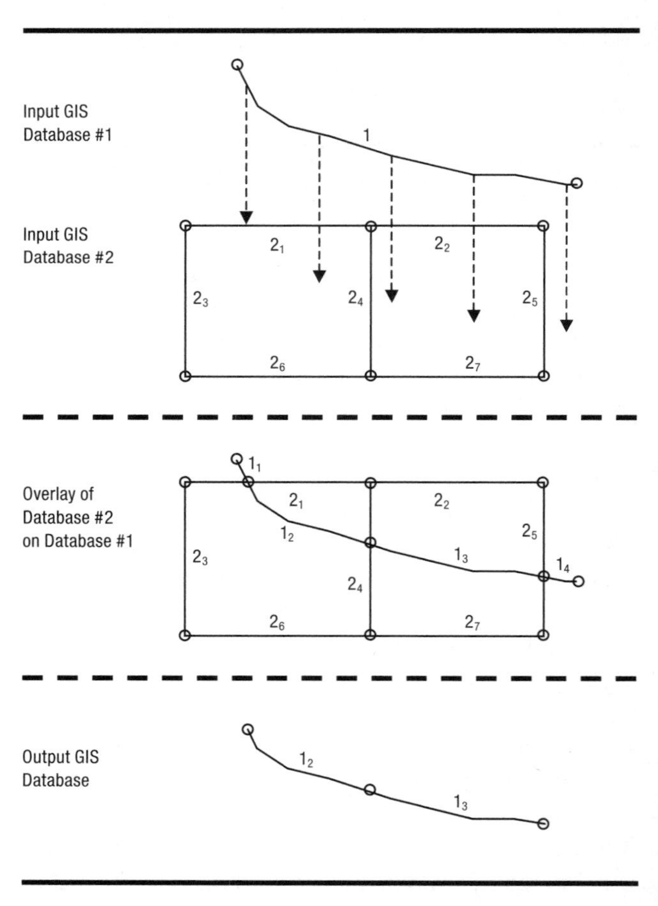

Figure 11.15 An example of the manipulation of landscape features during an intersect overlay of line and polygon databases.

and only lines that fall within the extent of the polygons will be retained in the output GIS database. While this represents a process similar to the clip process, the resulting lines contain the information (attributes) of the polygons within which they fell. In the example in Figure 11.15, a line (1) that represents a road (perhaps) is being overlaid by the two polygons from previous examples in this chapter. The line initially contains two nodes, but when the overlay process occurs, it is broken into four arcs (1_1, 1_2, 1_3, and 1_4) with 5 nodes. The pieces of the original road that fall outside the area covered by the polygons (1_1 and 1_4) are subsequently eliminated. The resulting GIS database contains a portion of the original line (sections 1_2 and 1_3), and each portion of the original line contains the attributes of the associated polygon within which it was contained.

An identity process using point or line GIS databases as the input database results in a GIS database that contains all of the original point or line features, yet the landscape features would contain the attribute information of the polygons within which they fell. Lines would also be split at polygon boundaries as in the previous example. The primary purpose of the identity process would be to distribute attribute data from an overlaid polygon GIS database to a point or line. The identity process with a point GIS database as the input GIS database is similar, in fact, to a point-in-polygon query, although the results here are not temporary.

Applying Overlay Techniques to Point and Line Databases

We present here some examples of applying overlay processes to point and line databases to demonstrate potential overlay applications. Our examples make use of the Brown Tract databases described earlier in the text. In the first example, consider the distribution of research plots as they relate to the Brown Tract forest stand boundaries (Figure 11.16). It might be of interest to determine the distribution of land allocation categories (described in the forest stand layer) that are associated with each research plot. A point in polygon intersect overlay is one approach an analyst could use to determine this information. The results of the intersect overlay in this case create a new point layer that contains all plot locations and an expanded list of data fields. Each of the research plot records would contain both the original data fields and the same data fields and values of the individual stand boundaries in which they were located. A statistical frequency could be generated from the new database and could demonstrate the distribution of land allocations contained in the research plot locations (Table 11.1). In this case, the majority (48 of 57) of research plots are located in even-aged stands. Given the geometry of the research plots and stand boundaries, an identity overlay process between these files should produce the same results as the intersect process.

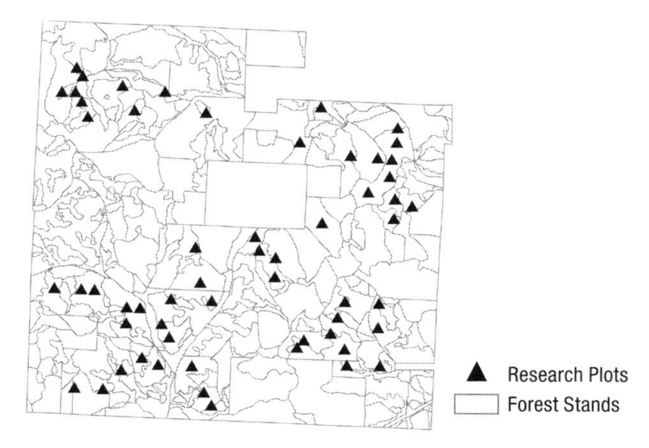

Figure 11.16 Research plots and forest stands on the Brown Tract.

As mentioned earlier, overlay operations involving line databases are also possible. As an example consider the arrangement of streams relative to stands in the Brown Tract (Figure 11.17). There are portions of the Brown Tract where the stream network extends beyond the forest boundaries. In this case, the intersect and identity overlay commands would lead to different output databases. The intersect overlay between these layers would result in an output stream system that would be reduced in extent from the original; only those stream segments that overlapped the stands would be retained in the output database. In addition, all data fields in both databases would be populated in the output database. The identity

TABLE 11.1 Frequency distribution of land allocation categories in research plot locations within the Brown Tract

Land allocation	Number of research plots
Even-aged	48
Research	6
Uneven-aged	3

Figure 11.17 Streams and forest stands on the Brown Tract.

overlay would retain all the stream segments, but we would only find stand data in a resulting database for a portion of the stream segments. The selection of which overlay to use—intersect or identity—will depend on the analysis goals. Let's assume that the analysis goals support retaining all stream segments in a final streams database. An identity output database between streams and stands contains 398 stream records since geometric intersections are created at all coincident locations in the input databases. The distribution of land allocation values for these streams is represented in Table 11.2. Note that there are 40 stream segments that do not have a land allocation value. These missing values represent streams outside the Brown Tract boundaries.

Although point and line spatial databases can be used with overlay processes to derive information from polygon databases, polygon databases are unable to serve as the output product of overlay analysis in these cases. In order to assimilate information from point and line databases into a polygon database, other approaches such as tabular or spatial joins must be considered.

Additional Overlay Considerations

Each of the three overlay processes discussed in this chapter are designed for different end products (Figure 11.18). Although different outputs will usually result depending on which overlay is chosen, there are situations where the output of two overlay processes will be the same, regardless of which process is used. For example, if two polygon databases have the same spatial extent, it will not matter whether an intersect, identify, or union process is used. In addition, if one spatial database is contained completely within the extent of another database and is used as the primary overlay database, an identity or intersect command should produce the same result.

TABLE 11.2 **Frequency distribution of land allocation categories in relation to stream segments within the Brown Tract**

Land allocation	Number of research plots
Even-aged	281
Meadow	2
Oak Woodland	3
Research	7
Uneven-aged	65
<No Value>	40

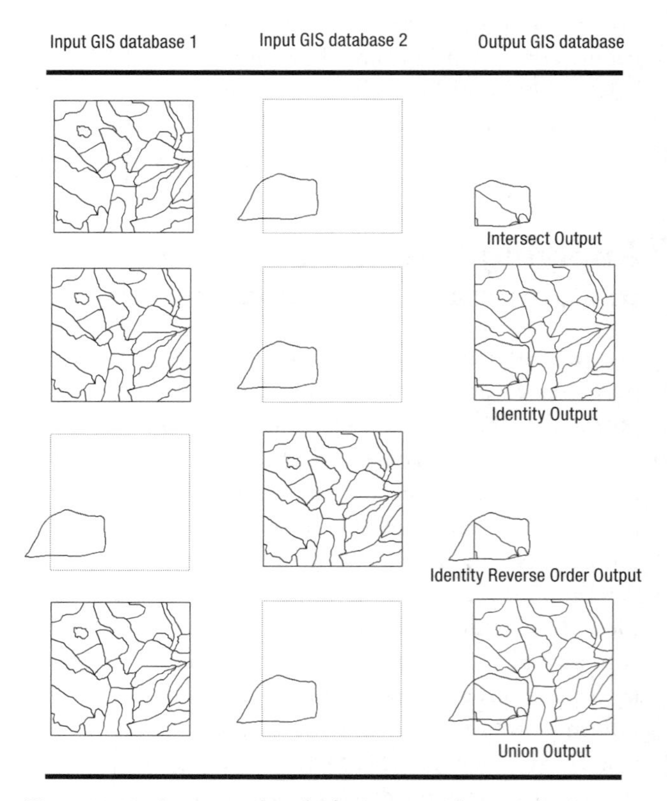

Figure 11.18 Summary of results for intersect, identity, identity reverse order, and union overlays for the stands and fire GIS databases.

The order in which GIS databases are selected for an overlay process may also have significance on the output database, depending on your GIS software. The identity overlay will take into account the spatial extent of one of your databases, and reduce other layers to the same spatial extent. For this reason it is important to recognize how your software selects the input database that is used to determine the identity spatial extent. The union and intersect overlays, on the other hand, are less discriminating as to the order in which spatial extents are considered. The union output will include the entire spatial extent of all layers while the intersect output will only include the common areas. Some GIS software, however, may only include by default the attributes of the initial input layer in the output database unless the user signifies otherwise.

Depending on your GIS software, you may also be able to involve more than two layers in an overlay process. Although this ability may be particularly useful for some applications, the complexity of the output product will increase with the number of layers that are involved. Overlay output databases take into account the topology and attributes of all input layers depending on operator choices. Bear in mind that overlay processes are complex operations, particularly when many features are present in input databases. As such, and depending on computing resources, some overlay output databases may take more time to complete than other, less intensive, GIS processes. Users may have to be patient in awaiting overlay output results. Finally, many GIS software programs allow users to select the attribute fields from both the input GIS database and the polygon overlay GIS database that are to be carried into the resulting output GIS database. Often you will find that selecting a subset of attribute fields for the desired analysis will reduce the time requirements related to interpreting and analyzing the results of overlay processes.

Given the considerations expressed above, you should always carefully consider the objectives of your analysis to determine which overlay process to use. In some cases, knowing whether input vector databases represent point, line, or polygon spatial features, or some combination thereof, will help lead to an appropriate overlay command choice.

Summary

With the conclusion of this chapter you have examined most of the common vector processes available within GIS software programs. As we have shown in this and previous chapters, to address natural resource management problems a number of different courses of action can be used, each utilizing different techniques or different sequences of GIS processes. For example, the following three processes might be employed to develop a summary of forest resources within owl buffer areas:

1. Buffer owl nest locations. Clip the owl buffer areas from a stands GIS database. Summarize appropriate statistics.
2. Buffer owl nest locations. Intersect a stands GIS database with the owl buffer GIS database. Summarize appropriate statistics.
3. Buffer owl nest locations. Overlay a stands GIS database on the owl buffer GIS database using an identity process. Summarize appropriate statistics.

The GIS toolbox available to readers of this text, as it relates to vector GIS processes, should now contain a number of useful tools. The challenge lies in deciding which tool(s) to use to address each natural resource management issue. The intersect, identity, and union processes each result in different outcomes, and you need to match the process to the type of information you desire. While the union process is restricted to polygon input and output databases, point and line vector databases can be used in intersect and identity process. The intersect process only provides information for features that overlap. An identity process breaks the features contained in one GIS database along the lines provided in a second GIS database, yet retains the full extent of features contained in the first database. In this case, some of the features in the first database will not contain information from the second where the features in the second are absent. The union process also breaks the features contained in one GIS database along the lines provided in a second GIS database, yet retains the full extent of features contained in both databases. Where features overlap, the attributes of both databases will be present in the attribute table. Where features did not overlap, only attributes from one of the GIS databases will be present.

Applications

11.1. Fertilization Plan. You have been asked to help develop a fertilization plan for the Brown Tract. In next year's plan of action, the managers of the forest may consider fertilizing some forest stands. However, the managers need to know how much area of forest could be fertilized, and what cost to anticipate. After a short discussion with the Brown Tract managers, the following criteria for identifying potential fertilization was determined:

- Only stands ≥ 25 years old and ≤ 35 years old should be considered.
- Only stands on soil types leading to a high fertilization response should be considered (soil types 'PR' and 'DN').
- Only stands outside of a 50-meter stream buffer (around all types of streams) should be considered.

The forest managers want to know the following:

a) How much area of land could be fertilized, given the criteria noted above?
b) Assuming the forest will develop a contract for the fertilization project, and assuming the contractor will use a helicopter to spread the fertilizer, how much area of land would you recommend, and why? Consider the following in your recommendation: (1) Are some of the stands of the appropriate age bisected by two or more soil types? (2) Would it be possible for the helicopter pilot to recognize these changes from the air? (3) Are some of the areas suggested by the GIS analysis for fertilization too small to be worth the effort? (4) Is there a minimum size assumption would you make regarding potential fertilization areas?
c) If the cost of fertilization was $250 per hectare, what might be the total cost of the project?
d) How much fertilizer is needed if you expect to use 450 kilograms of fertilizer per hectare (about 400 pounds per acre)?
e) Please provide the staff with a map of the potential areas that you are recommending to be fertilized (from part b above). Include the Brown Tract roads and streams on the map.

11.2. GIS processing (1). There are a number of processes that can be used to arrive at the information requested in Application 11.1. For example, you may have taken advantage of the intersect process, and obviously used a buffer process at some point. To illustrate the steps taken to accomplish the analysis associated with Application 11.1, develop a flow chart similar to the one below (Figure 11.19) that describes the process you used.

11.3. GIS processing (2). Use a flow chart to describe two other (different) processes that could also be used to generate the same results as those generated for Application 11.1. You may want to actually perform the process and check your results to see whether the results are identical to those obtained in Application 11.1.

11.4 Fire losses. How much forest area, by vegetation class, burned on the Daniel Pickett forest during the July 2, 2002 fire? Use the '070202_fire' GIS database to describe the boundary of the fire.

11.5 Operations around research plots. Assume that the managers of the Daniel Pickett forest have decided to clearcut stand 28. They noticed, just prior to allowing

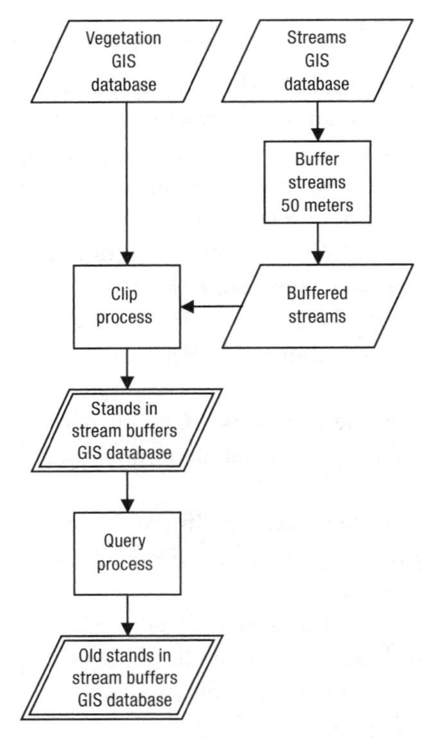

Figure 11.19 Hierarchy of intermediate and final GIS databases created in the development of a GIS database describing the older forest vegetation within 50 meters of all streams.

loggers to begin operations, that a research plot (number 3) was located in the stand. If a 100 meter buffer was left around the research plot,

a) How much forest area in stand 28 will actually be clearcut?
b) If timber values were $400 per thousand board feet (MBF), what is the value of timber that will remain around the research plot [hint: attribute 'Mbf' represents thousand board feet of timber volume per acre]?
c) What is the difference in timber volume and value between leaving the buffer around the research plot and not leaving the buffer?
d) Develop a map that describes the vegetation conditions on the forest after the harvesting operation has completed. Assume that the 100-meter buffer was maintained around the research plot. Illustrate on the map (either with annotation or with a shading scheme) the basal area of all of the stands.

11.6. Integrating streams and vegetation data. Develop a streams GIS database for the Brown Tract where each stream contains the data related to the vegetation polygon within which it is located. Using this data, develop a thematic map that illustrates one of the vegetation characteristics of the forest around each stream. In this exercise, change the appearance of the streams to illustrate the condition of the forest.

11.7. Fish bearing streams.

a) How many stands in the Brown Tract contain fish bearing streams?
b) How many stands are within 100 meters of fish bearing streams?

References

McHarg, I. (1969). *Design with nature*. Garden City, NY: Natural History Press.

Merriam-Webster. (2007). *Merriam-Webster online search*. Retrieved April 29, 2007, from http://www.m-w.com/cgi-bin/dictionary.

Chapter 12

Synthesis of Techniques Applied to Advanced Topics

Objectives

As you may have found during your various interactions with GIS, there are a number of spatial processes (or mixtures and arrangements of methods) that can be used to find the appropriate solution to a natural resource management problem. As we near the end of Part II of this book, this chapter seeks to integrate and synthesize the GIS processes introduced in previous chapters, and apply them to more complex natural resource management problems. At the conclusion of this chapter, you should be familiar with:

1. how a set of complex management rules or assumptions can be synthesized into quantitative information that can be used in a GIS analysis;
2. how a number of GIS processes can be integrated to allow you to develop a spatial representation of a value (ecological, economic, or social) that represents some aspect of a landscape; and
3. how ecological, economic, or social descriptions of a landscape can be developed to provide managers and other decision-makers with information regarding the current status of a landscape.

Many of the GIS processes presented in previous chapters can be integrated to allow users of GIS to perform complex landscape analyses. As we progressed through chapters 5 to 11, we built upon GIS processes to show how they may be complementary. For example, query processes allowed you to quantitatively examine the results of analyses that required buffering processes (chapter 7). In another example, clipping and erasing processes were used in tandem to create a new database of select features (chapter 8). Chapter 12 presents some examples of advanced topics in natural resource management for you to consider, and each requires an integration of the GIS techniques presented in earlier chapters. The tools acquired by working through the previous chapters should be more than adequate to address the problems introduced in this chapter; three types of advanced management problems are presented and the 'Applications' at the end of the chapter provide an opportunity for readers to perform similar analyses themselves. Once again, there are a number of paths you can take to approach each management problem presented in the applications section. You should approach each problem according to your preference of methods and techniques. However, only one set of answers to each natural resource management problem exists, and no matter what process you select, you should ultimately locate the appropriate answers.

We begin with land classification, where separate and distinct categories of land are delineated, each suggesting a different level of natural resource management activity will be allowed. Land classifications are, in fact, common starting points for the development of land management plans. You may need to use the querying, buffering, erasing, and other GIS processes to parse a land ownership into distinct non-overlapping classes that completely cover a landscape. A similar problem is then addressed, where a landscape is delineated into Recreation Opportunity Spectrum (ROS) classes. ROS classes are designed to describe and emphasize the potential recreation opportunities across a landscape, ranging from primitive recreation opportunities with few visitors to those where motorized vehicles and many visitors may be prevalent. Again, the querying, buffering, erasing, and other GIS processes may be used to delineate the non-overlapping ROS classes. Finally, wildlife habitat suitability index measures across a landscape are examined. Here, suitability is a function of the condition of the landscape, and how far each type of vegetation is from the road system. A hypothetical habitat suitability index is presented to provide you with a challenging spatial analysis. A number of GIS techniques can be used to develop the habitat suitability index values, including complex mathematical calculations within the attribute table of GIS databases.

In conjunction with these advanced natural resource management problems, we emphasize the use of flow charts to maintain order in the analysis process. The flowcharting process may be useful for your own GIS analyses, as a means of developing a logical approach to addressing management issues. Since many intermediate (temporary) GIS databases are created and used, tracking the process of an analysis with a flow chart may prove to be very useful, thus several examples are provided.

Land Classification

Land can be classified by vegetation, soils, range, habitat, landform (physiography), and other measurable physical or socio-economic characteristics. Any map that delineates unique pieces of land can be considered a land classification. Land classifications have many purposes in natural resource management, from serving as a basis for assessing the status of land resources to serving as a framework for assessing the local management opportunities (Frayer et al., 1978). Land classifications are necessary for providing both policy direction (knowing what types of resources are available) and for assisting with policy implementation (knowing where the resources are located). Land classification systems are generally based on landscape characteristics that can be seen and measured, and they ideally would be based on flexible, logical, general, and professionally credible concepts, and thus would be described with a quantifiable set of rules (Frayer et al., 1978).

Land classifications can be made from a management perspective where stratification of land is based on economic and management-related variables (roads, streams, etc.). The example we provide below captures the essence of this type of land classification. Alternatively, land classifications can be made purely from an ecological perspective, using vegetation, soils, water, climate, and other physical variables. The four ecological classifications used in Canada are a good example of these. They are designed in a hierarchy and range from broader ecozones to smaller ecodistricts. Each of the classifications have ecosystems that are predominantly woody vegetation, and are delineated without regard to commercial value (Canadian Forest Service, 2007).

Most natural resource management organizations establish their management plans within a land classification framework. Therefore, one of the initial steps in the development of a management plan is to describe the resources (the land) that are managed, and the type of management appropriate to each portion of the land base. After a land classification has been performed, the goals, strategies, and implementation of management can be planned and implemented accordingly. For example, after classifying the land base, an organization may decide to exclude certain management activities in some areas, limit the types of activities allowed in other areas, or allow full consideration of silvicultural or operational activities in other areas (Table 12.1). For example, the Washington State Parks and Recreation Commission (2006) uses a socio-economic, ecological land classification system that integrates physical land features with potential human

TABLE 12.1	An example of a management-related land classification system
Class 1 (Reserved)	
Administratively withdrawn areas (offices and other facilities related to resource management)	
Wilderness areas	
Areas of special concern	
Rock pits	
Ponds or lakes	
Viewsheds	
Other areas where management activities are precluded	
Class 2 (Limited management)	
Riparian areas	
Visual quality corridors around trails	
Areas designated as buffers around wildlife habitat	
Other areas where management is limited	
Class 3 (General management)	
Areas not classified as Class 1 or 2	

uses of the land. Associated with each land class is a description of the philosophy of each class, the appropriate physical features used to delineate the land class, and a matrix of allowed and prohibited activities within each land class area.

A land classification, in addition to guiding the development of a management plan, may also be a requirement for participation in voluntary stewardship programs. For example, organizations that need to comply with the Sustainable Forestry Initiative® (SFI) (American Forest & Paper Association, 2002) are required to classify their land according to the SFI land classification system. This requirement may be in addition to (and different than) the land classification a natural resource management organization may develop during their normal course of management planning.

The number of classes within a land classification may vary. For example, the Washington State Parks and Recreation Commission (2006) land classification system uses six classes, while the Oregon state forest land classification system (Oregon Department of Forestry, 2007) contains three: general stewardship, focused stewardship, and special stewardship. Each land class should be described by key quantitative rules that allow physical identification. There is an 'order of precedence' about a land classification: the highest (most restrictive) classes should be identified first, then the next most restrictive, and so on. Land classified into the highest class cannot subsequently be categorized as lower classes. This systematic approach to land classification helps avoid categorization errors.

Soils data are, in some cases, the drivers for land classification systems. In most cases, these are agricultural land classification systems. One such system (American Farmland Trust, 2006) uses soils characteristics, past irrigation practices, topography, and vegetation cover to describe the quality of land for farming practices. The US Bureau of Reclamation (1951) uses soil and topographic conditions along with economic opportunities to classify land. The main goal in this system is to express the anticipated influence of mappable physical features on the potential productivity of a farming enterprise, the cost of farm production, and the cost of land development. The USDA Natural Resources Conservation Services uses a land capability classification for interpreting soil groupings. The land capability classes for arable soils describe their potential for sustained production of cultivated crops that do not require specialized site preparation. The land capability classes for non-arable soils (those unsuitable for long-term cultivated crop management) are grouped according to their potential for production of permanent vegetation cover, and grouped according to the potential risk of soil damage (Klingebiel & Montgomery, 1973).

The Canada Land Inventory provides broad examples of land classifications related to forestry, agriculture, land use, recreation, and wildlife. The development of land capability classes related to forestry uses is based on a national classification system where land is rated according to its capability to grow trees for commercial uses. The rating system considers land that has not undergone improvements (such as fertilization or drainage activities), and focuses on seven tree productivity classes:

1. Land with no limitations on the growth of commercial forests, where soils are deep, have good water-holding capacity, and are high in fertility. Productivity of tree growth is greater than 7.77 m^3 per hectare per year (111 ft^3/ac/year).
2. Land with slight limitations on the growth of commercial forests, where soils are again deep and have good water-holding capacity, yet there is some limitation on growth (climate, rooting depth, low fertility, for example). Productivity of tree growth is between 6.37 and 7.76 m^3 per hectare per year (91–110 ft^3/ac/year).
3. Land with moderate limitations on the growth of commercial forests, where soils are shallow to deep and have good water-holding capacity, yet they may be slightly low in fertility, and have periodic water imbalances. Productivity of tree growth is between 4.97 and 6.36 m^3 per hectare per year (71–90 ft^3/ac/year).
4. Land with moderate to severe limitations on the growth of commercial forests, where soils are shallow to deep and have other highly variable characteristics. The main limitations are too much or too little moisture, restricted rooting depth, and low fertility, Productivity of tree growth is between 3.57 and 4.96 m^3 per hectare per year (51–70 ft^3/ac/year).
5. Land with severe limitations on the growth of commercial forests, where soils are shallow and poorly drained. The main limitations are too much or too little moisture, restricted rooting depth, low fertility, excessive rock content, and high levels of carbonates. Productivity of tree growth is between 2.17 and 3.56 m^3 per hectare per year (31–50 ft^3/ac/year).
6. Land with severe limitations on the growth of commercial forests, where soils are shallow, excessively

drained, and are low in fertility. The main limitations are restricted rooting depth, low fertility, excessive rock content, excessive soils moisture, and high levels of soluble salts and exposure. Productivity of tree growth is between 0.77 and 2.16 m^3 per hectare per year (11–30 ft^3/ac/year).

7. Land with severe limitations on the growth of commercial forests, where soils are shallow and may contain toxic levels of soluble salts. A large portion of these areas include poorly drained organic soils. The main limitations are restricted rooting depth, low fertility, excessive rock content, excessive soils moisture, and high levels of soluble salts, and exposure. Productivity of tree growth is less than 0.77 m^3 per hectare per year (11 ft^3/ac/year).

The land classification is based on potential natural tree growth and soil characteristics. The proximity of some areas of land to the ocean may also affect the classification of forest land. These seven main classes can be further subdivided into subclasses that are based on climate, soil moisture, rooting depth, and other soils characteristics. An example of the broad-based land classification for a portion of southern Ontario (using data obtained from Natural Resources Canada, 2000) is provided in Figure 12.1.

An important aspect to consider in any land classification process is that the sum of the area in the various classes should equal the sum of the area in the landscape being managed. For example, using the State of Oregon system, the sum of the area in the special, focused, and general stewardship land classes should equal the total area of the landscape being classified. If not, one or both of the following situations exist: (1) there is some overlap among the landscape features (polygons) in one or more of the land class GIS databases, or (2) some area of the landscape is not being represented by any of the stewardship classes.

As an example of developing a management-related land classification for a managed property, we will illustrate the application of the general classes represented in Table 12.1 to the Brown Tract. First, some quantitative assumptions about the three land classes need to be made to allow us to delineate the areas on a map. We will assume that class 1 (reserved) areas will contain meadows, research areas, rock pits, and oak woodlands. Class 2 (limited management) will be those areas of land that are within 50 m of streams, 100 m of hiking trails, 100 m of homes, and 300 m from any owl nest locations. Finally, class 3 (general management) areas are assumed to contain the land that remains after class 1 and class 2 management areas have been delineated. To delineate these three classes, a series of GIS techniques such as querying, buffering, clipping, and erasing processes may be required (Figure 12.2). However, other arrangements of GIS processes could have also resulted in the same solution. The polygons represented in the resulting three land classes (Figure 12.3) should not overlap, which means that no single unit of land will be counted twice. Put another way, each unit of land can only belong to a single land class. In this example of a management-related land classification, class 1 consists of 229 hectares (567 acres), class 2 consists of 671 hectares (1,657 acres), and class 3 consists of 1,222 hectares (3,020 acres). When all three of the land classes are added together, they equal the size of the Brown Tract.

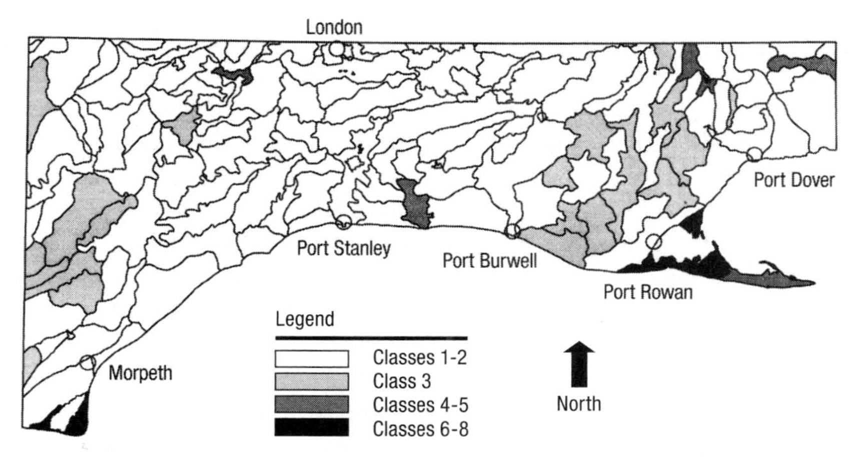

Figure 12.1 Land classification example for a portion of southern Ontario.

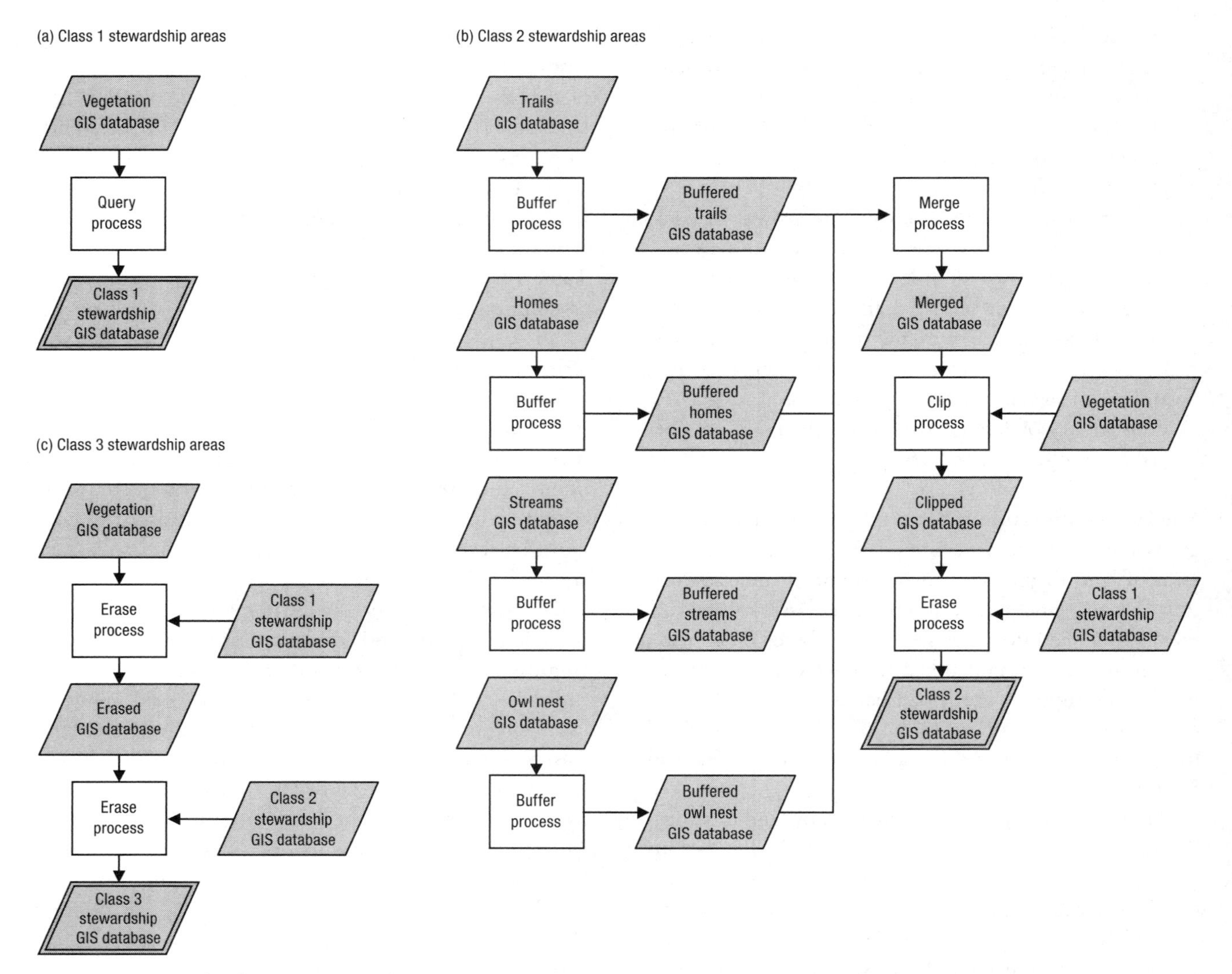

Figure 12.2 Hierarchy of intermediate and final GIS databases created in one process that facilitates the development of the Brown Tract land classification.

Recreation Opportunity Spectrum

A number of classification processes for outdoor recreation have been developed, including the Recreation Opportunity Spectrum (ROS), carrying capacity, limits of acceptable change, and the Tourism Opportunity Spectrum (Butler & Waldbrook, 1991). The latter system is based on the Recreation Opportunity Spectrum, and includes aspects of accessibility (i.e., transportation systems), tourism infrastructure, social interaction, and other non-adventure uses. The ROS was developed by the USDA Forest Service and the USDI Bureau of Land Management as a tool for managing recreation and tourism on federal land, and for integrating recreation and tourism with other land uses (Clark & Stankey, 1979). ROS is used to describe and identify recreational settings, and to illustrate the likelihood of recreational opportunities along a spectrum that is divided into several classes. This system combines the physical landscape characteristics of *location* and *access* to allow you to delineate areas of land that may be used for different recreational purposes. For example, in the most recent forest plan for the Hiawatha National Forest in northern Michigan (USDA Forest Service, 2006) it is suggested that recreation-related development, activities, management practices, and access will be consistent with the delineated

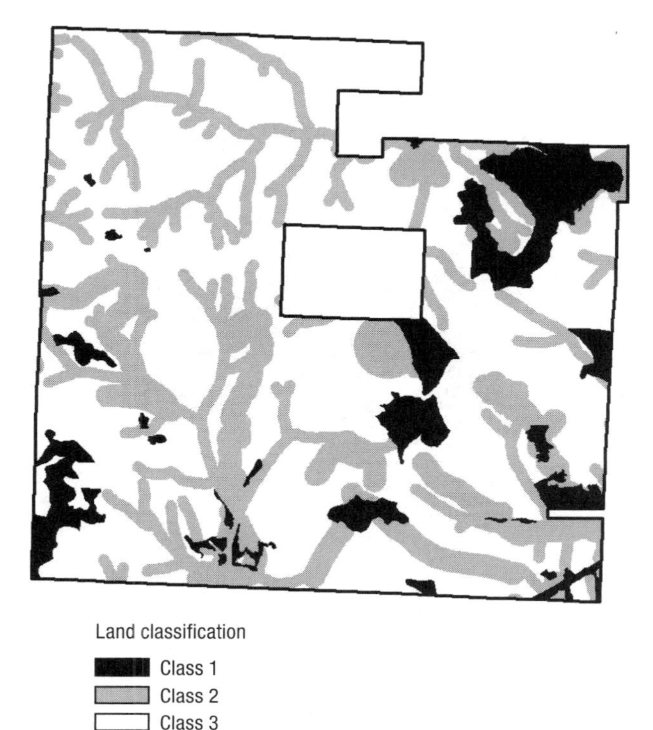

Figure 12.3 A land classification of the Brown tract.

ROS class for each area. Thus, the recreation opportunities will be provided in a manner consistent with the ROS designation for each management area. The ROS classes represent a wide range of recreational experiences, from those that include a high likelihood of self-reliance, solitude, challenge, and risk, to those that include a relatively high degree of resource development and interaction with other people. A recreation opportunity class, therefore, is an area of land that may yield certain experiences for recreationists in a specific landscape setting.

Consider an activity such as cross-country skiing, a popular recreational activity in western North America. Cross-country skiing experiences in and around cities, such as Bend, Oregon, are likely to result in experiences that are exercise-oriented yet include a high frequency of interaction with other people and developed resources. However, cross-country skiing experiences in the backcountry, such as the nearby Deschutes National Forest, while also exercise-oriented, are more likely to include elements of solitude, risk, personal challenge, and will likely have a lower frequency of interaction with people. Therefore, the same activity, cross-country skiing, can be associated with different experiences in different landscape settings. As a result, there is a need to delineate those areas spatially, so that management activities related to recreational activities (and other management objectives) can be planned accordingly.

The original version of the ROS classification (Clark & Stankey, 1979) divided land areas into six classes:

1. Wilderness (now called primitive)
2. Semi-primitive non-motorized
3. Semi-primitive motorized
4. Roaded natural
5. Rural
6. Urban

Category 4 has since been expanded to two classes, *roaded natural* and *roaded modified*, although the exact classes used seems to vary from one management situation to the next. In the most recent forest plan for the Hiawatha National Forest (USDA Forest Service, 2006), the original six classes are used.

The ROS classes suggest that specific kinds of recreation activities and experiences owing to certain physical (e.g., size), social (e.g., encounters with other people), and managerial (e.g., legally designated wilderness area) characteristics can be supported. The rules that define the ROS classes can include spatial relationships. For example, to be considered a primitive area, land must be more than 1.5 miles from *any* road (Table 12.2). Further, some spa-

TABLE 12.2 **A subset of rules with spatial considerations for delineating recreational opportunity spectrum (ROS) classes**

ROS class	Rule
Primitive (P)	Areas of land greater than 1.5 miles from a road.
Semi-primitive, non-motorized (SPNM)	Areas of land that are greater than 0.25 miles from a road, have forest stands ≥ 50 years of age, and are ≥ 202.3 hectares (500 acres) in aggregate size.
Semi-primitive, motorized (SPM)	Areas of land that are greater than 0.25 miles from a *paved* road, have forest stands ≥ 50 years of age, and are ≥ 202.3 hectares (500 acres) in aggregate size.
Roaded natural (RN)	Areas of land with stand ages ≥ 50 years, and ≥ 16.2 hectares (40 acres) in aggregate size.
Roaded, managed (RM)	Areas that do not fit into any of the other classes.

tial aggregation of polygons may be necessary. The semi-primitive, non-motorized ROS class suggests that there must be at least 500 contiguous acres (202.3 hectares) of land of certain forested conditions before land can be classed as such. Here, you may have to add together the area of several contiguous polygons to determine how much area they represent *in aggregate*.

As with the land classification example described earlier, the sum of the landscape area in the ROS GIS database(s) (Figure 12.4) should equal the total area of the landscape being classified. If not, one or both of the following situations exist: (1) there is some overlap among the landscape features in one or more ROS classes, or (2) some area of the landscape is not being represented by any of the ROS classes. A process you might use to create the primitive and semi-primitive portions of an ROS map might resemble the flow chart illustrated in Figure 12.5. As you may gather from this set of processing steps,

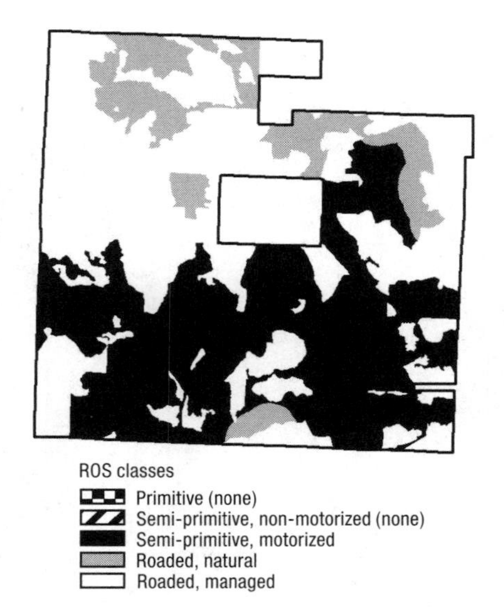

Figure 12.4 Recreation Opportunity Spectrum (ROS) classes for the Brown Tract.

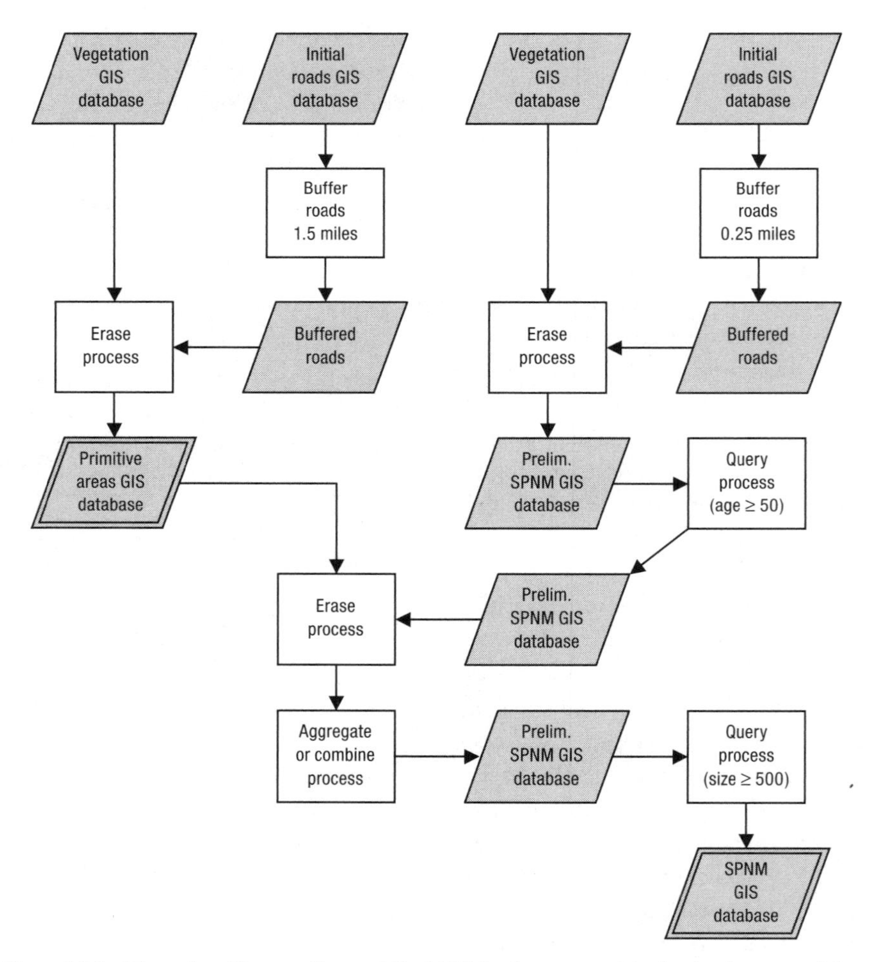

Figure 12.5 Hierarchy of intermediate and final GIS databases created in the development of the primitive and semi-primitive portions of a Recreation Opportunity Spectrum map.

when you consider the development of the full range of ROS classes, the set of processing steps may become cumbersome and confusing. Developing a flow chart to describe the processing steps you used, and to identify the intermediate and final GIS databases, will alleviate some of this confusion.

Habitat Suitability Model with a Road Edge Effect

Habitat suitability models provide natural resource managers with a glimpse into the potential of a landscape to support habitat for a specific species of wildlife, or group of wildlife species (Morrison et al., 1992). These models generally describe habitat suitability as the geometric mean of two or more variables that represent (or influence) the occurrence and abundance of a particular wildlife species. A geometric mean is calculated by taking the *nth* root of the product of a group of numbers, where *n* is equal to the number of observations. Rempel and Kaufmann (2003) define *habitat* as the set of forest structural conditions that provide some means (e.g., nesting, reproduction, foraging) for a species of wildlife during its life history. Each forest stand in a vegetation database is assigned a single habitat value based on the structural conditions that exist in (and perhaps around) the stand during some period of time, conditions which can change as management activities are implemented. The suitability of habitat is generally scaled between 0 and 1, and wildlife managers are called upon to determine what levels are appropriate to describe optimal habitat. While there is considerable debate concerning the usefulness and accuracy of habitat suitability models (see Brooks, 1997), when well-developed and validated, they do allow natural resource managers to examine the relative quality of one area versus another with respect to some species of interest.

To illustrate the development and display of a habitat suitability map, a hypothetical habitat suitability index (HSI) is developed for a fictional species of vole. The model will allow you to evaluate habitat suitability as a function of forest basal area, age, and the distance of habitat from roads.

HSI = *f* (basal area, age, distance from roads)

The HSI incorporates these three parameters into a single non-linear model that is used to calculate the

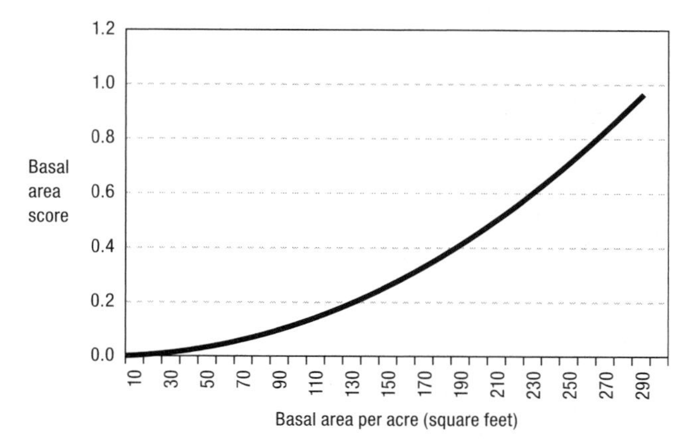

Figure 12.6 Basal area scores for a range of stand basal areas.

geometric mean of the scores of each variable. The purpose of this exercise is to obtain a graphical description of the landscape features that are important in describing the habitat of the vole, given an understanding of the vole's habitat requirements.

HSI calculation = (basal area score × stand age score × distance from road score)$^{1/3}$

In order to calculate each of the individual parameter scores, a set of quantitative rules is needed. These rules are generally developed through research, literature reviews, or perhaps are based on the advice of biologists who are experts on the life history of voles. Since the vole assumed here is a fictional species, the set of rules have been developed by the authors of this book, and are hypothetical. The basal area score, for example, is a function of the square of stand basal area (ft^2 per acre) multiplied by a constant, to indicate a non-linear positive response of vole abundance to more heavily stocked timber stands (Figure 12.6).

Basal area score = 0.0000115 × (basal area)^2
If Basal area score > 1.0, then Basal area score = 1.0

The stand age score is a linear function of stand age (Figure 12.7), where age is multiplied by a constant. When combined with the basal area score, the two portions of the HSI favor older stands that are well-stocked (where both the basal area score and stand age score are high), over older stands that are not well-stocked, or younger stands that are over-stocked (where one score is high and the other low).

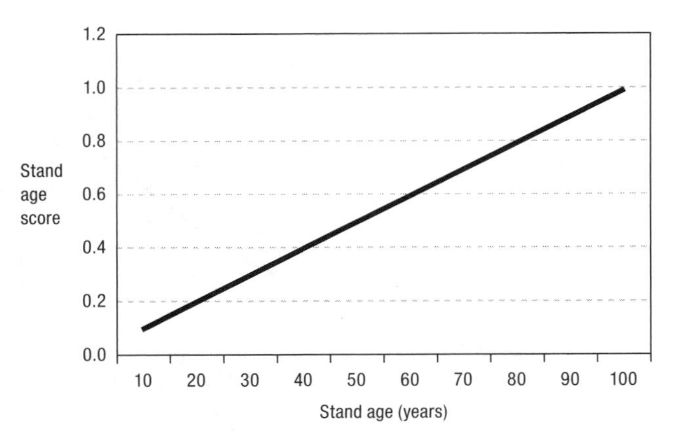

Figure 12.7 Stand age scores for a range of stand ages.

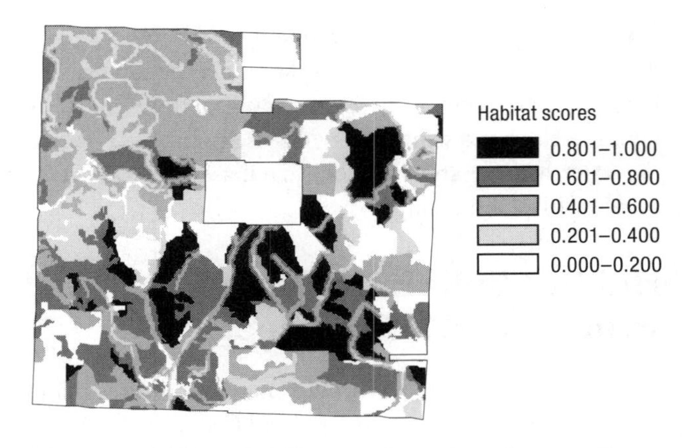

Figure 12.9 Habitat suitability scores for a vole on the Brown Tract.

Stand age score = 0.01 × (stand age)
If Stand age score > 1.0, then Stand age score = 1.0

While the basal area and stand age parameters may more likely describe the abundance of the vole, the distance from road parameter describes the potential occurrence of the vole species. In other words, the distance to road factor is used to represent the assumption that areas near roads will represent lower habitat quality than similar areas of land farther away from roads (Figure 12.8).

Distance from road score = 0.25 if within 15.24 m (50 feet) of any road, or
= 0.50 if 15.25 to 30.48 m (50.1 to 100 feet) of any road, or
= 0.75 if 30.49 to 45.72 m (100.1 to 150 feet) of any road, or
= 1.00 everywhere else

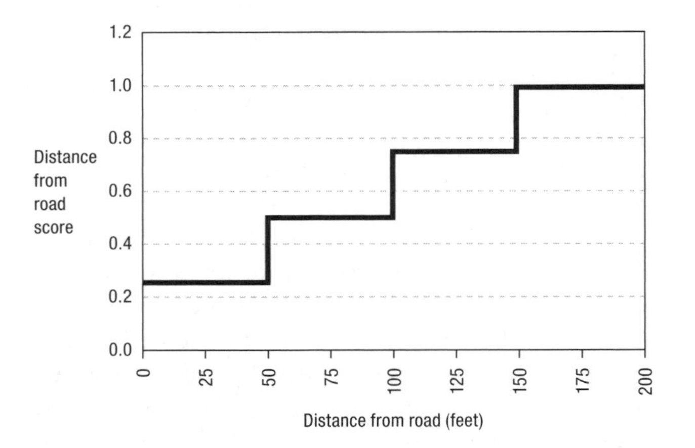

Figure 12.8 Distance from roads scores for a range of distances from the road network.

Graphically displayed, the map of HSI for the vole indicates the relative quality of vole habitat across the landscape (Figure 12.9). A score of 1.0 represents optimal habitat, a score of 0.0 represents the poorest quality habitat.

Somewhere along the 0.0–1.0 range, biologists will need to determine the threshold levels that separate good habitat from poor habitat. One process that can be used to arrive at these scores is presented in Figure 12.10. Here, the roads are buffered three times (15.24, 30.48, and 45.72 m). Two of the buffer GIS databases are then subjected to an erase process, resulting in a buffer band around each road (a 15.25–30.48 meter band and a 30.49–45.72 meter band). The 0–15.24 meter buffer is then combined with these two buffer bands to create a GIS database that represents three of the buffer distances by polygons, with no overlapping polygons present. The fourth buffer distance, as you can imagine, is everything not included in these three buffer distances. The buffers are then overlaid on the vegetation GIS database, breaking vegetation polygons at the buffer boundaries. The basal area, stand age, and distance from road scores can then be calculated in the tabular portion of the resulting GIS database. The final HSI score can be calculated as a function of the basal area, stand age, and distance from road scores, and a thematic map can be developed to illustrate the distribution of vole habitat across the landscape. In addition, the final GIS database can be queried to develop a table of area by habitat class.

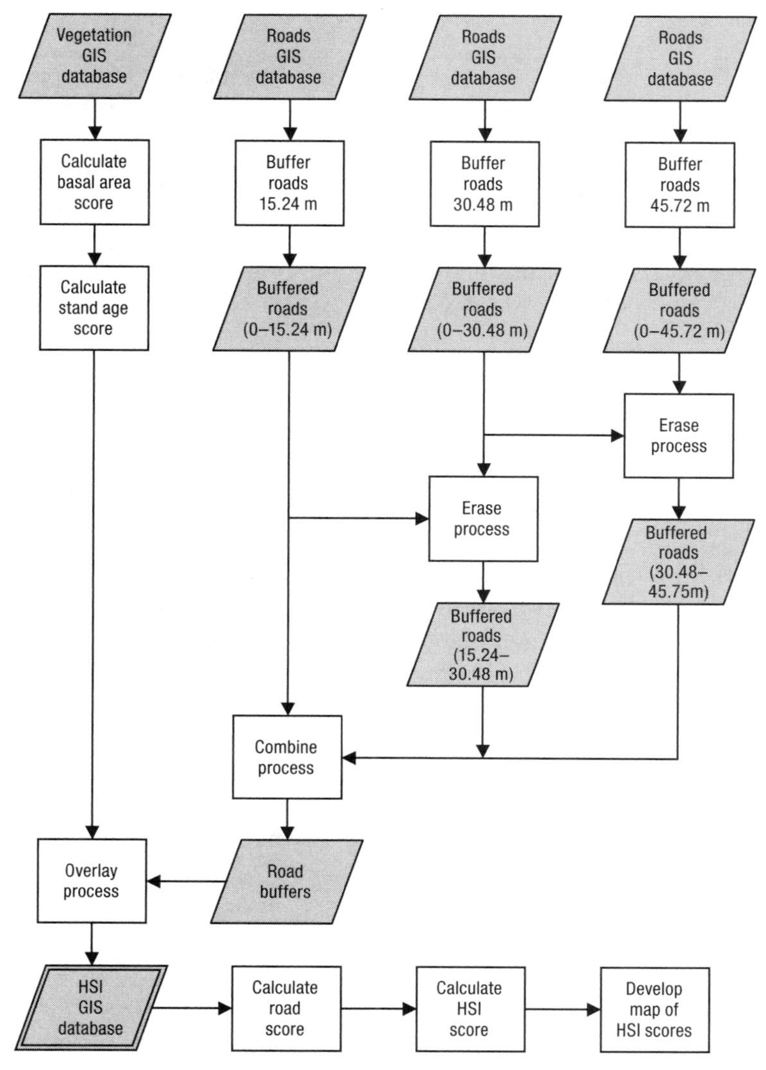

Figure 12.10 Hierarchy of intermediate and final GIS databases created in the development of an analysis of potential wildlife habitat suitability (HSI) areas for a vole on the Brown Tract.

Summary

This chapter illustrates just a few of the more complex spatial analyses that may be performed (or requested) by natural resource managers. The number and arrangement of GIS processes could vary in addressing analyses such as these, and may include buffering, clipping, erasing, and querying of landscape features. Therefore, the chapter represents a synthesis of the tools readers have acquired from previous chapters in this book. It should be apparent by now that it is important to explicitly define the quantitative rules and the GIS processes that might be used to address complex spatial analyses. For example, the need for quantitative rules and a logical set of GIS processes to separate one set of landscape features from another in an analysis of ROS classes is important because a single unit of land must be assigned only one ROS class, and all units of land must be assigned a class. The graphical display of the result of a complex GIS analysis, such as the ones illustrated in this chapter, is also important because land managers typically use these products to help them visualize and make decisions regarding the management of natural resources.

Applications

12.1 Land classification. Becky Blaylock, manager of the Brown Tract, wants you to develop a management-related land classification for the forest. She asks you to develop GIS databases for each of three classes using the following rules:

1. *Special stewardship areas* will consist of the following landscape features: Oak woodlands, Meadows, and Rock pits.
2. *Focused stewardship areas* will consist of the following landscape features:
   a. streams buffered according to the Oregon State Forest Practices Act (30.48 meters [100 feet] around large fish-bearing streams [Size = 'Large' and Fishbearing = 'Yes'], 21.34 meters [70 feet] around medium fish-bearing streams, 15.24 meters [50 feet] around small fish-bearing streams, 21.34 meters [70 feet] around large non fish-bearing streams, 15.24 meters [50 feet] around medium non fish-bearing streams, and 6.10 meters [20 feet] around small non fish-bearing streams);
   b. a buffer of 100 meters around all water sources that are not culvert spills or water towers;
   c. a buffer of 100 meters around all authorized trails; and
   d. research areas.
3. *General stewardship areas* will consist of whatever land remains.

Do the following:

a) Develop and illustrate a process (flow chart) for accomplishing the task of defining the land classifications of the Brown Tract forest according to the rules listed above.
b) Determine how much land area is contained in the special stewardship land classification.
c) Determine how much land area is contained in the focused stewardship land classification.
d) Determine how much land area is contained in the general stewardship land classification.
e) Produce a map of the entire Brown Tract, illustrating the three land classifications.

As a general strategy, you may want to follow this process:

- Develop a special stewardship GIS database.
- Develop a focused stewardship GIS database by performing the appropriate buffer and query processes, and then intersecting these GIS databases. (Why would you not use a merge process here?)
- Erase the special stewardship GIS database features from the focused stewardship GIS database features.
- Erase both the special stewardship GIS database features and the focused stewardship GIS database features from the stands GIS database, creating the general stewardship GIS database.

12.2 Recreation Opportunity Spectrum. The District Manager associated with the Brown Tract (Becky Blaylock) would like you to determine how much area might be classified in the five recreation opportunity spectrum (ROS) classes (see Table 12.2). Based on this subset of the ROS criteria,

a) How much land area is contained in the primitive class?
b) How much land area is contained in the semi-primitive, non-motorized class?
c) How much land area is contained in the semi-primitive, motorized class?
d) How much land area is contained in the roaded natural class?
e) How much land area is contained in the roaded managed class?
f) Develop a thematic map illustrating the five ROS classes on the Brown Tract.
g) Draw a flow chart to describe the processes used to develop the ROS classes, including the GIS operations and all GIS databases used (original, intermediate, and final GIS databases).

12.3. Visual quality buffers. You have been asked by the manager of the Daniel Pickett forest to evaluate the potential impact of two proposed organizational policies for the forest resources found there. It seems that the owners of the property are becoming very concerned with the public perception of management on the forest, thus they are interested in the trade-offs associated with alternative management policies.

Policy #1: Buffers next to neighboring landowners. Assume for this example that even-aged forest management is practiced across the property. This potential policy suggests that clearcut harvesting activities adjacent to neighboring landowners of the Daniel Pickett forest will be restricted.

a) If a 50-meter uncut buffer were to be left adjacent to all other property owners, how much land would this require as a voluntary contribution to the protection of adjacent landowners resources?
b) How much timber volume of vegetation class A would be found in the buffer (vegetation class A is the older timber class, perhaps that which can be harvested in the near-term), and what percentage of the total volume in this vegetation class would be affected?

Policy #2: Buffers next to paved public roads. This potential policy suggests that visual quality buffers may be maintained along paved roads within the Daniel Pickett forest. These buffers will not be managed, but rather treated as reserved areas, where harvesting is precluded.

a) If a 50-meter buffer were required around all paved roads, how much land area would this involve, and how much timber volume in vegetation class A would it affect?
b) If the State decided to convert the North-South paved road on the Daniel Pickett forest to a highway, and required a 100-meter wide corridor to be transferred to State ownership, how much land area would be affected?
c) If bare land values were assumed to be $200 per hectare, and timber volumes $400 per thousand board feet (MBF), how much would you ask the State to compensate the owners of the Daniel Pickett forest for the loss of this land?
d) If a 30-meter visual quality (i.e., uncut) buffer was then proposed around the 100-meter highway corridor, what is the total effect to the forest resource base, in terms of land area now affected in each vegetation type?

12.4. Habitat suitability index for a vole. The biologist associated with the Daniel Pickett forest, Will Edwards, has recently become aware of a vole habitat suitability model, and is interested in understanding the extent of vole habitat on the forest. Will asks you to apply the model described in the 'Habitat Suitability Model with a Road Edge Effect' section of this chapter to the Daniel Pickett forest, and to:

a) Calculate the amount of land area on the Daniel Pickett forest in the following habitat suitability classes: 0.000–0.200 (low quality), 0.201–0.400 (low/moderate quality), 0.401–0.600 (moderate quality), 0.601–0.800 (moderate/high quality), 0.801–1.000 (high quality).
b) Develop a map illustrating the habitat quality for the vole by suitability class.
c) Draw a flow chart of the process you used to develop the habitat suitability classes.

References

American Farmland Trust. (2006). *Land classification system*. Washington, DC: American Farmland Trust. Retrieved February 17, 2007, from http://www.farmland.org/resources/futureisnow/landclassification system.asp.

American Forest & Paper Association. (2002). *Sustainable Forestry Initiative (SFI)®*. Washington, DC: American Forest & Paper Association. Retrieved December 10, 2007, from http://www.afandpa.org/Content/Navigation Menu/Environment and Recycling/SFI/SFI.htm.

Brooks, R.P. (1997). Improving habitat suitability index models. *Wildlife Society Bulletin, 25*, 163–7.

Butler, R.W., & Waldbrook, L.A. (1991). A new planning tool: The tourism opportunity spectrum. *Journal of Tourism Studies, 2*(1), 2–14.

Canadian Forest Service. (2007). *Ecological land classifications*. Ottawa, ON: Canadian Forest Service, Natural Resources Canada. Retrieved February 17, 2007, from http://ecosys.cfl.scf.rncan.gc.ca/classif/intro_strat_e.asp.

Clark, R.N., & Stankey, G.H. (1979). *The recreation opportunity spectrum: A framework for planning, management, and research. General Technical Report PNW-98*. Portland, OR: Pacific Northwest Forest and Range Experiment Station, USDA Forest Service.

Frayer, W.E., Davis, L.S., & Risser, P.G. (1978). Uses of land classification. *Journal of Forestry, 76*, 647–9.

Klingebiel, A.A., & Montgomery, P.H. (1973). *Land capability classification. USDA agricultural handbook 210*. Washington, DC: US Government Printing Office. Retrieved February 17, 2007, from http://

soils.usda.gov/technical/handbook/contents/part622p2.html#ex2.

Morrison, M.L., Marcot, B.G., & Mannan, R.W. (1992). *Wildlife-habitat relationships: Concepts and applications*. Madison, WI: University of Wisconsin Press.

Natural Resources Canada. (2000). *Overview of classification methodology for determining land capability for forestry*. Ottawa, ON: GeoGratis Client Services, Natural Resources Canada. Retrieved August 10, 2007, from http://geogratis.cgdi.gc.ca/CLI/frames.html.

Oregon Department of Forestry. (2007). *Oregon administrative rules, department of forestry, Division 35, management of state forest lands*. Salem, OR: Oregon Department of Forestry. Retrieved March 12, 2007, from http://arcweb.sos.state.or.us/rules/OARS_600/OAR_629/629_035.html.

Rempel, R.S., & Kaufmann, C.K. (2003). Spatial modeling of harvest constraints on wood supply versus wildlife habitat objectives. *Environmental Management, 32*, 646–59.

US Bureau of Reclamation. (1951). *Land classification. Bureau of reclamation manual, vol. V, irrigated land use, part 2*. Denver, CO: US Bureau of Reclamation.

USDA Forest Service. (2006). *Hiawatha national forest, 2006 forest plan*. Milwaukee, WI: USDA Forest Service, Eastern Region. Retrieved February 18, 2007, from http://www.fs.fed.us/r9/hiawatha/revision/2006/ForPlan.pdf.

Washington State Parks and Recreation Commission. (2006). *WAC 352-16-020 land classification*. Olympia, WA: Washington State Parks and Recreation Commission. Retrieved February 17, 2007, from http://www.parks.wa.gov/plans/lowerhoodcanal/State%20Parks%20Land%20Classifications.pdf.

Chapter 13

Raster GIS Database Analysis

Objectives

The skills and techniques you'll learn in this chapter should provide insight into the examination and application of raster GIS databases for natural resources research, and how raster GIS databases might be included in supporting natural resource management decision-making. At the conclusion of this chapter, you should have an understanding of:

1. how landscape contour GIS databases are created from a DEM;
2. how landscape shaded relief GIS databases are created from a DEM;
3. how slope GIS databases are created from a DEM;
4. how to calculate slope gradients for a linear landscape feature, such as a road, trail, or stream;
5. how to conduct a viewshed analysis for a portion of a landscape; and
6. how to create a watershed boundary based on digital elevation data.

As mentioned in chapters 1 and 2, there are two general types of data structures used in GIS today: vector and raster. Until now, we have focused on vector GIS databases and the GIS operations related to the typical kind of applications performed in natural resource organization field offices. This chapter now delves into the use of raster GIS databases for natural resource applications, and a few of the GIS operations that can be performed using them. An emphasis is placed on how raster GIS databases might be used in field offices to support natural resource management decisions. Although we will turn our attention in the next chapter to other raster database applications, the primary raster GIS database that is considered in this chapter is a digital elevation model (DEM). Many different types of landscape information can be cultivated from a single DEM database.

Digital Elevation Models (DEMs)

As their name implies, DEMs contain information related to the elevation of a landscape above sea level or relative to some other datum point. They are different than the typical USGS Quadrangle maps discussed in chapter 4, in that they are in digital form. As with other raster GIS databases, each unit on the landscape is typically represented by a landscape-related value (or set to a null or 'no data' value), and each unit is exactly the same size and shape as the other units (Figure 13.1). The most prevalent DEM databases available in the US are the USGS 30 meter DEMs (US Department of Interior, US Geological Survey, 2007). Within Canada, Natural Resources Canada (2007) provides access to digital topographic data. Raster databases are often described in terms of their spatial resolution, as in the phrase '30 m DEM'. This infers that each grid cell in the DEM database is 30 m by 30 m in size in terms of on-the-ground area that it represents. Many regions in the US also have 10 m DEMs available for areas within federal and state agency administrative boundaries. In some cases DEMs for states, provinces, or other large regions can be purchased from commercial entities.

DEMs can be used for a variety of analytical purposes, but the most general of these purposes is simply to view the relief of a landscape. DEMs can use shades of color or gray tones to illustrate differences in elevation through a

Figure 13.1 Raster grid cells from a digital elevation mode (DEM).

separation and classification of elevation values. Figure 13.2 illustrates a gray tone color-shading scheme applied to a 10 m DEM of the Brown Tract, and uses a twelve-category equal-interval classification scheme to highlight the changes in elevation. The equal-interval classification takes the distribution of elevation data values found within the 10 m DEM and divides it equally into twelve sub-sets of elevation ranges. Most GIS software will allow users to define the number of elevation range categories shown visually, and offer choices for color and gray tone schemes to illustrate distinctions between elevation categories.

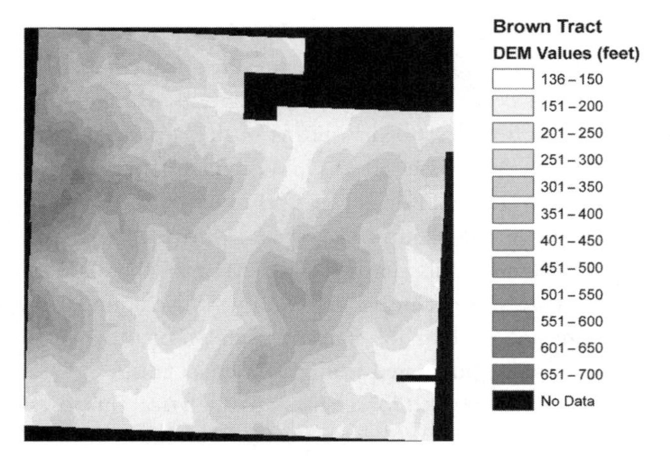

Figure 13.2 Elevation categories for the Brown Tract using a 10 m DEM.

One notable category represented in the legend in Figure 13.2 is the 'No Data' category. This category is necessary because raster GIS data must be stored as a set of grid cells that combine to form a rectangular or square shape—the width and height of the image is defined by the number of grid cells. Therefore, when landscape features of interest do not match a rectangular shape (e.g., the shape of the Brown Tract), the grid cells that are not associated with the landscape features of interest are given a null, or No Data, value. For example, all of the grid cells that represent areas outside the boundary of the Brown Tract contain no data. For mapping purposes, the symbolization used to display cells with a null value can be assigned a transparent shade. Almost all raster GIS software programs allow the recognition of a null value, yet are designed to ignore this value in analytical computations. When performing a multiple GIS database overlay analysis, some raster GIS software programs are also designed to ignore any cells that overlap null cells in any other GIS database being analyzed. For example, if two spatially coincident raster databases were overlaid on each other and any portion of either database contained cells with no data, any database resulting from an overlay or comparative analysis that involved both databases would also contain no data cells at the same locations. This result would occur even if actual values were present in a portion of one of the raster databases. While this functionality may or may not be appropriate for particular GIS analyses, users should be cognizant of how null values assigned to grid cells will be handled within their selected GIS software program(s).

Elevation Contours

DEMs can be used to create **elevation contours**, or lines that indicate a constant or nearly constant elevation across a landscape. Contour lines are created adjacent to each other such that elevation represented in even increments, such as every 30, 50, or 100 m. An elevation contour GIS database is usually represented through a vector data structure based on a user-defined elevation interval. Contour lines allow you to examine the relative relief of a landscape and to make inferences about landscape topography to support management decisions. For example, contour intervals can be used to delineate likely hydrologic drainage patterns and watershed boundaries. When designing road systems, engineers typically need to keep the slope of each road below some maximum gradient, since slopes too steep will either prevent the movement of certain types of vehicles (if the vehicle is travelling uphill),

In Depth

A contour, to most people, represents the outline of some figure or body. Contour intervals, as used on maps, represent the outline of all areas that have the same elevation. It would be as if you were to slice (horizontally) the landscape every 100 feet (or whatever interval was chosen) in vertical elevation. Contour plowing is a common practice in agriculture, where plow lines are laid parallel to the contour of the landscape to reduce the erosion potential of the agricultural practice. This also reduces strain on farm machinery; this practice encourages moving laterally rather than perpendicularly through elevation gradients. Each section of a plow line is, theoretically, at about the same elevation as every other section of the line.

or will be too dangerous to travel (if the vehicle is travelling downhill). With many raster GIS software programs, users have the ability to choose a contour interval and the starting elevation value at which contours will be created.

To create contour lines a process such as that described in Figure 13.3 can be used. In cases where elevation units (i.e. meters, feet) do not match horizontal mapping units, a unit conversion factor may be needed in order to bring the units into agreement. Within ArcGIS a DEM must be opened as a layer, the Spatial Analyst Extension must be activated, and the Spatial Analyst menu must be opened. From the Spatial Analyst menu, choose Surface Analysis, then choose the Contour option. This should open the Contour dialog box, which will prompt you for the Input Surface, contour dimensions, and an output database name. Using the Brown Tract 10 m DEM, a contour interval of 50 feet with a base contour elevation of 100 feet was chosen (Figure 13.4). Using the 100-foot base elevation should result in contour lines that originate from 100 feet and increment in 50-foot steps. The contour line GIS database that is created is a vector GIS database, and each line contains an attribute describing the elevation. Users can then modify this vector GIS database to display different color shades or line thickness for different contour lines of interest.

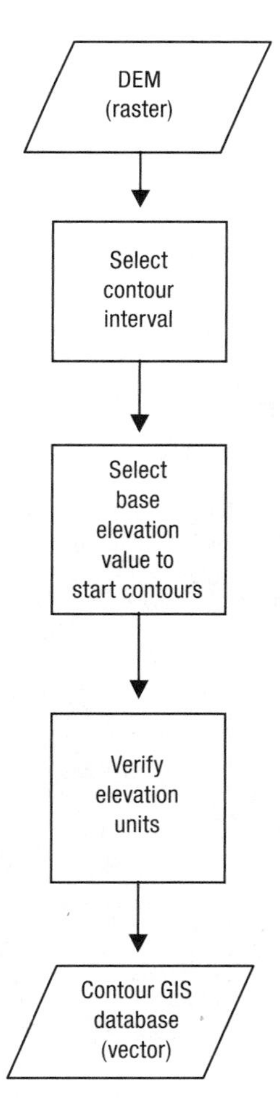

Figure 13.3 A general process for the development of a contour line GIS database from a DEM.

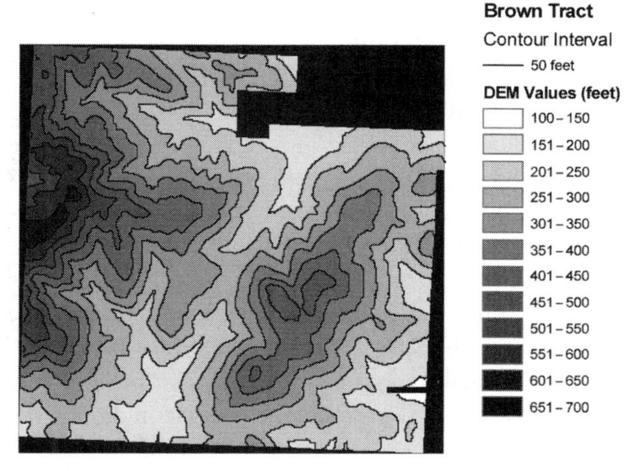

Figure 13.4 A contour line GIS database for the Brown Tract displayed on top of the Brown Tract 10 m DEM.

Whenever both horizontal coordinate positions and vertical elevations are processed simultaneously, as is the case in the creation of contour lines discussed above, it should be ascertained that both types of measurements use the same units. Typically coordinates and elevations will be recorded using meters, international feet, US survey feet, or some combination of these units. It is not uncommon within the US, however, to discover DEMs that have coordinate values in meters but that store elevation values in survey feet. A GIS analyst might mistake the resulting contour line values to represent meters, thus over-representing the elevations along contours. The Spatial Analyst Contour option dialog box contains an input box where users can specify whether elevation units differ from coordinate units within a DEM.

Shaded Relief Maps

Another product that can be derived from a DEM is a **shaded relief map**. Shaded relief maps are intended to simulate the sun-lit and shaded areas of a landscape when assuming that the sun is positioned at some location in the sky. Landscape features that face toward the sun will appear more brightly lit than objects facing away. For raster GIS software programs that provide the ability to create a shaded relief map, the result of performing a shaded relief map process is a raster GIS database, and each grid cell typically contains an attribute value describing a gray tone ranging from light (facing towards the sun) to dark (facing away from the sun). The shaded relief map is useful for illustrating the topography and provides a three-dimensional perspective of the landscape.

Shaded relief maps can be created with the general process described in Figure 13.5. The azimuth selected specifies the direction from which the sun is shining. An azimuth of 90°, for example, indicates that the sun is positioned in the east, and an azimuth of 180° indicates that the sun is positioned in the south. The altitude defines the angle of the sun above the landscape. An altitude of 0° typically indicates that the sun is located directly overhead, whereas an altitude of 90° would indicate that the sun is at the horizon. Within ArcGIS a DEM must be opened as a layer, the Spatial Analyst Extension must be activated, and the Spatial Analyst menu must be opened. From the Spatial Analyst menu, choose Surface Analysis, then choose the Hillshade option. This should open the Hillshade dialog box, which will prompt you for the Input Surface, Azimuth, and Altitude amounts. Options are also provided for vertical unit conversion to horizontal map unit (Z factor), an output cell size, and an output database name. Using the Brown Tract 10 m DEM, an azimuth of 210°, and an altitude of 45°, a shaded relief map is created (Figure 13.6) that shows (relatively speaking) how much sunlight reaches each part of the landscape in the late afternoon (the sun azimuth of 210° indicates that the sun is located directly to the southwest of the

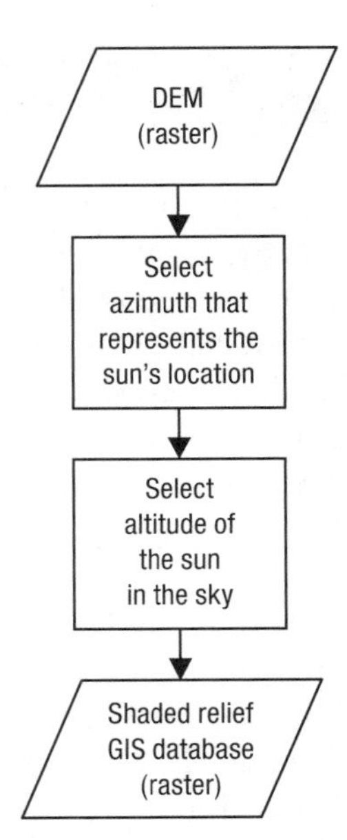

Figure 13.5 A general process for the development of a shaded relief GIS database from a DEM.

Figure 13.6 Shaded relief map of the Brown Tract using a 10 m DEM, an illumination azimuth of 210°, and an illumination altitude of 45°.

In Depth

The orientation and presentation of 'direction' has not been discussed to great extent in this book, however, it is important for readers to know the difference between an **azimuth** and a **bearing**. Why? Because compasses used in fieldwork either represent direction as azimuths or bearings. In some cases, both types of measurements will be represented. Azimuths are degrees of a circle, with North being 0° (or 360°), East being 90°, South being 180°, and West being 270°. A compass line indicating an azimuth of 353°, therefore, indicates a direction of almost due North. A bearing is represented as any angle of 90° or less from either the North or South (and directed towards the East or West). Thus an azimuth of 353° represents a bearing of N7°W, since the angle would arise from the North half of a compass, and is directed towards the West 7°. Similarly, an azimuth of 89° represents a bearing of N89°E (the angle arises from the North half of the compass and is directed towards the East 89°), and an azimuth of 190° represents a bearing of S10°W (the angle arises from the South half of the compass and is directed towards the West 10°). Property deeds, commonly used within North America to legally state ownership of a land area, often use bearings to describe the land boundary locations.

landscape, and the 45° altitude indicates a sun position halfway between 'directly overhead' and 'setting'). With this shaded relief analysis, you can obtain a sense of the varied topography of the Brown Tract. Other landscape features, such as study areas, roads, and streams, might then be displayed on top of the shaded relief GIS database to allow an examination of how these resources might be influenced by landscape topography.

Slope Class Maps

A third product that can be derived from an analysis of DEMs are GIS databases that represent the **slope class**, or gradient, of each portion of a landscape. Slope class values are measurements that indicate the steepness of a landscape, and provide insight into the rate at which other resources, such as water, vehicles, or people, are likely to travel over those portions of the landscape. Since each grid cell in a DEM contains both horizontal (e.g., latitude and longitude) and vertical (elevation) measurements, the slope of each grid cell can be computed based on the position and height of the neighboring grid cells. Most raster GIS software programs have the ability to compute slope classes, and are able to express slope class as an angle (degrees) or as a percentage of the difference in elevation of each grid cell as compared to the neighboring grid cells.

It is important to understand that there are a number of different methods used in choosing the values for slope class calculations. In a raster GIS database, each raster grid cell will have eight neighbors that share a portion (a side or a point) of its boundary: four neighboring cells will share a corner point and four neighboring cells will share a side. Theoretically, eight possible grid cell values can be used to calculate the slope class. Many raster GIS software programs use a formula that takes into account the values of these neighboring grid cells in calculating the average slope class of a single grid cell (Burrough & McDonnell, 1998, pp. 190–3). In Figure 13.7, the weighted average slope gradient between the cell of interest (the center cell with a 293 m elevation) and the eight neighbors can be calculated to determine the slope class change by computing the elevation change among the cells. You can imagine, however, that perhaps only the cells that share a side might be used to calculate the slope class for the cell of interest, or a broader window can be used (e.g., one more ring of cells around the cell of interest, or 24 neighboring cells).

To create a layer representing slopes within ArcGIS a DEM must be available, the Spatial Analyst Extension must be activated, and the Spatial Analyst menu must be opened. From the Spatial Analyst menu, choose Surface Analysis, then choose the Slope option. This should open the Slope dialog box, which will prompt you for the Input Surface

1 302 m	2 300 m	3 298 m
4 290 m	293 m	5 295 m
6 287 m	7 288 m	8 290 m

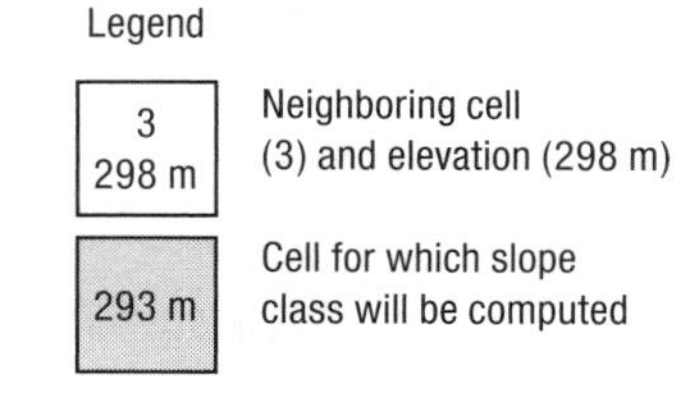

Figure 13.7 Slope class computation within a raster GIS environment.

and whether degrees or percent slope is desired. Additional options include vertical unit conversion to horizontal map unit (Z factor), output cell size, and output database name. Given that slope is a direct function of distance and elevation comparisons, it is imperative with the slope process that users know whether the measurement units of coordinates and elevations are the same. If the measurement units are not the same (e.g. meters for coordinates and feet for elevations), the Z factor input can be used to reconcile differences. The slope class GIS database created from the Brown Tract 10 m DEM (Figure 13.8) shows that slopes are reported in degrees, and are divided into nine categories. The darker-shaded slope class categories represent areas where slopes are steep, and the lighter-shaded slope class categories represent areas where slopes are gentle.

Many natural resource management organizations prefer to work with slope classes expressed as a percentage, and thus it may be important for to know how to perform this conversion:

slope class (percent) = tan (α) × 100

where

tan = tangent trigonometric function
α = slope in degrees

Brown Tract slope (degrees)

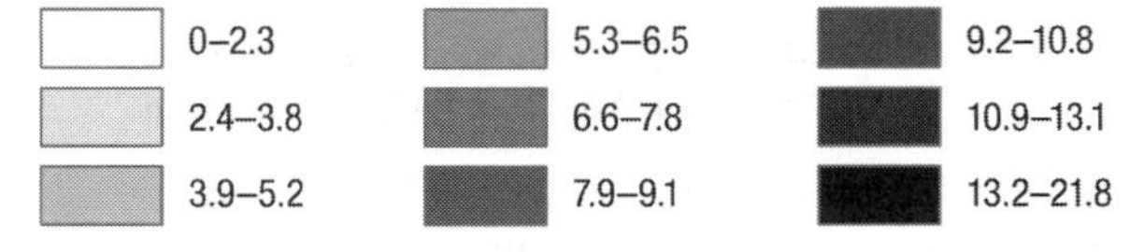

Figure 13.8 Brown Tract slope class GIS database created from a 10 m DEM.

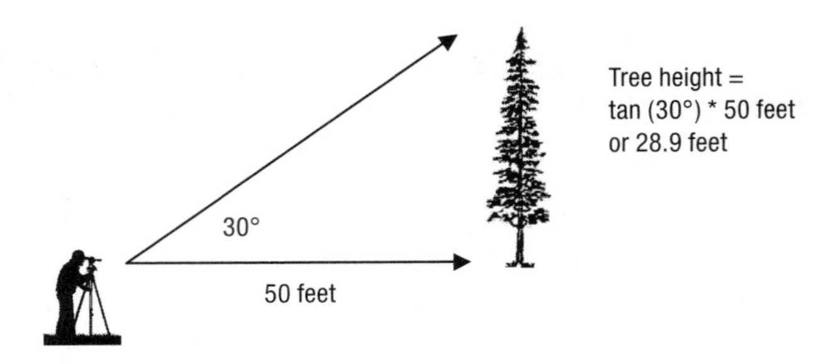

Angle (degrees) = 30°
Angle (percent) = (28.9 feet / 50 feet) = 57.7%

tan (30°) * 100 = 57.7, providing a quick conversion from degrees to percent slope

Figure 13.9 A simple example of the conversion process from degrees to percent slope.

To prove this rather simple conversion from degrees to percent slope, assume that a person was standing on flat ground (Figure 13.9) and needed to determine the percent slope from their location to the top of a tree. By knowing the angle (30°) from their position to the top of the tree, and the distance from their location to the tree, the person can calculate the height of the tree (28.9 feet). The slope from the person to the top of the tree, as expressed in percentage terms, is then the rise (the height of the tree, or 28.9 feet) divided by the run (the distance the person is from the tree, or 50 feet), or 57.7 per cent. And, by simply inserting the angle into the equation noted above,

slope class (percent) = tan(30°) × 100

you can arrive at the same conclusion. Depending on the GIS software program being used, you may need to convert between degrees and radians, since the angle reported after a slope class calculation may be reported as a radian. An examination of the software documentation will reveal whether this consideration is necessary.

Interaction with Vector GIS Databases

There are a number of methods by which you can perform a GIS analysis using both vector and raster GIS databases simultaneously. This ability has traditionally been uncommon in many desktop GIS software programs, but as technology progresses you will see the expansion of these capabilities, and field personnel (those with access primarily to desktop GIS software programs) will be able to perform more complex analyses. Two types of analyses that combine vector and raster GIS databases will now be explored: an examination of the slope class characteristics of land management units, and an examination of the slope class characteristics of streams.

Suppose you were assigned the task of developing a management plan for an area the size of the Brown Tract, and one where there was significant amount of relief associated with the landscape. The set of management activities appropriate to each management unit defined on the landscape may vary based on the slope class within each unit. For example, if you were to consider planning a forest thinning operation on the Brown Tract, it would be useful to know the locations of areas where thinning operations should use a ground-based logging system (e.g., fell-bunchers skidders, harvesters, forwarders, etc.) and the locations of areas where the thinning operations should use a cable-based logging system. Since ground-based logging systems are appropriate for the gentler slopes, slope class measurements will help identify those management units that have the steeper slopes more appropriate for cable logging systems. The slope class condition of a management unit can be measured in the field with clinometers or other surveying instruments or hypsometers, or the slope class condition can be computed using a DEM in conjunction with the vector GIS database that describes the management units. In the case of the Brown Tract, rather than having field crews spend several days collecting slope measurements, the average slope class of each management unit can be calculated with GIS

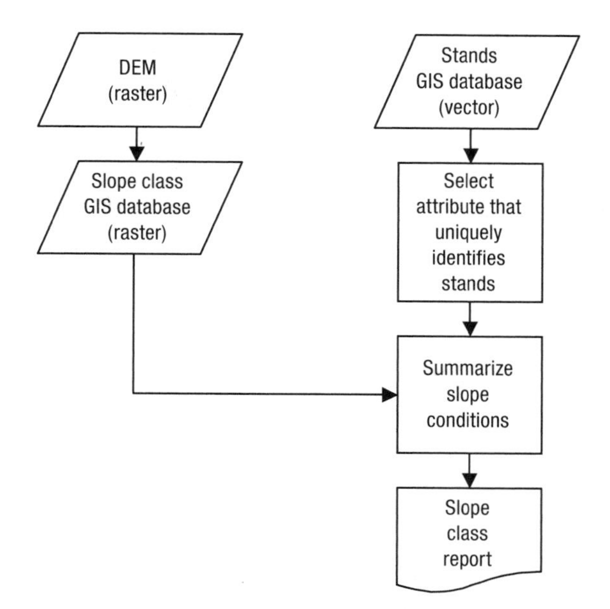

Figure 13.10 A general process for the development of a slope class condition information for each stand (management unit) on a landscape.

using a process similar to that described in Figure 13.10.

A tabular report is generated by the process described in Figure 13.10, and provides a summary of the slope class condition for each of the management units. An annotated version of the output, showing information for the first ten stands of the Brown Tract, is provided in Table 13.1. The first variable in the table represents the attribute that was selected to uniquely identify each stand (the stand

TABLE 13.1 Output of percent slope values for management units

Stand	Count	Area	Min	Max	Range	Mean	Std	Sum
1	319	343603	0.11	15.44	15.33	5.31	3.81	1692.78
2	2186	2354595	0.34	23.55	23.21	9.41	3.76	20564.20
3	770	829386	0.44	22.46	22.02	10.22	4.15	7866.61
4	2884	3106428	0.28	23.01	22.73	9.54	3.66	27521.07
5	533	574107	1.71	19.80	18.09	8.34	3.14	4446.68
6	1195	1287164	0.44	23.72	23.28	8.51	4.24	10168.51
7	338	364068	0.20	15.15	14.95	6.20	3.52	2096.76
8	2494	2686349	0.15	26.11	25.95	13.65	4.27	34040.15
9	337	362991	3.20	25.41	22.21	15.03	3.91	5066.74
10	2395	2579714	1.55	24.25	22.70	11.52	3.90	27591.07

Count = number units (10 m grid cells) in the database
Area = square feet
Min = minimum value in the database
Max = maximum value in the database
Range = (maximum value – minimum value)
Mean = average slope
Std = standard deviation of values in the database
Sum = sum of the slope for all units

number). With this value, you could join the tabular data to the stands GIS database, using a one-to-one join process (see chapter 9 for a review of join processes), and facilitate a graphical display of the slope class for each management unit. The variables 'Count' and 'Area' list the number of grid cells from the slope class GIS database that fall within each management unit, and the area that the grid cells represent. The 'Min', 'Max', and 'Range' provide the minimum slope class value within each management unit, the maximum slope class value, and the difference between these two values for each management unit. The 'Mean' is the average percent slope (what was hoped to be obtained for the thinning opportunity analysis) and the 'Std' variable is the standard deviation of the slope class values of the grid cells located within each management unit. The standard deviation provides information on the distribution and variation of slopes classes within each management unit. Large standard deviations indicate a wide variation of slope class values whereas small standard deviations indicate a narrow variation.

In the management of natural resources, the condition of a stream system may also be important to know from both a hydrologic and fisheries perspective. For example, you might need to understand the ability of streams to support fish populations, or to understand the potential water runoff implications from extreme rainfall events. Stream slope class (gradient) is one common measure of the condition of a stream system. Stream slope class can be calculated by field personnel using clinometers or other surveying instruments or hypsometers, yet this requires a visit to each stream to provide measures for the entire landscape, a very costly and time consuming proposition. The slope class conditions of streams across a landscape can, alternatively, be estimated rather quickly if a DEM and a streams GIS database is available for the landscape.

The straightforward approach to calculating slope values for streams would be to follow the previous example of supporting a thinning operation, and use the slope class GIS database for the entire Brown Tract. However, since the slope class GIS database was created for the entire landscape and only a small portion of the landscape is of interest (the streams), a different approach might be appropriate (Figure 13.11). One solution would be to create a raster GIS database of the streams. A raster database of

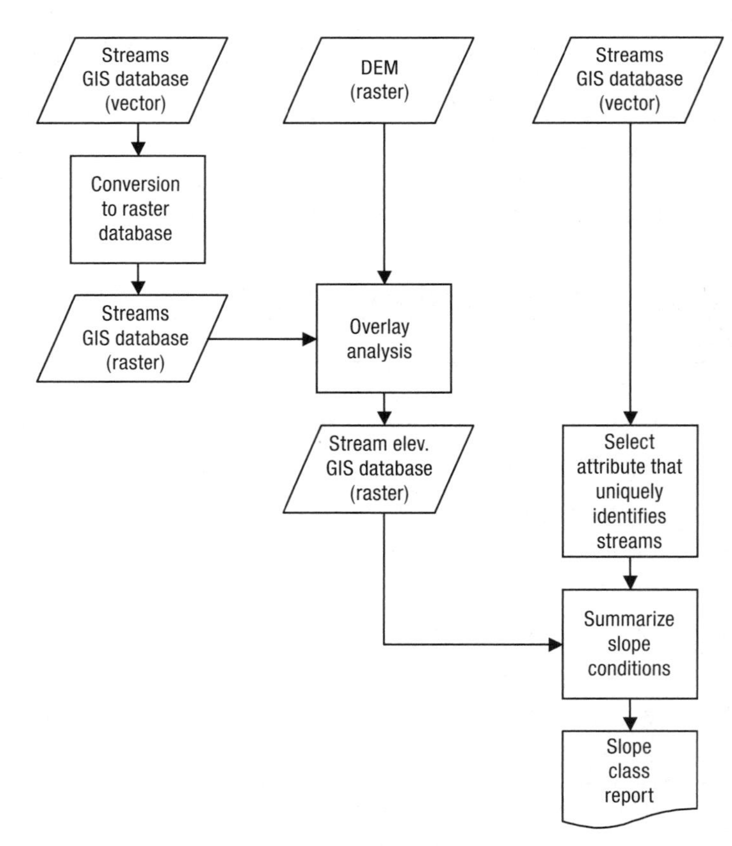

Figure 13.11 A general process for the development of slope information for each individual stream on a landscape.

In Depth

Intervisibility is a term used to describe the number of viewpoints with a view of each unit of land. For example, a number of homes may be within the viewshed of a property being managed by a natural resource organization. These homes could be considered viewpoints. By performing the viewshed analysis described above, however, you do not necessarily understand the intervisibility of the landscape, or the number of homes that actually have a view of each unit of land. In fact, some areas of the viewshed may be visible to a single home. From a management perspective, this information may be important, and allow natural resource managers to focus their public relations efforts to the directly affected homeowners. If you were to develop a GIS database that describes the intervisibility of a landscape, the phrase 'cumulative viewshed map' (rather than 'viewshed map') might be used, to illustrate the number of homes viewable from each land unit.

streams that matches the spatial extent and resolution of other raster GIS databases previously developed for this landscape would facilitate future overlay analyses. The grid cells that contain actual values (other than 'no data') should only be those that overlap a stream in the vector streams GIS database. An overlay analysis is then performed in conjunction with a DEM to enable the creation of a raster GIS database that describes the elevation of each grid cell representing a stream. Now that only those raster grid cells that touch a stream have been identified, and the elevation of each is available, you can calculate the average slope class values for each of the streams.

Similar to the previous analysis, the result of this analysis is also a tabular database. By using the attribute that uniquely identifies each stream, you could join this tabular data with the streams GIS database to facilitate the display of the slope classes for each stream. A similar processing approach could be used to identify slope or elevation characteristics for any vector GIS database that contains lines (e.g., roads, trails, facility corridors, etc.).

Viewshed Analysis

The maintenance or enhancement of aesthetic values is becoming increasingly important in natural resource management. Research has shown that the experiences of recreational visitors are influenced by the visual appearance of the landscape, and thus can be negatively affected by management activities that leave visible impacts (Ribe, 1989). A key to reducing (or heading off) potential public relations problems in natural resource management is to ascertain what portions of a landscape are visible to recreational visitors, and to adjust the management plans associated with those areas accordingly (Wing & Johnson, 2001).

Viewshed analysis can facilitate an understanding of the portions of a landscape that are visible from specific landscape features of interest. For example, observation or viewing sites (overlooks) may be represented in a GIS database by points, and the location and elevation of the points would be used to determine which other parts of the landscape would be visible. If the viewing sites were described by lines, the vertices (points where the line direction changes) of each line would be used for the viewshed analysis. For example, lines that represent a road or trail through a landscape, could be used to determine which other parts of the landscape can be seen from those features.

The landscape in a viewshed analysis is represented by a DEM or a TIN (see chapter 2 for a description of a TIN). The objective of a viewshed analysis is to calculate the line of sight between the viewing sites (e.g., observation points, homes) and other landscape features (Figure 13.12).

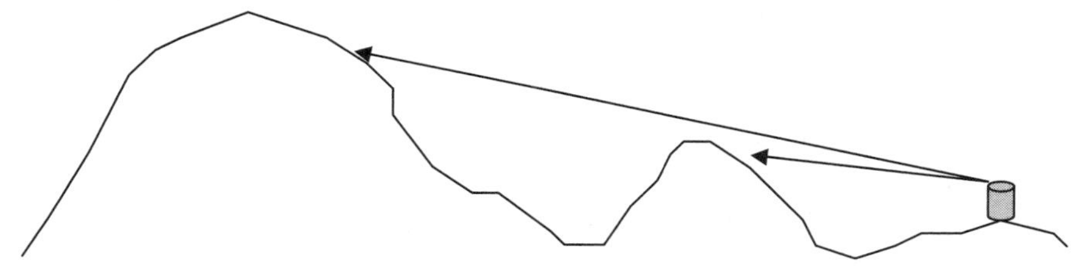

Figure 13.12 Line of sight from a viewing site to the surrounding landscape.

Features that are identified as being in the line of sight of the viewing sites are considered visible, whereas all other landscape features are not considered visible.

A number of considerations must be taken into account when conducting a viewshed analysis. Although a DEM associated with a landscape may have been acquired (perhaps from the US Geological Survey), an assessment of its fitness to represent the landscape in a viewshed analysis should be performed. In heavily forested landscapes, the DEM may not represent the effect of tree heights on your view, as tree canopies extend above the elevation surface (the ground, as represented in the DEM). The current management of the landscape will also affect a viewshed analysis because different aged stands have different heights, and therefore the top of the current canopy (which is what is usually seen in a scenic view) may be misrepresented. One solution to this problem is to acquire a vegetation GIS database that contains tree height information, and to incorporate these measurements into the DEM elevations. Another consideration is to make sure that the DEM surface covers the entire landscape area between a viewing site and the landscape being analyzed. In some applications, a manager might be interested in determining the visibility of a resource from surrounding viewing sites (homes or roadways) but only has access to a DEM that describes the area managed (such as the DEM for the Brown Tract, as shown in Figure 13.2). To perform a viewshed analysis, it will be necessary to acquire a DEM that covers not only the landscape of interest (the land being managed), but also contains the areas where the viewing sites are located.

One consideration in viewshed analyses often overlooked is that GIS users can generally modify the height of the viewing site. For example, the height of the viewing site is generally considered to be the actual ground-level elevation at the location of the viewing site. If the viewing sites were meant to represent people who were standing and viewing the landscape, then an observation height of five or six feet higher than the ground elevation at the viewing site might be more realistic than assuming a ground level view of 0 feet. This relatively modest change in viewing elevation can have significant impacts on viewshed results when large land areas are involved. As viewing elevation rises, the amount of land returned by a viewshed analysis will typically also increase. Other considerations in performing a viewshed analysis might include setting limits as to how far viewing sites are allowed to 'see' across the landscape, and limits on viewing angles.

To illustrate the development of a viewshed analysis, a GIS database representing home locations surrounding the Brown Tract will be used as the viewing sites. A cursory investigation of the homes GIS database reveals that there are multiple homes along the eastern half of the Brown Tract (Figure 13.13). A viewshed analysis will allow you to determine what portions of the Brown Tract are visible by residents of the nearby homes. One of the first steps in the viewshed analysis is to define the observer height. You might assume that the average person has a view height (above ground level) of 5.5 feet, and that owners of the homes around the Brown Tract could likely view the forest from their second floor windows (add 10 feet), resulting in an adjusted observer height of 15.5 feet.

To perform the viewshed analysis, you might pursue a process similar to that described in Figure 13.14. Since the original DEM for the Brown Tract was clipped to the ownership boundary of the tract, you could acquire USGS DEMs and create a new DEM GIS database that includes the Brown Tract and the areas that cover the homes. As mentioned earlier, addressing vegetation height in the viewshed analysis might be appropriate, since much of the Brown Tract is forested. The stands GIS database contains a variable named 'Height' that provides the average height of each of the management units. By converting this vector GIS database to a raster GIS database format, height information can be added to the Brown Tract

Figure 13.13 Locations of homes near the Brown Tract.

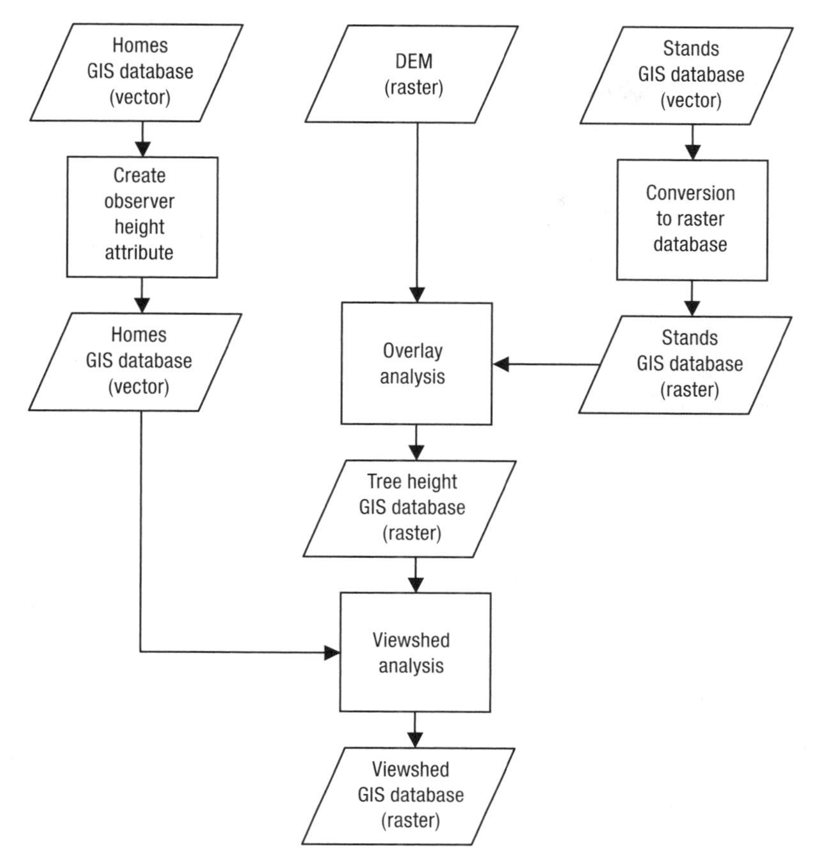

Figure 13.14 A general process for the development of a viewshed analysis for the Brown Tract based on a set of viewshed sites (nearby homes).

DEM values to get an adjusted elevation that includes the average tree heights for each management unit contained in the stands GIS database (Figure 13.15). The Spatial Analyst extension within ArcGIS contains a convert menu that offers a 'features to raster' utility for this purpose. The dialog box for this option will prompt for the database to be converted and a variable to be represented in the raster (height in this case). Once converted the stand height raster can be added to the DEM to create a new elevation surface that contains updated elevation data for the Brown Tract and surrounding areas (Figure 13.16).

After performing the preliminary database development steps, the actual viewshed analysis can be performed by computing the areas of the landscape that are visible from each home. Within ArcGIS, this can be accomplished by using the Viewshed command, under the Surface Analysis tools within the Spatial Analyst extension. A dialog box will open and offer prompts for the input surface (DEM or other modified elevation surface) and observer points (locations from which the input sur-

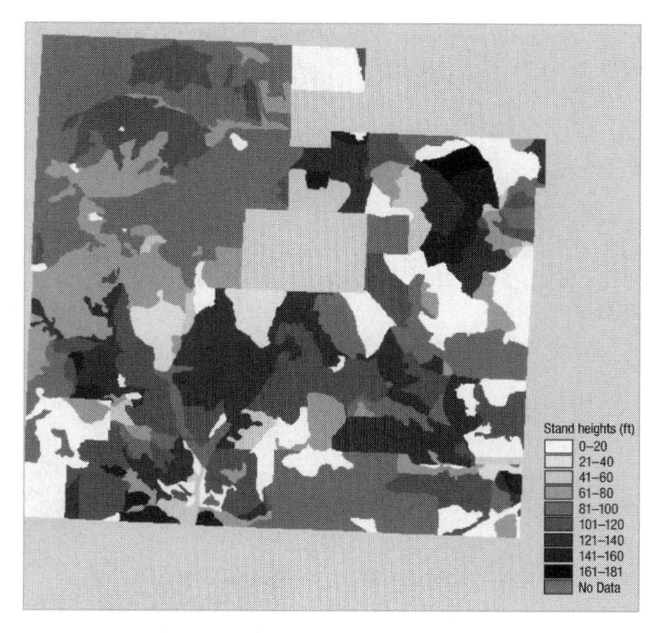

Figure 13.15 Average tree heights for management units contained in the Brown Tract stands GIS database.

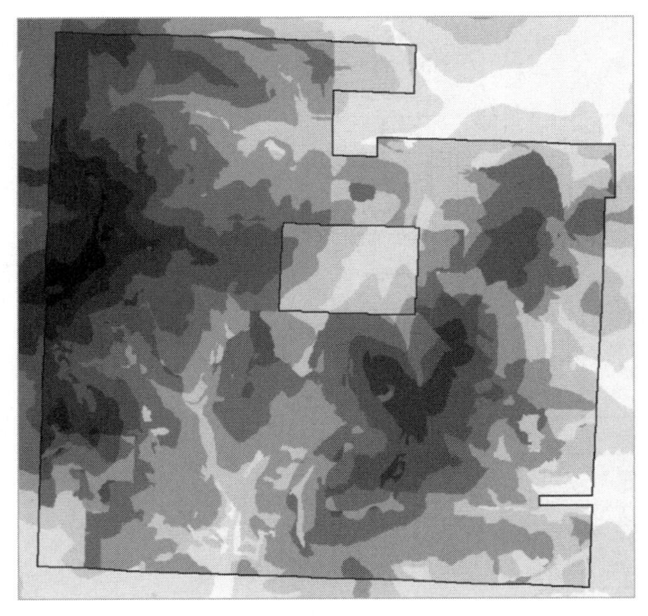

Figure 13.16 DEM GIS database for the Brown Tract and surrounding areas that includes the average tree heights of the management units contained in the stands GIS database.

face will be analyzed). Depending on the size of the DEM, this processing time may be lengthy, however, after processing is complete, a raster GIS database is created that illustrates the areas within and around the Brown Tract that are visible from the nearby homes (Figure 13.17). The areas within the Brown Tract that are visible from

Figure 13.17 Areas (dark shade) within and surrounding the Brown Tract that are visible from the nearby homes.

the nearby homes are located mainly along the tract borders, however a significant amount of land within the tract boundary is also visible from one or more homes (although you cannot, from this analysis determine how many homes have a view of each piece of land).

Watershed Delineation

GIS has come to be a standard tool for hydrologic research and water resource modelling. GIS applications for water resources are common and vary from simply mapping water resources to sophisticated analyses that consider water quality influences and future water availability (Wilson et al., 2000; Martin et al., 2005). A primary use of GIS for water resource management is that of **watershed delineation**. A watershed can be defined as a landscape area that shares a common drainage. This definition assumes that if water were able to flow freely over a landscape area, it would follow a downhill trajectory and exit the landscape area at a common point. Watershed boundaries are commonly used by federal, state, and provincial organizations to separate a landscape into smaller management areas. In addition, watershed councils—groups that help to determine management activities within a specific watershed—are becoming more common in North America. Some refer to watersheds as 'catchments' but the intent is the same: to describe an area according to the flow of water over its surface. The delineation of a watershed boundary is a function of topography and changes in landscape relief. You must decide on the scale of the watershed to be created and have access to topographic information in order to begin delineation.

Previous to digital GIS and the availability of DEMs, watersheds were created primarily by using available contour maps and using the visual clues provided by contour shapes to draw or digitize a watershed boundary. Clues from contour shapes included saddles, peaks as indicated by closed contours, and the direction of funnel-like contour shapes that represent water flow paths or ridges (Figure 13.18). If existing stream networks were present on the contour map or on additional maps, these features also provided clues as to watershed boundary locations. The delineation process was usually begun from the perceived lowest elevation point of the watershed and then continued uphill, with the goal of identifying the ridge lines that surrounded a watershed and separated it from other watersheds. Contour line funnels indicated either water flow paths or ridges, with water flowing from the bottom of the funnel (u- or v-shaped end) to the open

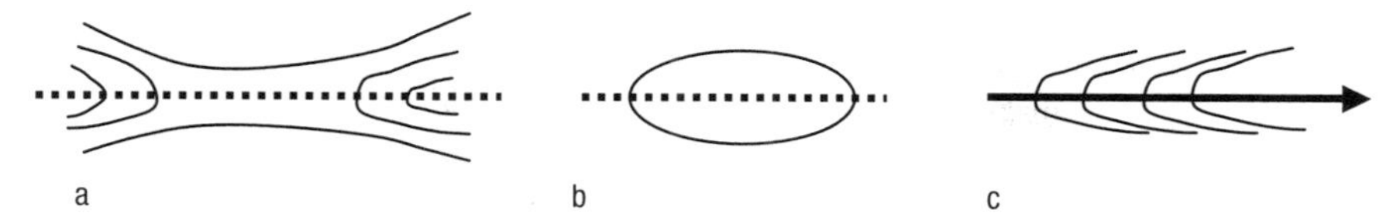

Figure 13.18 Watershed boundary location and stream flow patterns derived from contour interval shapes: (a) watershed boundary parallels contour saddles and (b) splits peaks as indicated by a closed contour homes, and (c) water flows from the bottom of contour funnels through the top.

end. Closed contour shapes were to be split through the middle as they indicated a peak, and boundaries should split contour saddles in a parallel orientation. This required those involved in the delineating watersheds to continually ask themselves 'which way would water flow if a bucket of water were dumped at a particular location?' The delineation process could be drawn directly on the contour map or accomplished through heads-up digitizing on a monitor with background databases of streams and contours being displayed. Nonetheless, manual watershed delineation is a tedious process with significant opportunities for human error. If a common drainage point, line, or area can be digitally represented in a GIS and a DEM exists of the area, many GIS can automatically generate a watershed area boundary given some basic data manipulations.

In addition to a feature that represents the drainage location(s) and DEM, a database representing flow directions is usually required for watershed delineation. A flow direction database is represented through a raster database structure and assigns an expected direction of water flow to each raster cell. Fortunately, a flow direction layer can usually be created from a DEM through an algorithm that evaluates the elevations of each raster cell and all neighboring cells. The possible directions include the path to each potential neighboring raster cell (eight directions) as determined by the lowest elevation among the neighbor cells. In addition, a 'no direction' choice is possible and is referred to as a 'sink'. A sink occurs when a raster cell has an elevation value that is less than all surrounding cells.

Any portion of a landscape can be analyzed for its watershed boundary given that a DEM is available of the entire potential watershed extent. If the DEM does not cover the entire potential watershed area, an incomplete watershed boundary may result without any warning from the supporting GIS software as to its incompleteness. As an example of GIS-based watershed delineation, a portion of the stream network within the Brown Tract that falls within the smaller interior private ownership area is considered (Figure 13.19). A logical sequence of steps for the watershed creation process is represented in Figure 13.20. A separate database must first be created that represents only this stream network portion, or at least the lowest elevation point within the portion. This lowest elevation point in this database will serve as the watershed source, sometimes referred to as a pour point. Rather than risking a misidentification of the lowest point, it's probably more expedient and reliable to clip, or otherwise separate, the entire portion of the stream network into a new GIS database. Within ArcGIS, you can use the Select Features tool to select all of the stream features and then export the selected features into a new database (right click on the layer and choose Data, then Export Data). Some GIS software will require that line and polygon features intended to serve as a watershed source must be converted into a raster database. Within ArcGIS, once the selected streams database is ready, the Spatial Analyst Convert menu contains a 'features to raster' command that will make this conversion.

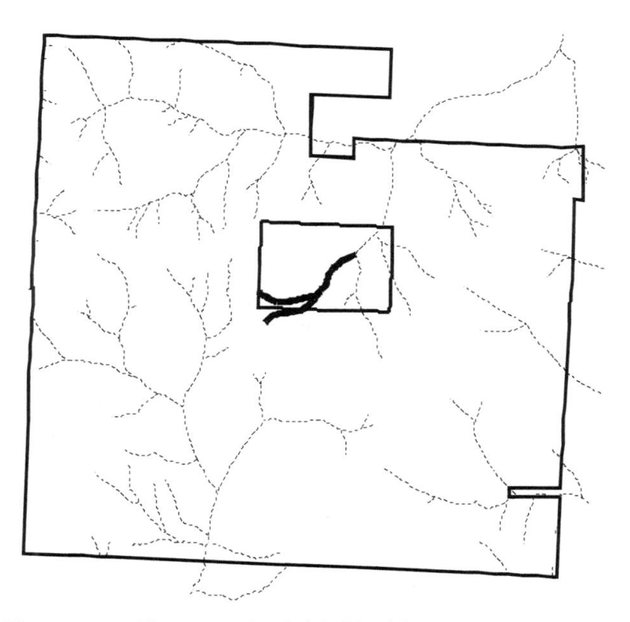

Figure 13.19 The portion (in dark bold) of the stream network within the Brown Tract for which a watershed area is to be created.

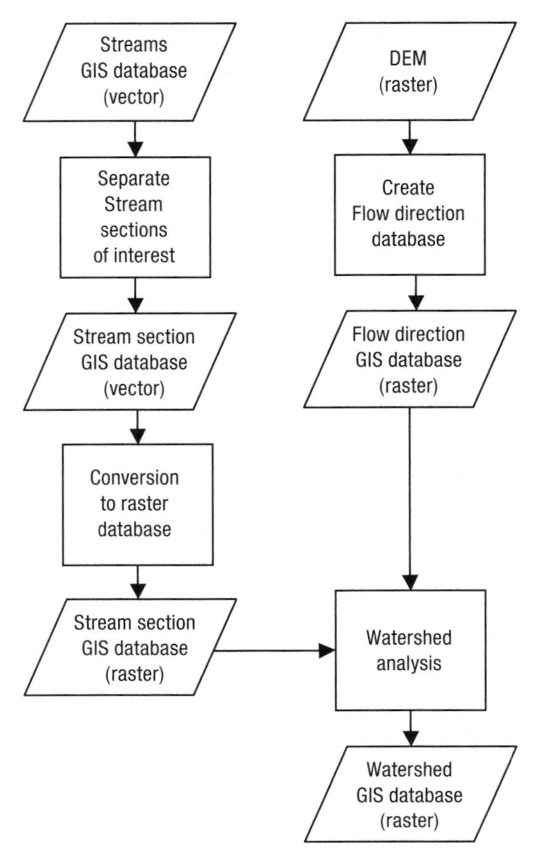

Figure 13.20 A general process for the delineation of a watershed area for a portion of the stream network in the Brown Tract.

The next step is to create a flow direction raster based on the Brown Tract DEM. This process can be accomplished within ArcGIS by using the raster calculator and flowdirection command. Assuming that the extended DEM for the Brown Tract is used, the raster calculator syntax would be 'flowdirection ([browndemex])'. The result of this command should be a temporary raster file with cells assigned to one of nine possible values: one for each direction to the eight neighboring cells and a sink or 'no lower elevation' value. A sink can be caused by a natural feature, such as a lake or other water body with no surface outlet, or can be the result of a DEM error. Regardless, it may be necessary to detect and eliminate sinks within the DEM before flow directions can be calculated. Within ArcGIS, the fill command can be used to manipulate potential sinks but must be run through the Spatial Analyst Hydrology tools in the ArcToolbox or through the command line interface. The fill command will create a new raster with elevation values for sink locations raised to the minimum elevation found in surrounding cells. The modified ('filled') raster can then be used to calculate flow directions. Once the flow direction raster is complete, the watershed area can be derived. Within ArcGIS the raster calculator can be used with the Watershed command to create watershed boundaries. The syntax is 'watershed ([flowraster, streamsection])' where flowraster is the name of the flowdirection raster and streamsection is the name of the raster version of the portion of the Brown Tract stream network for which a watershed will be created. The watershed created should not be very large in comparison to the Brown Tract boundary (Figure 13.21).

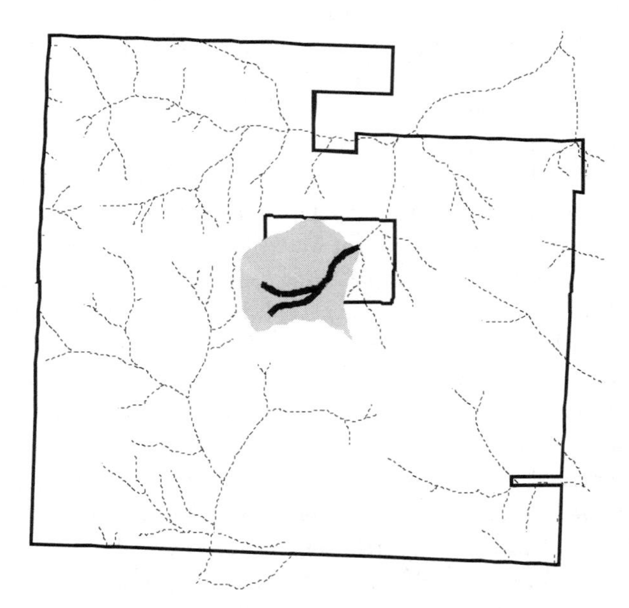

Figure 13.21 The watershed (in gray) for a portion of the Brown Tract stream network.

Summary

The examples in this chapter have demonstrated several GIS operations that are possible when using raster GIS databases, and have concentrated on the wealth of opportunities associated with using a DEM. From the basic elevation information contained within a DEM, you can generate contour lines, a shaded relief image, and a slope classification map. Raster GIS databases can be integrated in spatial analyses to calculate slope classes for management units as well as gradients of streams, and to create a database of the viewshed related to nearby landscape features, such as the homes located around the Brown Tract. In addition, DEMs can be used to analyze landscape topography and can create additional raster databases to support watershed delineation. These examples are but a small sample of the types of applications that are possible with a single raster GIS database in terms of examining landscape topography. We turn our attention in the following chapter to exploring other potential raster data applications.

Applications

13.1 Shaded relief map. You have been asked by the manager of the Brown Tract to create a shaded relief map that simulates the sun's light on the landscape in early morning.

a) What values would you use to approximate the sun's azimuth and altitude?
b) Develop a map that illustrates the shaded relief.

13.2 Slope map of the Brown Tract. During one of your monthly planning meetings, it was suggested that a slope map would be of value to the foresters who plan timber sales on the Brown Tract. This map would be of equal value to the recreation manager, who is interested in potential trail systems throughout the forest.

a) Create a slope GIS database of the Brown Tract that has 80 ft contour lines that originate from a 300 ft base elevation.
b) Develop a map illustrating the slope classes.

13.3 Viewshed analysis planning. Your supervisor has approached you and asked that you assist in a viewshed analysis for part of your resource management area. She is interested in knowing what portions of your resource management area are visible from a nearby road. She needs the information quickly and has asked you to scope out the project with regard to how this might be accomplished in GIS (e.g., what GIS databases are necessary, what processes are necessary, etc.). Provide a brief description that lists required GIS databases, techniques, and potential pitfalls of conducting the visibility analysis.

13.4 Road gradient analysis planning. Your supervisor is interested in knowing the average slope gradients of a set of roads within your organization's management area. She has asked you for a one-page description of how this might be accomplished in GIS. Prepare a report for her that describes the necessary GIS databases, potential techniques, and considerations that would be involved in using GIS for this purpose.

13.5 Road gradients on the Brown Tract. The forest engineer associated with the Brown Tract has asked you to calculate road slope gradients (percent slope) for the roads in the Brown Tract using two GIS processing methods. For the first method, the slopes should be calculated using the entire DEM available for the landscape. For the second method, the slopes should be calculated using only those raster grid cells that overlap the road network. The engineer wants to understand the difference, if any, that would be observed in the average slopes gradients when using these two methods. Prepare a short report to convey this information to the forest engineer.

13.6 Brown Tract viewshed analysis. The manager of the Brown Tract would like you to develop a viewshed analysis using only those homes located on the north side of the Brown Tract. Use the assumptions and processes described in the example provided earlier in this chapter. How do the viewshed results differ from those of the example presented earlier in this chapter?

13.7 Brown Tract watershed analysis. The hydrologist for the Brown Tract, Samantha Wasser, has asked for your help in identifying a watershed for a portion of the Brown Tract stream network. The lowest elevation stream segment in the portion is identified through a numeric value of 123 in the stream variable within the streams database. Develop a watershed for this stream network portion and report the resulting watershed area in both acres and hectares.

13.8 Brown Tract beaver watershed analysis. The Brown Tract manager has asked you to delineate the watershed area that flows into a large beaver pond. The beaver pond location is described in the water sources database as a point location. Create a watershed for this point and report the resulting watershed area in both acres and hectares.

13.9 Brown Tract stand watershed analysis. The Brown Tract manager has asked you to delineate the watershed area that flows into a stand inside the forest. The stand is identified as number 36 with the stand field in the Brown Tract stands database. Create a watershed for this stand and report the resulting watershed area in both acres and hectares.

13.10 Basic characteristics of the Brown Tract. Becky Blaylock, the manager of the Brown Tract, is interested in some basic knowledge of the landscape. Specifically, she wants to know the following:

a) What is the minimum, maximum, average, and standard deviation of slope (degrees) for the Brown Tract?
b) What is the minimum, maximum, average, and standard deviation of slope (percent) for the Brown Tract?

13.11 Combining raster GIS databases. What is the result when two raster GIS databases are combined, yet 'no data' values are present in some portions of the landscape in one of the two raster GIS databases?

13.12 Highest and lowest elevation stands. Which stand in the Brown Tract is located at the lowest average elevation? Which stand is located at the highest average elevation?

13.13 Elevation information for a single stand. What are the average, minimum, and maximum elevations of the stand described in question 13.9 (stand 36)?

13.14 Water source elevations. What is the average elevation for each of the points described in the water sources database?

References

Burrough, P.A., & McDonnell, R.A. (1998). *Principles of geographical information systems*. Oxford: Oxford University Press.

Martin, P.H., LeBoeuf, E.J., Dobbins, J.P., Daniel, E.B., & Abkowitz, M.D. (2005). Interfacing GIS with water resource models: A state-of-the-art review. *Journal of the American Water Resources Association, 41*, 1471–87.

Natural Resources Canada. (2007). *Mapping*. Retrieved May 26, 2007, from http://www.nrcan-rncan.gc.ca/com/subsuj/mapcar-eng.php.

Ribe, R.G. (1989). The aesthetics of forestry: What has empirical preference research taught us? *Environmental Management, 13*, 55–74.

US Department of Interior, US Geological Survey. (2007). *USGS Geographic data download*. Retrieved May 26, 2007, from http://edc2.usgs.gov/geodata/index.php.

Wilson, J.P., Mitasova, H., & Wright, D.J. (2000). Water resource applications of geographical information systems. *URISA Journal, 12*(2), 61–79.

Wing, M.G., & Johnson, R. (2001). Quantifying forest visibility with spatial data. *Environmental Management, 27*, 411–20.

Chapter 14

Raster GIS Database Analysis II

Objectives

Chapter 14 builds on the previous chapter and further explores raster data for spatial analysis. Like chapter 13, this chapter also involves topographic applications of raster data but broadens the treatment of raster data applications. In addition, more technical detail is provided for raster data processing techniques and analysis. At the conclusion of this chapter, readers should understand and be able to converse about:

1. the potential applications of raster data for natural resource problem analysis,
2. how distance functions can be applied to raster data,
3. the types of statistical summary search functions for raster data,
4. the capabilities and applications of density operations,
5. raster data reclassification and map algebra processes, and
6. data structure conversion considerations.

Raster Data Analysis

We further develop raster data analysis in this chapter by describing some general considerations for raster analysis working parameters, examples of raster analysis functions, and applications of processing commands for manipulating raster data. We then describe some application examples that make use of select raster functions. The raster analysis functions we discuss include distance, statistical summary search, and density functions. In some cases, the functions can accommodate both vector and raster databases as input. General processing commands covered in this chapter involve raster reclassification, raster map algebra, and data structure conversions. We also provide more detailed information about the procedures within the world's most popular GIS software—ArcGIS—that can be used for raster analysis. An assumption for these procedures is that the Spatial Analyst software extension for ArcGIS is available. This extension is specifically designed for raster data analysis and provides a set of menu choices and commands that make some of primary raster capabilities more readily available. If readers are using different raster software (i.e., Imagine, GRASS, Idrisi, etc.) for their applications, the procedures described in this chapter should provide a template for applying the functions and analyses that we present.

Raster Analysis Software Parameters

Some raster software will allow you to set working environment parameters that the software will observe during use. These conditions establish resolution, analysis, output, and other conventions that affect output raster databases that are created during processing. One of the distinguishing characteristics of raster databases in comparison to vector data is that they are typically larger in size in terms of digital storage space. Raster working environment parameters can help constrain raster databases to specific study areas and resolutions so that output files do not become overly large in size. Some raster software will only create temporary raster databases unless otherwise specified by a user. These temporary raster databases will be removed from the hard drive once the GIS software is closed, which can cause some distress for unsuspecting users the next time the software is restarted. Depending on the raster software, working environment parameters may need to be reestablished at the beginning of every analysis session.

An important raster parameter to consider is an analysis mask. An analysis mask, as established by the dimensions of another GIS layer, will restrict processing to only those areas coincident within the layer. The extent options allow you to further restrict the areas that will be considered when new raster output is created. Depending on your software, possible choices include existing layers, spatial combinations of existing layers, or a bounding set of coordinates. In addition to raster databases, vector databases can usually also be used to create an analysis mask.

Another critical choice for processing output raster databases is cell resolution. Typically, working parameter choices will allow users to set a designated raster cell resolution for output databases. This can typically be set to match other analysis layers, combinations of other layers, user-specified cell resolutions, or a user-specified set of columns and rows. Setting a cell size and analysis extent to match an existing layer will ensure that output raster cells are registered to the existing layer, and that output raster cells from both layers will be coincident and not offset from one another. These potential parameters described here, and others that may be available depending on software, can help make raster analysis more efficient. These choices can help establish a common resolution, analysis area, and output location for raster processing results. The appropriate options should be set prior to any raster analysis session.

Distance Functions

Distance functions calculate the measured distance to features of interest. There are various ways of computing distances, ranging from a straight line measurement to more involved approaches (i.e., shortest paths) that use constraints of moving across a landscape due to changes in relief or other impediments. Distance functions can be used to help determine the next closest feature, areas that might be ecologically sensitive because of their proximity to natural features, or the most expedient route to take from one feature to another. Typical distance functions include the straight line, allocation distance, cost weighted distance, and shortest path (Theobald, 2003).

The straight line distance function will create a raster output with direct (also called Euclidean) distances to the closest feature. The function can be applied to either vector or raster databases. As an example, consider the water sources point locations within and around the Brown Tract (Figure 14.1). It might be beneficial for emergency response units combating wildfire to know how far away the closest water sources are located, particularly if helicopters will be used to carry water from sources to a forest fire and straight line distances were of interest. A distance function could demonstrate relative distances to the nearest water source (Figure 14.2).

An allocation distance function assigns areas to the closest feature. In this case, raster cells will be assigned a pixel value that recognizes the nearest zone of influence, similar to the vector representation of a Thiessen polygon. A Thiessen polygon is created for each feature in a spatial database and represents the area that each feature is nearest. In the case of the nine potential water sources, nine allocation zones are identified with each water source having its own zone of influence (Figure 14.3).

A cost weighted distance function allows for assigning different weighting to raster cells that take into account what is required for a pathway to cross through the cell (Chrisman, 1997). This function might be used to assess the amount of time needed for water to flow through one end of a stream network to another, the relative costs of materials to construct or resurface a trail, or resources necessary to develop access to areas in mountainous terrain. The cost weighted function requires that in addition to

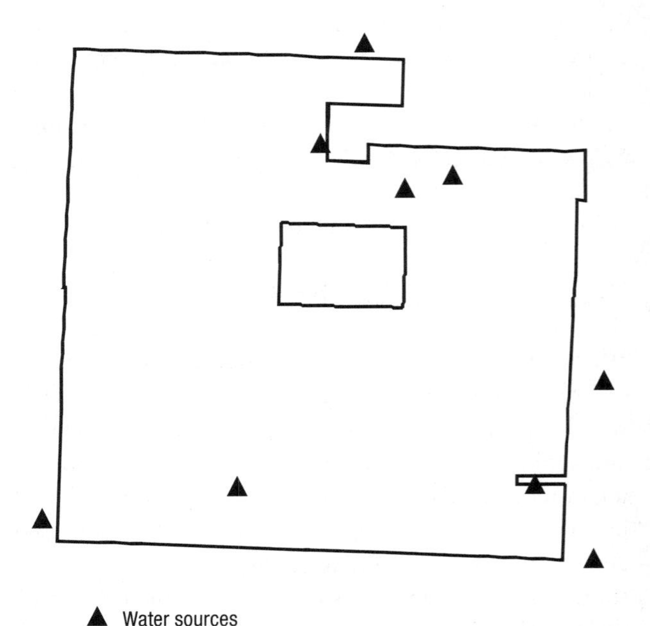

Figure 14.1 Water sources in and around the Brown Tract.

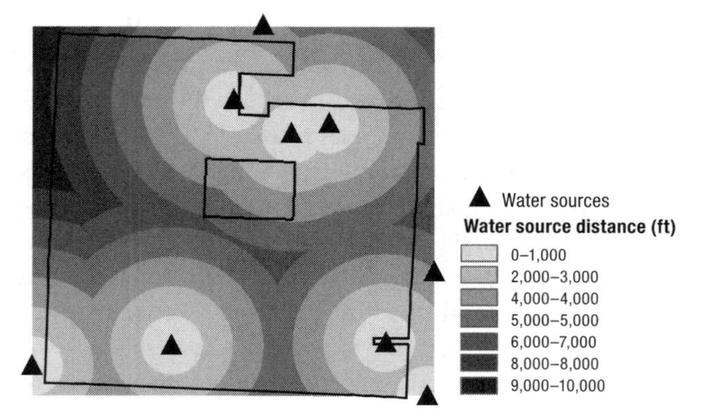

Figure 14.2 Straight line distance categories to the nearest water source in the Brown Tract.

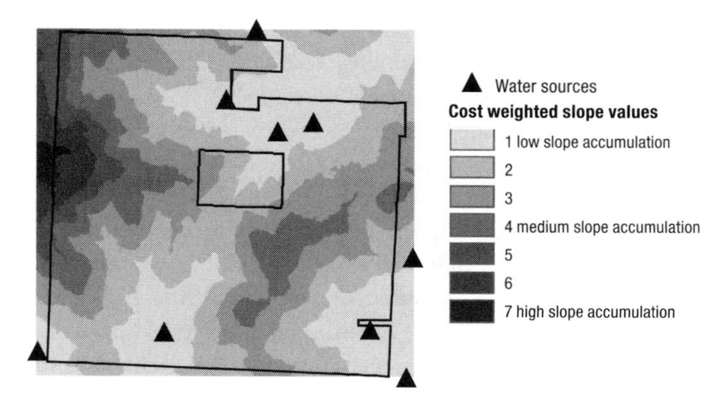

Figure 14.4 Cost weighted slope values to water sources in the Brown Tract.

the raster database representing desired destinations, the cells contain values representing the weights. The weights represent the cost inherent in accessing the cell, be it symbolic of steepness, required construction materials, or some other value that has significance. The costs are then added together, starting at the destination(s) and moving outward, to form an output raster that symbolizes the entire costs or resources from each cell to the desired destination(s). A cost weighted distance surface with the water sources as destinations and the Brown Tract slope values as weights is shown in Figure 14.4.

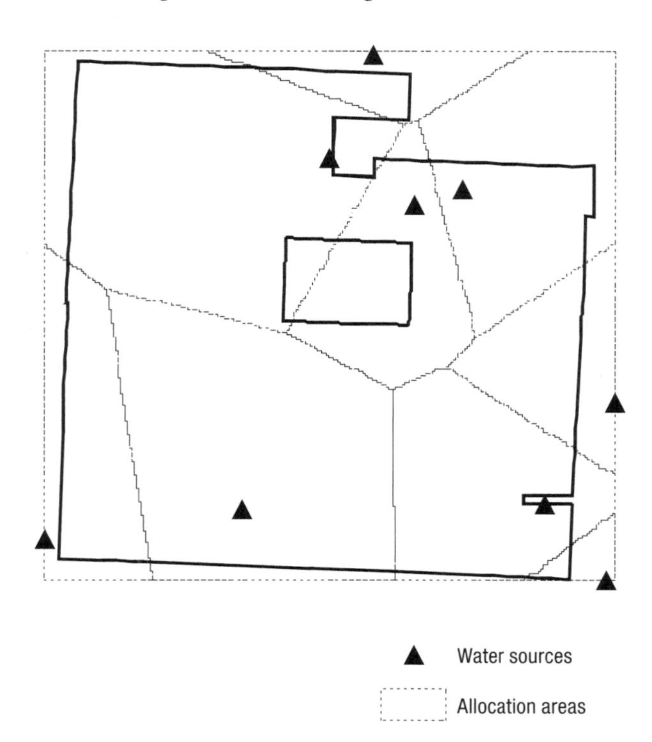

Figure 14.3 Allocation areas for water sources in the Brown Tract.

The shortest path function is intended to do what the name implies: identify the path of least distance or resistance to a desired destination. The shortest path function requires supporting databases representing cost weighted distances and least cost directions. The least cost direction raster is generated for each raster cell and is used to designate which of the eight neighboring cells is upon the least cost path.

Statistical Summary Search Functions

Statistical summaries can be generated from individual raster databases, multiple raster databases, and combinations of raster and vector databases using raster-based search functions. These functions search within a database and return summary values based on search criteria provided by the user. The search criteria may be based on areas within a second coincident database, a search area within a given pattern and size, or values contained within multiple raster files. Three general types of raster statistical summary search functions include cell, neighborhood, and zonal statistics options (DeMers, 2002). Cell and neighborhood search functions are sometimes called local and focal searches, respectively. Each search function requires user input to direct the search extent and the content of statistical summary information that is returned.

Cell, or cellular, statistics allow multiple raster databases to be evaluated in the creation of a new raster database. Each coincident raster cell in all input databases is considered and a statistical summary of cell values is selected for output into the new database. Statistical summaries can include average, summary, minimum, maximum, standard deviation, most common value, or other possibilities. Values within input databases are also restricted to numerical formats. The cellular statistics capability represents a type of raster overlay operation but all values in all input databases are treated equally according to the statistical summary selected. Potential applications of cellular statistics involving multiple raster databases include adding ground and structure elevations to create a surface elevation layer, developing a composite fire risk index by summarizing potential fuel and landscape qualities, and developing an average temperature given a set of databases that contain annual temperature measurements.

Neighborhood statistics functions are for single raster or vector files and allow for statistical summaries based on area searches within a single database. The search will consider the entire extent of the database but will consider each feature (point, line, or polygon) or raster cell and apply the search area parameters for each. Regardless of raster or vector input, a new output raster is created that contains the results of area evaluations in a summary value within each output raster cell.

Neighborhood search functions can have several different shapes including rectangles, circles, and wedges. In addition, an annulus shape is possible which has the functional appearance of a donut, with an inside area (donut hole) that is ignored in searches and an outside area (donut) that is evaluated. The user designates the size of the neighborhood search area shapes. Within each area, a statistical summary is possible of any single numeric field in the input database. Statistical numeric summaries include the average, summary, minimum, maximum, standard deviation, most common value, and other frequency evaluations. Possible uses of neighborhood search functions include identifying the number of available water sources occurring within a 100 km radius or any feature, the number of nest locations at least 50 m away but less than 2,000 m distant, and the maximum tree height of all trees contained within a 20 m by 20 m plot.

Zonal statistics, applied in examples in the previous chapter, are similar to neighborhood search functions in that a location or area is being evaluated. In contrast to neighborhood functions, zonal statistics functions offer the advantage of allowing two databases to be evaluated simultaneously. A new output raster database is created that provides a numerical summary of values in one database that are coincident with areas in a second database. The areas in the second database are considered to be analysis or summary zones for which values in coincident locations in the other database will be analyzed. The zones can be designated by an attribute value in either numeric or categorical format, such as county names, land cover categories, or road numbers. Output includes a statistical summary in tabular format for each identified zone and contains the number of coincident raster cells and associated area. In addition, numerical summaries include the minimum, maximum, range, average, standard deviation, sum, variety, majority, minority, and median values found within each zone. Zonal statistics offer the advantage of allowing raster databases to be summarized in relation to zones described within a vector or raster database in a single analysis. The vector database can be point, line, or polygon format. Examples in the previous chapter included calculating slopes for forest stands and streams. Other potential applications include determining maximum elevations of precipitation gauges, average aspect of a stream, and the average elevation in the home range of a wildlife species.

Density Functions

The intensity or frequency with which something occurs across a landscape or portion of a landscape can be demonstrated through a **density function**. Implicit in the requirements of density calculation is the tabulation of some resource in magnitude or location relative to some area quantity (Chang, 2002). Within many raster-based GIS programs, density can be calculated for point and line vector layers with output results being written to a raster database. The creation of smoothed density surfaces is also possible for point and line features (Silverman, 1986).

Density estimates are useful for describing the road system within a forest, assessing the relative quality or suitability of habitat areas given the number of wildlife or other features present in an area, and for demonstrating conglomerations or 'hot spots' of an activity or resource condition when many features are present in a database. Hot spots are apparent locations or areas where stronger concentrations of some condition are observed. When thousands of locations are present in a database, it may be challenging to map and determine locations that have heavier concentrations than others (Wing & Tynon, 2006). This is particularly true when point features are of interest and many occur in the same location. A single point in this situation can obscure other points when plotted on top of one another on a map or computer monitor. In these situations, density functions can be applied to create shades that demonstrate concentration

In Depth

What is a hot spot? In terms of GIS analysis, hot spots are locations where some feature of interest is occurring with greater frequency. A hot spot can be indicated through increased density of point, line, or polygon features in a vector database. Groupings of raster cells can also mark hot spots. Beyond the density of spatial features, an attribute field that contains one or more features with increased or heightened attribute values in comparison to other feature values can also create a hotspot. In other words, a subset of attribute values might differ markedly from other attribute values. The locations of the features would then be used to identify the hot spot's location. Hot spots are frequently mentioned in spatially-based crime research and a number of analytical techniques have been developed to identify hot spot locations. These techniques can also be applied to natural resource applications.

intensities and can more quickly draw ones attention to likely hot spots.

Density functions within GIS typically allow a user to select an attribute within a layer to serve as the population quantity for density calculation. If no attribute is selected, it is then the number of points or length of line within a search distance that is quantified. Two types of densities are usually available: simple and smoothed. For each type, a search radius and area unit must be specified and output results are written to a raster database. The search radius determines how far from each raster cell in the output database in which to search for features. The area unit will be the size of the landscape unit area in which frequencies are assessed: per square meter, square kilometer, or other area unit. The simple density method will result in a raster output in which each cell shows the number of features per unit area. If the user chooses a population field, the field quantity is used as the number of times in which to count each feature in the density summation. The smoothed density also uses the same approaches but stretches or 'smoothes' the results such that density will be highest at each feature location but reduced gradually to 0 at the outside radius of the designated search distance. The output of the smoothed density is typically more aesthetically pleasing than the simple density output, but it is less precise in its demonstration of density. Some refer to the smoothed density approach as the kernel density method.

An example of a simple density raster is shown in Figure 14.5. In this example, the roads in the Brown Tract have been selected as the source to create for the density raster. A 2,000 ft radius was selected as the search distance and areas with greater road density are displayed using darker shades.

Raster Reclassification

Raster cell values can represent almost any numerical or categorical value, ranging from reflected electromagnetic energy to descriptions of land cover categories. It may be necessary to recode raster values so that they represent a modified range or more representative range of values, given an analysis objective. It may also be necessary that raster values need to be aggregated to form a smaller set of values. The reasons for needing to reclassify raster values vary and include:

1. Values within a raster may have been updated through additional data collection.
2. Numerical values are needed instead of current values that are described using categorical or nominal values.
3. A more detailed description of categorical raster values may be desired.

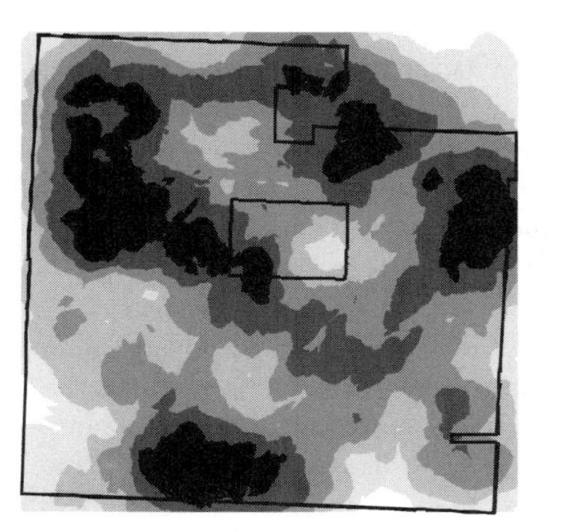

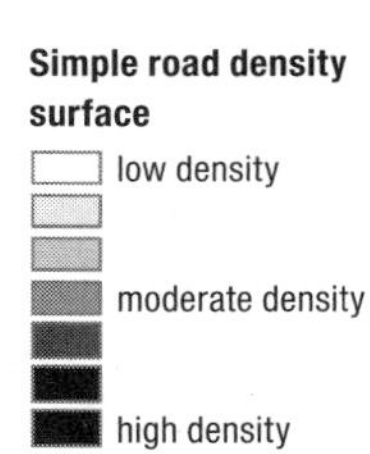

Figure 14.5 Simple density surface for roads in the Brown Tract using a 2,000 ft search radius.

4. Raster values may need to be rescaled in order to support a single raster analysis.
5. Raster values may need to be rescaled in order to support a multiple raster analysis.

Reclassification is different from altering the symbology or legend of a raster database for display purposes in that it goes beyond simply altering the appearance of a raster database. Regardless of the reasons for reclassifying a raster database, the reclassification process leads to a new raster database with pixel values reflecting the recoded values. The new raster database and its reclassified values can then be used for mapping and analysis purposes.

Within the ArcGIS Spatial Analyst, a reclassify command is available. Selecting the reclassify command will open a dialog box that prompts the selection of a raster database, reclassification field, and input cells for reclassification values. By default, this interface will use the existing symbolization used to display the raster. For example, should current raster values be displayed using five ranges or categories, these same five classifications would be displayed on the reclassify cell options. To select different classifications from which to begin the recoding, the 'classify' button can be chosen.

Raster Map Algebra

The ability to access and evaluate values within multiple raster databases and output results to a new raster database presents powerful analysis opportunities. Raster map algebra involves a mathematical evaluation that is applied to one or more raster databases to create a new database (DeMers, 2002). Mathematical evaluations for single raster databases may include adding or multiplying raster cells by a constant value, or applying trigonometric, logarithmic, and other mathematical transformations to raster values. A common example of single raster map algebra might be taking an elevation-based raster and multiplying the elevation values by a conversion factor to move from one measurement unit to another, such as converting elevation values from meters to feet. Mathematical evaluations for multiple raster databases might involve adding or multiplying raster values, comparing multiple raster databases and returning the largest value for all coincidental cells, or using raster database values within a formula that captures landscape processes such as that for water velocity and discharge rates.

Whether one or more raster databases are used, raster map algebra results in a new raster database that contains the results of the mathematical evaluations specified by the user. Some forms of map algebra can be accomplished through techniques discussed earlier, such as cellular statistics that allow for statistical summaries to be calculated from multiple raster databases. Operations that require mathematical manipulations of single or multiple raster databases, however, require more direct approaches. Within the Spatial Analyst function of ArcGIS, the raster calculator is provided in support of raster map algebra, and also provides access to additional raster related functions.

Database Structure Conversions

It may be necessary to convert a spatial database from a vector to raster or from a raster to vector data structure for a variety of reasons. Potential reasons include:

1. supporting a GIS process or analysis that only accommodates vector data,
2. supporting a GIS process or analysis that only accommodates raster data,
3. sharing data with a colleague who can only access one data structure type,
4. meeting the data requirements of a client or funding organization, and
5. database storage size considerations.

As your work with GIS continues, it is likely that you will have to convert spatial data from one structure to another. Regardless of which structure you are converting to, there are decisions that must be made for either transformation. The good news is that a second database is typically created through the conversion process. This usually ensures that if the transformation is unsuccessful, subsequent attempts can be made until a satisfactory product is created.

The creation of point, line, or polygon features is of interest in converting vector databases to raster. Usually an attribute field is selected as the information to be carried into the new raster database. Depending on the software you use, the output raster format might be integer or floating point. Integer and floating point are the two primary types of raster data formats supported within typical GIS software. A key distinction between these two types is that integer raster databases will usually accommodate multiple fields or variables, and can accommodate non-numeric data. Floating point raster databases are typically used for numeric data that includes precision, as evidenced by decimal values. Floating point raster databases are usually restricted to one attribute value, rather than multiple, being associated with each raster cell. In some cases, it may be possible to convert a floating point raster database to an integer format database, or vice versa, but you must consider what types and formats of data are intended for the final product.

In addition to the attribute field to carry into the new raster database, a raster cell resolution must also be selected. This is a critical choice and it will have a large influence on the specificity of the representation of vector features in the raster database and the scale at which subsequent analysis can occur (Mitchell, 2005). Equally important is that this choice will also impact the amount of digital storage space required by the resulting raster database. A smaller cell resolution will represent vector features more precisely, but will require more storage space. If a study area is large in size, and analysis goals are oriented more towards a landscape or regional scale, then a larger cell resolution may be an appropriate choice to consider. If a study area is small in size, or if more detail is desired in the representation of landscape features, then a smaller cell resolution should be chosen. Increasing a cell resolution by half, however, results in a four-fold increase in the number of raster cells. Keep in mind that many raster analysis processes will either require or depend on raster databases having the same cell resolution in order to produce reliable output results. **Raster resampling** involves changing the resolution of an existing raster layer and is a commonly used process among raster-based GIS users. However, decisions about how cell values are to be treated in the resampled product must be made. When the resampling of a raster database increases the spatial resolution, such as moving from a 30 m to a 10 m spatial resolution, the value for each resulting cell can be taken from the 'parent' cell directly. When the resampling of a raster database decreases spatial resolution, such as moving from a 30 m to a 1 km resolution, the value for each cell will require some statistical summation of input raster cells that contain numeric values. The statistical summation might be the average or highest value of input cells. For raster databases with categorical values, such as land cover or tree species, the choices will be more limited in how values are resampled. The value in the nearest or most completely coincident cell, or the most commonly occurring value in cells that will be combined in the output product, might be chosen.

In moving to a vector structure from a raster database, a raster value will need to be selected for the transformation product if the raster supports multiple fields. A key choice will be the vector feature type for the output database. Users will usually have to choose from point, line, or polygon feature types. The choice of one feature type over another will be a function of the input raster database and analysis objectives.

- Rasterization is the process of creating a raster database from a vector database.
- Vectorization is the process of creating a vector database from a raster database.

The spatial resolution of the raster database in either process may influence the quality of the resulting GIS database (Bettinger et al., 1996).

Getting Started with the ArcGIS Spatial Analyst

The Spatial Analyst extension software for use with ArcGIS must be purchased in addition to the base software, and will not work independently of the base software. The Spatial Analyst extension must be enabled within an open ArcGIS session in order for the extension software to work. As with all ArcGIS extensions, the Spatial Analyst is enabled by selecting the Tools menu, selecting the Extensions option, and selecting the check box next to the extension. The Spatial Analyst toolbar must then be enabled through either the Toolbar choice under the View menu, or by right clicking on an open location on one of the menus, then selecting the Spatial Analyst. Once the Spatial Analyst toolbar is available, session parameters can be established by accessing the options choice at the bottom of the Spatial Analyst menu. The general options allow you to set your working directory, where newly created raster databases will be saved, and also the parameters of an analysis mask. The analysis mask can be set to the spatial dimensions of another GIS layer and processing will only occur in coincident areas. The extent options, available in the second tab of the options dialog box, allow you to further customize the designation of areas considered for raster output. Choices include existing layers, spatial combinations of existing layers, or a bounding set of coordinates. Both vector and raster databases can be used. The cell size options are provided in the third tab of the options dialog box, and can be used to establish the resolution of output raster databases. Settings include other analysis layers, minimum or maximum area covered by all input databases, and user-specified resolutions. The number of columns and rows in the raster can also be chosen. Setting a cell size to match an existing layer will lead to spatial agreement between the layer and any output raster databases. This is a key choice that can help ensure consistent spatial registration of raster databases, an attribute that helps support reliable analysis output.

The default raster format within the Spatial Analyst extension is an ESRI grid. The grid data structure bears similarities to that of ESRI coverages in terms of transport, storage, referencing, and naming conventions. These conventions are not very forgiving but are manageable given a few ground rules. ESRI grids should be named using no more than 13 characters and should not start with a number, should not contain spaces, and should not use unusual characters such as an ampersand (&) or dollar symbol. These same rules should be considered for the naming conventions of folders under which grids are stored. An underscore can be used in place of a space. Just as with an ESRI coverage, an ESRI grid is actually composed of two folders and all information in both folders must be stored under the same directory. For this reason, an ArcInfo interchange file with an .e00 file extension is often used to transport ESRI grids and coverages, much as a common zip file format performs when it is used to compress and transport multiple files.

The majority of the raster analysis and processing functions described are available under the primary Spatial Analyst menu or within sub-menus. The previous chapter explored the majority of functions available in the surface analysis sub-group. We turn our attention to two examples that involve raster data processing and analysis functions.

Determining the Most Efficient Route to a Destination

Let's assume that the Brown Tract staff would like to take rocky fill material placed in one of the rock pits (described in the Brown Tract stands GIS database as stand 282) and transport it to another location near the southeast entrance of the Brown Tract boundary (Figure 14.6). The material will be used to create a new trailhead to accommodate the growing use within the forest. Multiple trips with a dump truck will be required and the staff would like to minimize the impact on the forest road system during the fill hauling process. A potential logical sequence for the GIS operations to support the shortest path creation is represented in Figure 14.7.

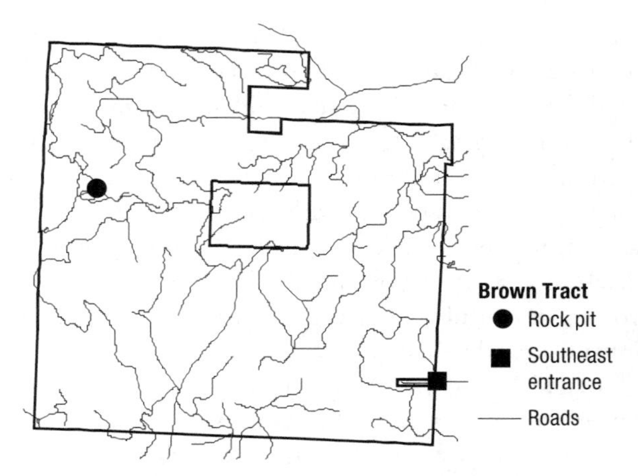

Figure 14.6 Rock pit, southeast entrance, and road system in the Brown Tract.

The roads database is in a vector data structure and describes the road surface type according to one of three possible values: paved, rocked, or dirt (unpaved). These values will be used to assign a cost to each of the Brown Tract roads. The identification of the most efficient route will assign costs of 1, 5, and 10 to the three road types, respectively. The slopes of the Brown Tract roads will also be considered in the analysis. Due to the relatively coarse resolution of the Brown Tract DEM (10 m), road slopes will be divided into three broad categories with cost values of 1 for mild slopes (< 5 per cent), 5 for moderate slopes (5–10 per cent), and 10 for greater slopes (> 10 per cent). Although the choice of a range between 1 and 10 is somewhat arbitrary, choosing values from a larger range will help differentiate possible routes more distinctly than will values from a smaller range. The two layers are then added to each other through raster map algebra to form a single database with cost values for each cell. Cost weighting and direction are then calculated for the Brown Tract road network in order to reach the rock pit. After the supporting databases have been developed a shortest path function can be used to identify a preferred transportation route for the fill material (Figure 14.8). Although the road system represented in the Brown Tract database is not substantially large in extent, many forest and other natural systems have extensive transportation networks.

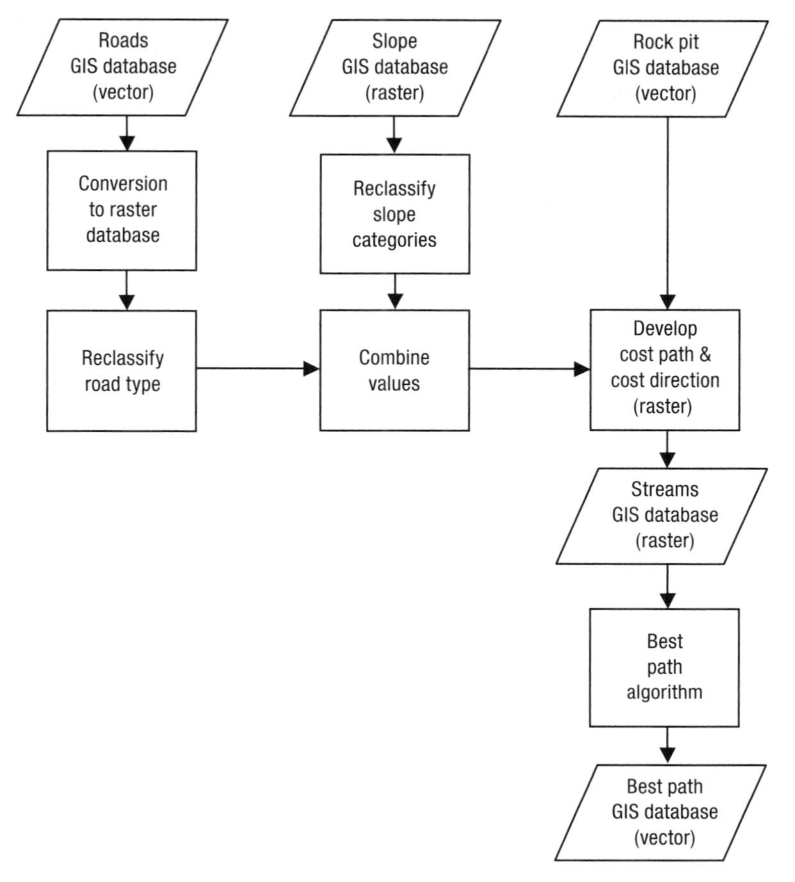

Figure 14.7 A general process to identify the shortest path between two locations on the Brown Tract.

In these situations, the number of potential routes can surpass the ability of transportation planners to systematically evaluate and select from a full range of options.

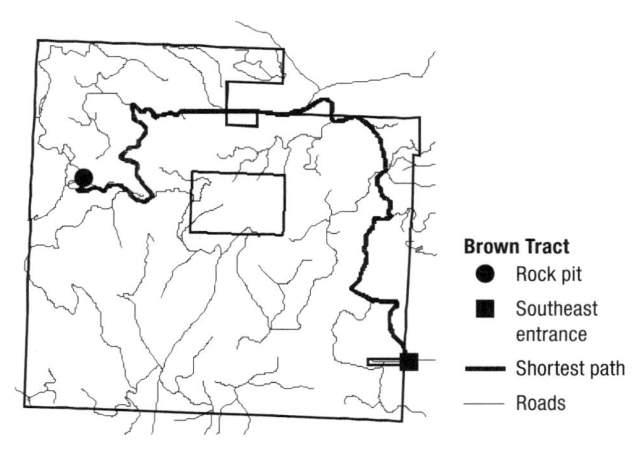

Figure 14.8 Shortest path between rock pit and southeast entrance of the Brown Tract, given cost weights for road surface and road slope.

Shortest path algorithms can assist planners and managers in making sound decisions.

Creating a Density Surface for the Number of Trees Per Acre

Density functions can be used to demonstrate the relative abundance or strength of the locations of features and attributes. The stands database for the Brown Tract contains an attribute named 'trees_acre'. This attribute has a relative weighting of the trees that you would expect to find in each stand within the Brown Tract on a per acre basis. It may be of interest to determine the areas in the forest where this field is strongest, indicating where higher numbers of trees are more likely to be found. Determining this information could be done through plotting the polygons and using shaded symbols to demonstrate intensity values of individual polygons. This approach, however, would neglect the influence of neighboring

In Depth

What is a centroid? A centroid is a coordinate pair that is intended to represent the mid-point of a feature or group of features. A centroid could be created to represent the center of a group of points by taking the average of all the longitude and latitude coordinates. In terms of a line feature, a centroid position is easily determined by dividing the total length of the line in half and using a coordinate pair to represent the half-way point. A polygon centroid can be more difficult to determine if the polygon shape is irregular or non-homogenous (is not round, square, rectangular, triangular, etc.), particularly if some or all of the boundary that make up a polygon contains curves, as is often the case when describing natural features. The centroid of such a polygon is determined through mathematical integration (calculus) with the goal of determining where the center of gravity of the polygon is located. The center of gravity can be thought of as the point at which the polygon would balance if set flat upon a pole.

polygons in its representation. A more helpful approach might be to create a density surface which would search surrounding areas and determine intensities that take into account the number of trees for each stand while considering the number of trees in neighboring stands.

A density surface must be created from a point or line feature type. In order to apply the density function to the stands layer in the Brown Tract, we'll need to convert the stands polygon feature type. A point representation is probably preferred over a line feature type for the converted stands. A common method for representing polygons as points is to calculate the centroid, or middle of a polygon's extent, which is determined geometrically if the shape is basic, such as that described by round, square, or rectangular features. For irregular polygon shapes, centroid determination must be accomplished with more rigorous mathematical techniques. Most GIS software systems will offer routines for centroid determination and can quickly create a centroid representation of a polygon or line feature with one output point created for each input feature. In addition, all of the attribute values will be carried into the point attribute table. Within the ArcGIS software, the ArcToolbox has a 'Feature to Point' conversion command that will accomplish this transformation. The XTools extension software, a popular low-cost ArcGIS extension program, has commands that also support centroid creation.

After the stand polygons have been converted to points, the density surface can be created. Figure 14.9 shows the result of a simple density surface based on the number of trees per acre. The darker shaded areas highlight the areas where greater numbers of trees would be expected. The search radius was set to 1,000 ft and density circles were created for each stand centroid to demonstrate the detected densities. Figure 14.10 shows the output that results for a smoothed density surface using the same stand centroids and a 1,000 ft search radius for trees per acre.

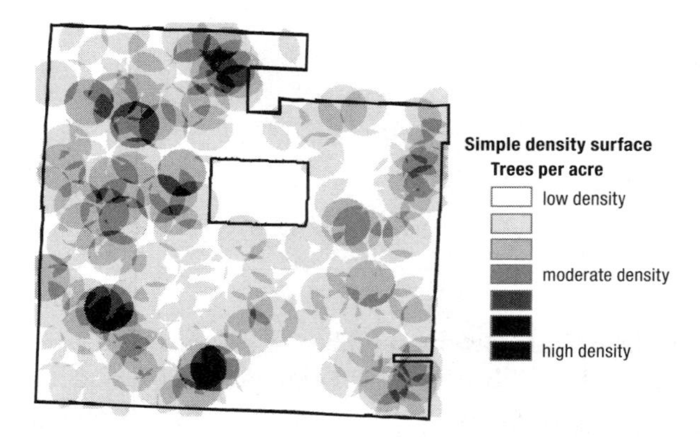

Figure 14.9 Simple density surface for number of trees per acre based on a 1,000 ft search radius.

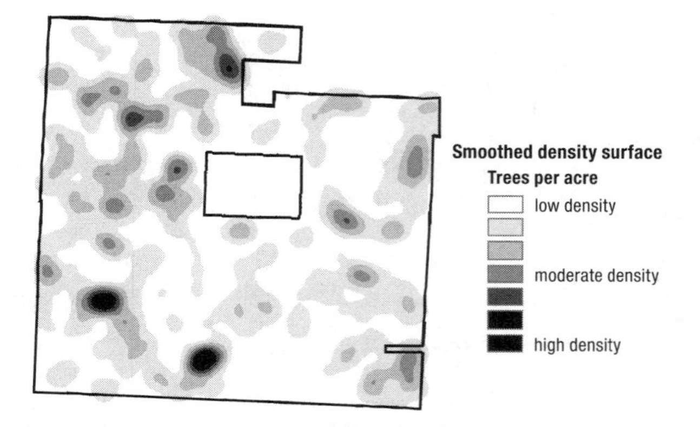

Figure 14.10 Smoothed density surface for number of trees per acre based on a 1,000 ft search radius.

Summary

We demonstrated in this chapter how raster databases can be manipulated and analyzed to solve questions related to natural resource applications. A host of functions are available to support analysis including distance, statistical search summary, and density functions. The functions differ in their application and in the types of database structures that can be used for analysis. Output may be tabular, vector, or raster depending on the function. Several common procedures within most raster-based software include raster reclassification, raster map algebra, and conversion routines between vector and raster databases. The procedures facilitate spatial analysis and support the ability to prepare spatial databases so that they are suited for specific analytical purposes. In addition, we presented two potential applications in which some of the functions and procedures discussed earlier in the chapter were applied. The raster analysis processes and examples we presented by no means represent the extent of the potential of raster analysis. Rather, these processes and functions describe some of the more useful and common commands and processes for natural resource analysis with raster data.

Applications

14.1 Straight line distance function for points. As part of your position as a natural resource manager, you manage the research areas on the Brown Tract. Concerned about their distribution across the landscape, you decide that a simple raster analysis might shed light on their spatial arrangement. Create a straight line distance raster database for research plot points contained in the Brown Tract.

14.2 Straight line distance function for lines. The density of stream systems can be used to define their character. Given the streams GIS database for the Brown Tract, create a straight line distance raster database for streams contained in the Brown Tract.

14.3 Straight line distance function for polygons. As another analysis related to the distribution of research areas, create a straight line distance raster database for the stand polygons in the Brown Tract where the LANDALLOC field is designated as 'Research'.

14.4 Allocation distance. As we mentioned earlier, in developing an allocation distance, raster cells are assigned a pixel value that recognize the nearest zone of influence, similar to the vector representation of a Thiessen polygon. Create an allocation raster database for research plot points contained in the Brown Tract.

14.5 Cell statistics. To further your understanding of the spatial distribution of research plots on the Brown Tract, you decide to embark on a series of raster analyses to determine their vertical distribution.

a) What is the average elevation of the research plots?
b) Which research plot (use the numbers in the field 'PLOT' to describe) has the highest elevation and what is the elevation?
c) Which research plot (use the numbers in the field 'PLOT' to describe) has the lowest elevation and what is the elevation?

14.6 Neighborhood statistics. Using the elevation layer for the Brown Tract as an analysis extent and a template for output cell resolution, what is the longitude and latitude of the center of the three by three grid cell neighborhood with the highest elevation?

14.7 Zonal statistics. Assume that a fire has started in stand 140 of the Brown Tract. If hand crews cannot control the fire, water must be acquired from nearby sources to help extinguish the flames. What is the average distance to the nearest water source for stand 140?

14.8 Zonal statistics. What is the average distance to the nearest water source for soils polygon 166? Use the SOILS_ field to determine where this stand is located within the soils database in the Brown Tract?

14.9 Density surface for basal area. Create both a simple and smoothed density surface for the Brown Tract stands. Base the density upon the basal area field contained with the stands attribute table.

14.10 Density surface for research plots. Create both a simple and smoothed density surface for the Brown Tract research plots. Base the density upon the field named 'SI' which contains a numeric value for the site index of each plot.

References

Bettinger, P., G.A. Bradshaw, & Weaver, G.W. (1996). Effects of geographic information system vector-raster-vector data conversion on landscape indices. *Canadian Journal of Forest Research, 26*, 1416–25.

Chang, K. (2002). *Introduction to geographic information systems*. New York: McGraw-Hill.

Chrisman, N. (1997). *Exploring geographic information systems*. New York: John Wiley & Sons, Inc.

DeMers, M.N. (2002). *GIS modeling in raster*. New York: John Wiley & Sons, Inc.

Mitchell, A. (2005). *The ERSI guide to GIS analysis. Volume 2: Spatial measurements and statistics*. Redlands, CA: ERSI Press.

Silverman, B.W. (1986). *Density estimation for statistics and data analysis*. New York: Chapman and Hall.

Theobald, D.M. (2003). *GIS concepts and ArcGIS methods*. Fort Collins, CO: Conservation Planning Technologies.

Wing, M.G., & Tynon, J.F. (2006). Crime mapping and spatial analysis in National Forests. *Journal of Forestry, 104*(6), 293–8.

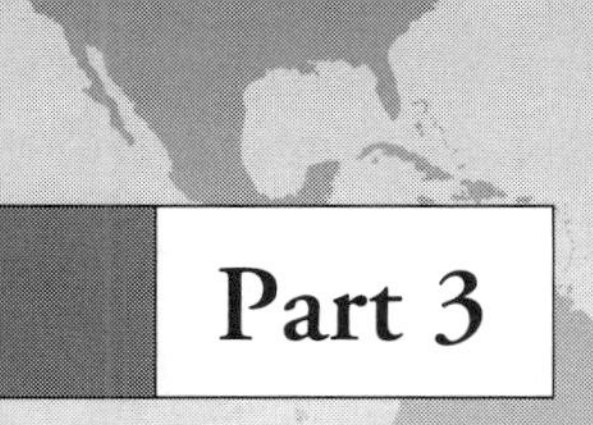

Part 3

Contemporary Issues in GIS

Geographic Information Systems: Applications in Natural Resources Management focused on the background and development of GIS in Part 1 and delved into GIS applications in Part 2. In Part 3, we try to provide a glimpse of where GIS use may be heading in the near future. Trying to look ahead and predict what may happen is a difficult task because both GIS-related technology and the society that surrounds it are changing rapidly. Nonetheless, in chapter 15 we discuss some trends that are associated with GIS in natural resource management. These trends are related to technological developments, the handling and sharing of spatial data, and the legal issues that may impact organizations that use GIS. Chapter 16 makes note of how the increased availability of GIS has transformed the delivery and structure of GIS operations in many organizations. Also important in chapter 16 is the discussion of possible barriers to successful GIS implementation and how implementation effectiveness can be assessed.

We also consider some of the current challenges within the GIS community in Part 3. The final chapter, chapter 17, examines the on-going and sometimes contentious discussion of how the GIS profession should be defined and recognized. There are a number of other established professions that are also involved with measuring and mapping features and there has been friction at times in agreeing on the capacity in which certain professions should apply GIS. Such discussions are evidence that GIS and GIS professionals have an important and necessary role in today's society. While many other professions have well-defined activities, competency standards, and governing bodies that describe and guide its members, the GIS community has only recently developed some initial pathways for certifying GIS competency. Criticism has been leveled toward these initial efforts on the grounds that sufficiently rigorous processes to establish competency have not been developed. The discussion of GIS competency arose initially from concerns voiced from the land surveying and engineering communities about the potential for GIS users to perform traditional surveying and measurement activities while not actually having professional license as a surveyor or engineer. The final chapter probes the issue of whether GIS users should be licensed, a hot topic of conversation particularly when you try to define what a 'professional' GIS user is and what sorts of activities that person is qualified to perform.

Chapter 15

Trends in GIS Technology

Objectives

GIS technology is constantly evolving, adapting, and changing according to the needs and capabilities of GIS users, particularly those within the field of natural resource management. While natural resource managers may only represent a portion of the total population of GIS users, natural resource management also benefits from the influence other fields (transportation, utility management, public planning, etc.) have on the evolution of GIS technology. This chapter provides a discussion of some of the current trends associated with GIS technology and use. When we initially developed this chapter in 2004 (Bettinger & Wing, 2004), we concluded that forecasting the direction and success of trends was challenging, but allowed you to consider what potential applications might exist within the area of natural resource management. As you will see, some of the trends in GIS technology have remained over the past three years, while others have recently appeared.

After considering the topics presented in this chapter, readers should have a reasonably firm understanding of a number of issues related to the trends in GIS technology and use. As a result, readers should be able to describe and debate the associated strengths and weaknesses of:

1. the common trends related to GIS technology, and how these might be applied in natural resource management,
2. the opportunities for strengthening GIS technology and applications within natural resource management organizations, and
3. the current and potential technological developments that might promote or hinder the advancement of GIS as an effective problem-solving tool.

Integrated Raster/Vector Software

For many years, GIS and other spatial software systems have been defined by their ability to work with either raster or vector data. In fact, many GIS software programs either restricted users to working with one data structure or the other, or allowed users to conduct analyses with one data structure and limited the use of the other data structure to rudimentary purposes (e.g., viewing only). Recently, however, almost all traditionally raster-based GIS software programs have begun to include algorithms and techniques to allow the capability of managing vector GIS databases, and similarly, almost all traditionally vector-based GIS software programs have begun to include algorithms and techniques to allow the capability of managing raster GIS databases (Faust, 1998).

The primary hindrances to providing the capability to use both data structures for spatial analyses were the marked differences between the two data structures and in particular, how they were stored. To further complicate matters, software manufacturers created their own proprietary formats for raster and vector data structures that were best suited to their product. In addition, each data structure could also be described by more than one format. The different structures (and formats of structures) overwhelmed the computing capabilities and software design efforts of earlier GIS software companies. As computer technology and software programming languages evolve, a totally integrated system, one that would be able to incorporate both vector and raster GIS data structures simultaneously in the spatial analysis of natural resource issues, is a trend in software development that will continue to drive the direction of future GIS software programs. For example, such a system would allow the use of vector GIS databases to assist in image classification, whereas previously only raster-based GIS databases would allow a system to perform GIS operations such as buffering, overlays, and proximity operations, with both raster and vector processes in a seamless and efficient manner. In a totally integrated GIS, processes such as vector-to-raster or raster-to-vector conversion (as described in chapter 3) and analyses that use both raster and vector data simultaneously (as described in chapter 13) would therefore be transparent to users of GIS (Faust, 1998).

Software such as ENVI (ITT Corporation, 2007) and Erdas Imagine (Leica Geosystems, LLC, 2007) not only provide a vast suite of raster-based analytical tools (e.g., image classification, terrain analysis), but also allow you to integrate vector data with raster data and perform the buffering, digitizing, and editing functions that were discussed earlier in this book. Erdas Imagine also allows you to clean and build the topology of vector GIS databases, which is useful when editing vector GIS data. Google Earth (Google, Inc., 2007) is a similar system but it may be more appropriately considered as a geospatial exploration program at this point in time. The Google Earth system has the ability to integrate vector and raster data to a limited extent, but its real value lies in allowing users to easily visualize landscapes through an Internet browser.

Linkage of GIS Databases with Auxiliary Digital Data

While we think of GIS as a system for displaying and manipulating geo-referenced maps and images, we have the ability in some GIS software programs to associate spatial data with other non-spatial data. Of course, data in an attribute table, or data from a non-spatial joined table can fall into this category as well, but what we refer to here is the association of an image that is not georeferenced with some spatially-referenced data. For example, in the field of urban forestry, you might capture the spatial position of trees within a city as a set of vector points. These points may be attributed with tree characteristics (species, height, etc.) and other local landscape variables. The points that represent the trees can also, in some GIS software programs, be attributed with a link to a picture and, when you select a point representing the tree, the picture of the tree is presented (Figure 15.1).

The linkage of GIS databases to this type of auxiliary data is generally made using a hyperlink. A hyperlink is a navigation element that allows, when selected, the viewing of the referenced information associated with the link. Hyperlinks are used widely on the Internet for navigation purposes, but they are not limited to Internet usage, obviously. They were designed as a way for you to link to specific portions of related documents without having to open each new document at its beginning and search for the desired page. Hyperlinking is a useful way to associate pictures, documents, videos, or any other relevant data to a mapped feature, thus allowing for a more comprehensive use of information systems. The urban tree example is but one of many logical and valuable uses of hyperlink-

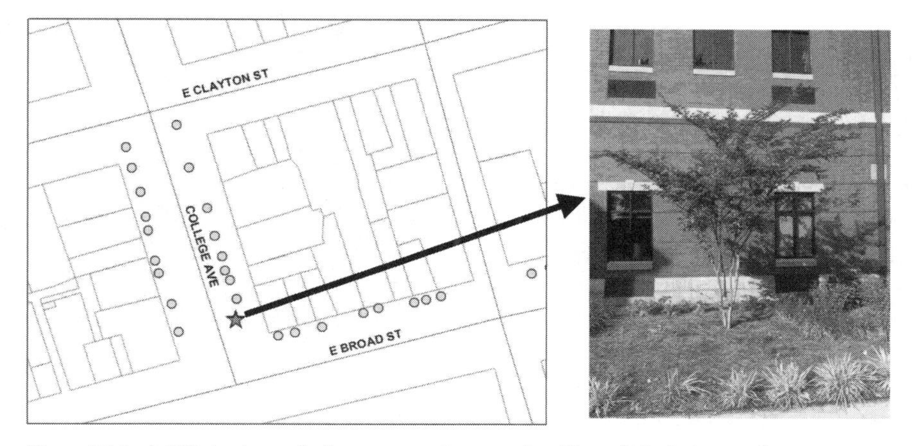

Figure 15.1 A GIS database of urban trees, and an associated hyperlinked picture of a tree (Courtesy of Andrew Saunders).

ing non-spatial data to GIS databases. Having the ability to view (with actual photos) oblique perspectives of landscapes from various vistas or overlooks represents another valuable use of hyperlinking data for natural resource management purposes.

High Resolution GIS Databases

New areas of research and development, called 'precision forestry' or 'precision agriculture', have recently been introduced in natural resource management. These areas of research and development seek to use digital technologies for improving, and making more efficient, natural resource management activities. 'Precision techniques' might include using GPS as a navigational aid for farm or forestry equipment, capturing remotely-sensed imagery to describe the status of soil properties (e.g., the need for fertilizer or pesticides), or using digital aerial photography to record crop plantings and outcomes. Precision agriculture techniques have been actively used and recognized as a discipline for at least a decade. In contrast, the first formal recognition of precision forestry occurred in June 2001 at the University of Washington's Precision Forestry Symposium. High among the list of goals of precision forestry is the identification of methods for the collection, analysis, and use of highly accurate and precise data from the Earth's surface to facilitate better management of natural resources. Examples of precision forestry techniques might include using electronic distance measuring (EDM) tools to capture the precise spatial position of forest landscape features, capturing precise and timely satellite imagery to assist in monitoring threats to forest health (e.g., fire or disease), or developing precise, fine-scale DEMs to identify steep forested areas susceptible to potential landslide activity.

The main obstacle to implementing precision forestry or agriculture techniques in natural resource management remains that of obtaining accurate, precise, and timely spatial data of landscapes. Some applications of precision technology may be implemented more easily in differing land uses, however. For example, in contrast to many agricultural applications, forests are characterized by a dense canopy cover and sometimes by mountainous terrain, which can limit the types of technologies that can be used to collect precise spatial data. A thick canopy cover, for instance, often hinders GPS reception, and hillsides can prevent satellite signals from reaching a GPS receiver. There are ways to avoid some of these problems, such as by collecting GPS data during the winter months when the tree canopy is least dense, or by scheduling GPS missions during periods when an increased number of GPS satellites will be available.

Spatial data collection technology continues to evolve, and natural resource managers are likely to see new techniques that improve upon present GPS, satellite imagery, and LiDAR data collection methods in the upcoming years. One of the most promising sectors of improved data collection technology is related to high spatial resolution GIS data. GIS databases developed from the rasterization of color aerial photography and developed from satellites such as IKONOS™ (GeoEye, 2007) are becoming available at 1 m to 4 m spatial resolutions (Figure 15.2). While geo-registered color aerial photography of large land areas can now be collected, processed, and made available to clients the following day, acquiring satellite imagery at 1 m resolution requires a longer time period (generally 10 days or more) and depends on the area and time frame of interest. Another promising area of improvement in data is high spectral resolution raster databases. Normally, aerial photographs that are converted into digital orthophotographs cover a 0.4 to 0.9 micrometer range of the electromagnetic spectrum. Some satellites systems capture energy in longer wavelengths, but usually within 10 distinct bands (ranges of energy) or less. Higher spectral resolution data implies

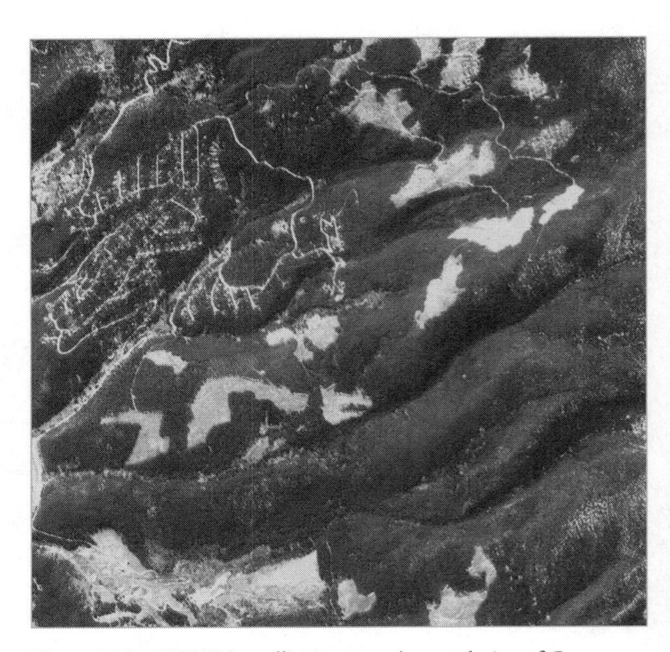

Figure 15.2 IKONOS satellite image at 4 m resolution of Copper Mountain located in the Colorado Rocky Mountains (Images courtesy of GeoEye).

that many more bands of energy have been captured, and can be used to monitor and evaluate the Earth's surface. AVIRIS (National Aeronautics and Space Administration, 2007) is an example of one such high spectral resolution system, although it has been available for almost 15 years. With this system, 224 bands of data can be captured for a single landscape at one point in time, allowing scientists and managers to use the appropriate spectral reflectances for analyzing various natural resource management issues.

Still among the primary concerns of most natural resource management organizations is the cost related to the acquisition of high-resolution spatial data. While initially relatively high, as new technologies and data sources become available the cost and availability of high-resolution GIS databases might be as low as $0.03 per acre. Another issue of concern is the data storage requirements. High-resolution images can require massive amounts of computer hard drive space. Although large hard drives (100 GB and above) have become the norm, storage space can be filled quickly as images are gathered and stored digitally.

Managed and integrated properly, high-resolution GIS databases will help facilitate the creation and maintenance of databases that have tremendous potential for organizations that manage large land areas. These GIS databases can be associated with methods that keep land cover information current, and allow you to conduct temporal analysis of land cover conditions. Two of the challenges to using high-resolution GIS databases will be in deciding how often to acquire new data, and how to integrate new GIS databases with existing GIS databases. These challenges are markedly different from those experienced in the recent past, where most natural resource management organizations struggled due to a lack of data. Now, the amount of data available to an organization can become overwhelming. The high resolution of these GIS databases will also help provide a more accurate representation of natural resources, which has typically been one of the drawbacks of using raster GIS databases (Faust, 1998).

Distribution of GIS Capabilities to Field Offices

There are a number of reasons why the use of GIS has, and continues to, spread from a centralized organizational office to field offices: more and more people are becoming comfortable using GIS, colleges and universities are educating students in the use and application of GIS in natural resources, and natural resource organizations are recognizing that more timely analysis and map products can be obtained if the work is more closely situated to the end user (Bettinger, 1999; Wing & Bettinger, 2003). In addition, the increased power of computer technologies (speed and memory) and the advancements made in the operating systems of personal computers have both allowed a wider user base to use GIS technology (Faust, 1998). Since computer systems are relatively inexpensive (a GIS workstation can now be purchased for under $2,500), and GIS software has been developed with end users (e.g., field personnel) in mind, the trend is toward a GIS system where, in larger organizations, data development and maintenance tasks are performed at a central office, and data analysis and map production tasks are performed at remote field offices. In smaller organizations, the distribution of processes may be less clear, because the distinction between a central office and field offices may be blurred (or non-existent).

In some organizations, approved personnel who work in the field perform the data maintenance tasks. Changes to GIS databases can be made at the field office, sent electronically to the central office for verification and integration, and eventually passed back to field offices. In systems such as these, only one person can be making changes at each point in time. As a result, data being edited is 'checked out' (like a library book) until the editing has been completed. The transfer of updated information to the field offices would ideally be instantaneous, but it is not. A delay of 15–30 minutes is required for centralized systems to complete their tasks.

The benefits of a distributed GIS system are aimed at enhancing local or field office productivity and decision-making. Two of the main benefits include a more timely response to analysis and map production needs of field offices, and a decreased work load on a centralized GIS office (allowing more time and effort to be devoted to GIS database quality and maintenance). Within a distributed GIS system, clearer channels of communication should exist, since generally speaking, the customers (those requesting maps or analysis) and suppliers (those performing the analysis or making maps) are in the same office (or perhaps are the same person). This face-to-face communication is often more effective in meeting the goals of a map or analysis request than communication processes that rely on e-mail or phone calls. In addition, field personnel involved in GIS analysis and map production are likely to feel as if they have a greater investment in the GIS program, and perhaps will develop a greater sense of responsibility for maintaining accurate GIS databases (Bettinger, 1999).

Given the ongoing technological (software and hardware) advancements related to GIS, and the proliferation of GIS training, it is highly likely that the distributed GIS model of capabilities will continue to grow, and become more prevalent than the centralized model. At some point, the distributed model may replace the centralized model completely in many organizations. The challenge in managing this paradigm shift will be to ensure that organizational protocols and monitoring are in place to protect distributed users from using spatial data improperly.

Web-based Geographic Information Systems

The widespread use of Google Earth has suggested to many that GIS can be both affordable and easy to use across the Internet. Some organizations have recognized the need to provide GIS data management services over the Internet as a way to more rapidly update databases and provide information to users in the field. Ideally, a natural resource organization would maintain a system where data being updated by an employee in a field office can be 'checked out' over the Internet and managed (updated or modified). While the data is being modified remotely, other users would have the ability to view the data, but not simultaneously modify the data. Once the data has been checked back in, other users in the organization can then use the modified data. Overall, this type of process has the potential for substantially reducing the time required to traditionally update GIS databases (see chapter 10). However, there may be some data quality issues related to remotely-performed modifications of databases that are not consistent with organizational standards.

As an example of a system such as this, the Virginia Department of Forestry has recently implemented the Integrated Forest Resource Information System (IFRIS). The benefits of the system include reduced paper-related processes (e.g., transfer of maps and data to a centralized office), and an empowerment of people to adopt and use new technology at the field level. Users in field offices can update GIS databases using heads-up digitizing, and submit data and reports on progress in managing the state's forests. Future endeavors include extending the real-time capability to hand-held data recorders equipped with GPS technology to allow an immediate data capture and update to occur, which can increase the efficiency of mapping and reporting wildfires, water quality problems, and insect and disease outbreaks. Although the intent of the system is to make administrative functions of state land managers more efficient, as with the previous discussion, there may arise some data quality issues related to remotely-performed modifications of databases that are not consistent with organizational standards. Only time will tell whether these advances will result in cost- and time-savings and in increased productivity with no substantial reduction in data quality.

Data Retrieval via the Internet

As we discussed in chapter 3, the Internet is becoming a common source for acquiring GIS databases, GIS metadata, and other information regarding the acquisition of GIS databases. In fact, the current popularity and prevalence of GIS can at least be partly attributed to the Internet. As public agencies began to produce and make GIS databases available, customers who wanted the data were often required to pay for the storage medium (tape, CD, etc.) and for the time required to place the GIS database(s) on the medium. Since the media had to be mailed to the customer, this process also required a period of several days (to weeks) before the user could actually use the GIS databases. Presently, most public agencies offer their

In Depth

Many organizations have developed their own **Intranet**. These are private, networked computer environments in which only members within the organization have access. Intranets look and act like the Internet, and can range in complexity from the very simple (a set of folders or subdirectories containing data or other organization information) to the sophisticated (those that offer graphical interfaces for access to data, software, or links to other organizational services). Each of these services is provided through an internal (to the organization) website. Intranets are a method that can facilitate a distributed GIS system within an organization.

non-sensitive GIS databases over the Internet, allowing GIS users the opportunity to quickly acquire the data at no cost. In cases where GIS databases are very large and therefore not practical for Internet transfer (for example, as with DOQs or other high-resolution raster imagery), public agencies may still require that consumers pay for data transfer costs. However, data compression software technology continues to improve, which reduces the need for non-Internet transfers. Private organizations that market and sell GIS databases also allow customers to download the databases from the Internet. In these cases, customers usually need to be registered with the private organization because access to the data is restricted.

Portable Devices to Capture, Display, and Update GIS Data

In the past decade the use of hand-held data collectors and personal digital assistants (PDAs) has become quite common for collecting forest inventory data and other attributes of landscape features. GPS receivers, in fact, use hand-held data collectors to allow you to capture the spatial attributes of landscape features. Integration of the two philosophies, allowing you to collect spatial locational information about landscape features and to collect attribute data, results in positive benefits to a natural resource management organization. Traditionally, data collected for natural resource inventories would be recorded in a field notebook or on a map, and would require manual data entry into a spreadsheet or GIS database (through attributing spatial landscape features). This manual process is time consuming and presents several opportunities for human error to be introduced into a GIS database. The integration of digital technologies allows information to be recorded in a computer database while a person is in the field. Hand-held data collectors and PDAs are able to connect to wired or wireless computer systems, allowing the data to be transferred to GIS. This greatly expedites the transfer of field-collected data to a GIS database where the data can be analyzed and mapped. This process also removes some of the error opportunities that might occur through the manual coding and inputting of data. Data collectors also offer users the ability to examine maps and images of landscape features as they are being measured. Field personnel can use this heads-up display to visually determine whether their measurements are in agreement with the landscape features being measured. Digital orthophotoquads or digital raster graphics are two common GIS databases that can also be used to visually check measurements, as well as to facilitate a traverse of the landscape.

Hand-held data collectors are moderately expensive ($1,000 to $5,000), depending on the quality of the instrument and the functions they allow. PDAs are less expensive (around $500), and can be used as data collectors, but they are generally less rugged and more prone to damage from environmental factors (e.g., rain) and human error (e.g., dropping the device). Some GIS consultants have developed software that will run on PDAs. ArcPad (Environmental Systems Research Institute, Inc., 2006) is perhaps the most widely known product. A growing list of accessories can also be purchased to make PDAs more durable and useful in inclement weather and under other conditions.

Standards for the Exchange of GIS Databases

The development and use of standards for exchanging GIS databases may seem like a trivial exercise for governmental employees and university researchers, since theoretically data transferred among US federal government organizations (for example) must adhere to federal data standards (http://www.fgdc.gov/standards). These standards specify data formats that are intended to facilitate the sharing of spatial data among organizations. Many university researchers also utilize this protocol (or something very similar) in some cases because they interact with federal granting agencies during the course of research. However, most private natural resource management organizations are not bound by these data standards. Thus acquisition and modification of GIS databases by private natural resource management organizations proceeds undocumented; transformations and re-projections regularly occur to allow an integration of the acquired GIS databases into the organization's system, since the type and format of data exchanged can vary considerably (Figure 15.3). Moving to a standard data exchange format usually suggests that one of two organizational policies will be used: (1) organizations convert all of the GIS databases currently in use to a standard format, thus avoiding the need to convert GIS databases when data exchange processes occur, or (2) organizations convert GIS databases to a standard exchange format only when data exchange processes occur. There is a cost associated with both policies, and it is a function of how often

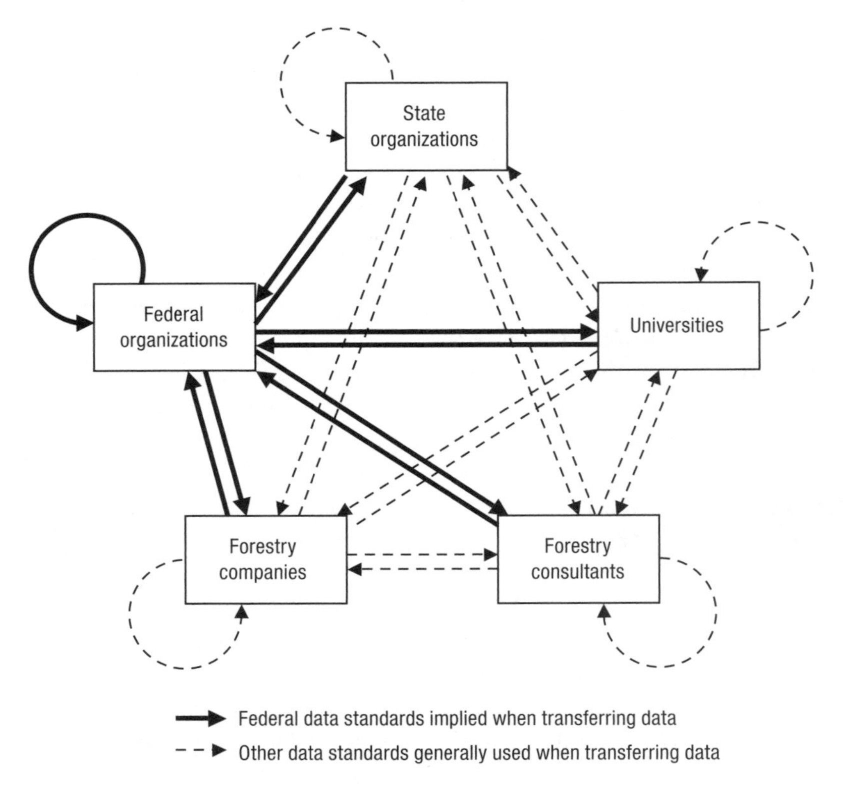

Figure 15.3 Database transfer and implied standards among organizations.

an organization perceives that it might exchange data with other organizations. If an organization positions its internal data standard closely to that of the data exchange standard, the cost will be minimized when acquiring GIS databases from, or sharing GIS databases with, federal agencies. If an organization does not plan to acquire GIS databases from, or share them with, federal agencies, the cost of deviating from the federal standard may be minimal.

Even if a natural resource management organization avoids implementing a data exchange standard a few common data exchange formats are used by other organizations. For example, it is not uncommon for public agencies to make GIS databases available in both ArcInfo export file format (e00) and ArcView shapefile format. Among computer-assisted design (CAD) software programs, the DXF (drawing exchange format) format is commonly used to exchange files. Most GIS software manufacturers recognize that users will need to accommodate data formats designed by other software vendors. For this reason, it is not unusual for a GIS software program to make available conversion and integration processes that make it possible to view other GIS database formats, particularly those that are used by the most common GIS software programs.

Legal Issues Related to GIS

Legal issues confront the GIS community on several fronts including issues related to privacy, liability, accessibility, and licensing. Some of these issues are relatively new, while others have been associated with GIS since its inception. In either case, the issues will continue to evolve as GIS software becomes more widely used. Licensing and certification of GIS professionals is an issue of current concern to many GIS users and other professionals, and will be discussed in more detail in chapter 17. Therefore issues related to privacy, liability, and accessibility are presented here.

GIS data is being collected at an ever-increasing pace, and used in novel ways as people begin to understand the power of connecting information to spatial position. For example, some organizations now rely on the ability to relate data about purchasing decisions with demographic and location information (Clarke, 2001). This information is used by businesses to direct mass mailings, to suggest the location of new facilities, and to place phone calls to potential customers in the evening to inquire whether they may be interested a purchasing a product or service. Although federal and state legislation exists to protect the privacy of information collected from individuals by public organizations, very little legislation currently exists to prevent non-public organizations from selling or sharing the information that is gathered during regular consumer transactions. GIS has thus enabled organizations to cultivate business using spatial analyses. In this way, GIS has become a business tool, and like it or not, an effective one. As private organizations continue to forge new ground in the collection, sale, and exchange of spatial data that describes the economic and social behavior of individuals, society will be challenged in establishing the laws and regulations that relate to privacy standards.

In the US, the federal government and state agencies have spent millions of dollars of public funds collecting and processing spatial data. The Freedom of Information Act (FOIA) was authorized in 1966 to grant taxpayers the right to access information related to the functioning of the government (Korte, 1997). Certain types of information, such as that related to security and law enforcement investigations, among others, are exempt from the FOIA. Other types of information, however, must be provided to the person making the request, usually at some minimal cost to cover the cost of processing the data and providing the media upon which the data is exchanged. Most states in the US have developed laws based on the FOIA that also require state governmental agencies to make government information available. Unfortunately, new threats to public safety and national security have emerged in recent years, and have necessitated a closer scrutiny of the types of government information made available to the public. More than likely, access to certain GIS databases describing such landscape features as water supplies and power facilities will be curtailed in the future, and access to other GIS databases will be delayed due to new security protocols.

Legal liability issues are associated with circumstances where a service or product provided by a producer is not satisfactory to the customer receiving the service or product. Onsrud (1999) identifies two types of liability that are pertinent to GIS: contractual and tort liability. Contractual liability issues arise when a contract between two parties has been breached. For a private organization that provides GIS products or services, this might involve a software product not behaving as advertised or a GIS database that does not adhere to a data accuracy standard. Tort liability issues arise when a party (person or organization) becomes injured (sustains a physical injury, loses business, etc.) as the result of another party's actions or products. An example might be an accident at sea as a result of using an inaccurate GIS database for navigation.

Private organizations that provide GIS products and services are responsible for adhering to, and demonstrating, a level of competency associated with their discipline. When others are injured as the result of incompetence, the organization providing the service or product may be liable for damages. In determining incompetence or negligence, a private organization may be responsible for producing inaccurate or insufficient data, but this alone does not prove incompetence. Rather, the courts have sought to establish incompetence by comparing services or products to those that would be produced from an organization that is acting 'reasonably' (Onsrud, 1999). Government agencies have typically been immune to litigation or responsibility for providing inaccurate spatial data due to sovereign immunity. An exception is made among agencies that produce goods or services that are considered discretionary. Discretionary services or products have resulted in government agencies being held liable for damages (Korte, 1997).

Both private and public organizations that are involved in providing GIS products and services can act to limit their liability risk. One method for limiting risk is to include information or disclaimers that accompany a product in order to describe its intended use, data accuracy, data reliability, and a warning that there may be errors in the data (as described in chapter 4). Organizations can further protect themselves by ensuring that all relevant parties have signed a clearly defined contract for products and services, and that the organization performs the specifics of the contract competently. If project requirements necessitate actions (for example, the development of other products or services) other than what is contained in the original contract, the organization providing the products or services should contact the other parties involved immediately to reach agreement on the specifics of the additional products and services (costs, time frame, etc.) before beginning to develop those products and services (Beardslee, 2002).

Licensing of GIS data products is another legal issue, and is tied directly to digital rights management (Cary, 2006). Many organizations require payment for use of data and software that they produced and, without payment, organizations using the data risk violating the terms of agreements that may have been implicit when the information was shared. While the appropriate model for licensing GIS data is currently being debated, the problem lies with the ease of copying digital data and sharing it with others in the absence of an agreement (similar to sharing music files). New types of data sharing arrangements will likely be formulated that are based on limited data sharing licenses. Cary (2006) suggests a system where users of GIS are granted access to certain GIS data based on the locations of actual landscape features or proximity to other features. This type of proximal data sharing could balance the need for openness (as desired by the user) with the need for confidentiality (as desired by the producer).

GIS Interoperability and Open Internet Access

Interoperability in terms of GIS refers to ability of different geospatial instruments, databases, and techniques to work together on applications. Interoperability involves creating standard terminology, data formats, and software interfaces that are both recognized and used by organizations involved in geospatial applications. The need for interoperability should not be surprising for any discipline that becomes popular among a wide number of potential users, such as has been witnessed by the rapid growth of GIS applications over the past two decades. The rapid growth of GIS gave rise to a number of GIS software interfaces and data formats that were proprietary and therefore not designed so that others could easily and freely access and exchange data with the proprietary formats. This inability led to frustration among GIS users and gave rise to the need for GIS interoperability. The Open Geospatial Consortium was founded in 1994 (Open Geospatial Consortium, Inc., 2007) and has 341 member organizations as of 2007. The OGC represents a coalition of both private and public organizations. The goals of the OGC are to promote public accessibility to geoprocessing tools and other location-based services. Significant accomplishments of the OGC include the standardization of terms for GIS features (points, polylines, polygons), the creation of the Geography Markup Language (GML) that provides an open source language for describing spatial data, and the development of standards for how geographic data can be requested and accessed from Internet servers (Longley et al., 2005). The OGC has had a profound influence on making geospatial tools and services available to Internet users and continues to work on promoting geospatial software accessibility. The OGC will likely continue to develop innovations that increase public access to location-based information services.

GIS Education

Attention is likely to increase around the methods and approaches instructors use to teach geospatial skills to students, not only at the university level, but also in high school and within the professional workforce. GIS capabilities are now essential for natural resource organizations, as well as for other disciplines throughout society (Wing & Sessions, 2007). The primary training ground for GIS skills is currently within the university system but geospatial skills are also being taught to students during elementary school years. In addition, training opportunities for in-career professionals appear to be growing.

Many disciplines, such as forestry, engineering, and surveying have accreditation bodies that review and appraise the curriculums of universities and colleges that offer related degrees. The accreditation bodies either approve the curriculum or report why it cannot be accredited and what steps are needed to gain accreditation. The accreditation process helps ensure that instruction supporting the necessary knowledge and skills to enter into professional disciplines is being delivered appropriately. The accreditation process helps educational programs become recognized and can be a significant incentive for drawing students. No such accreditation process exists specifically for geospatial technology instruction per se, although some engineering and surveying programs focus heavily on measurements and have accreditation through the American Board of Engineering Technology (ABET). Currently, a wide variety of methods and approaches are used to teach geospatial skills. This results in considerable variation in achieved learning outcomes (Longley et al., 2005). A recent publication titled *The Geographic Information Science and Technology Body of Knowledge* (DiBiase et al., 2006) has attempted to define critical concepts and skills that relate to geographic information science and technology. Written through extensive collaboration among GIScience researchers and educators, this work represents an initial attempt to provide a unified description of topics important for establishing geospatial skill competency. A second edition is planned that will provide additional detail and instruction that supports key concepts and skills.

Summary

GIS technology is evolving almost as quickly as general computer systems (hardware and software) evolve. People are thinking about natural resource management issues in ways unimagined just a few years ago, and spatial data is facilitating these efforts. Natural resource management organizations are actively engaged in testing and implementing new tools for collecting and analyzing spatial data. Organizations are also making GIS technology available to a large portion of their workforce, and the Internet has created an efficient avenue to make GIS databases available in a timely manner. These developments have had a positive effect on the acceptance and use of GIS, and encourage further GIS technology development. As GIS technology and use evolves, however, other issues (privacy, accessibility, liability, etc.) arise that must be addressed. These issues may require a close examination of policies and practices related to the use of GIS in natural resource management.

Applications

14.1 Local regulations regarding GIS distribution. Select an agency in your area (state, province, city, county, township, etc.) that uses and might distribute GIS databases.

a) What types of GIS databases are publicly available?
b) Are the GIS databases available for download over the Internet?
c) What data exchange formats are available?
d) Is there a cost related to acquiring the GIS databases?
e) Are there specific laws that relate to GIS database distribution, and how do they affect your ability to acquire GIS data from this agency?

14.2 Product liability (1). You work for a private organization that makes GIS databases available to customers. How might you help protect your organization from liability that could arise from customers that purchase your services and products?

14.3 Product liability (2). You work for a public agency that makes spatial data available to customers. How might you help protect your organization from liability that could arise from customers that use your services and products?

14.4 Precision forestry and agriculture. You work for the Bureau of Land Management (BLM) in central Colorado, and your supervisor, Mark Miller, has recently learned of precision forestry and agriculture. He has asked you how precision techniques might benefit the management of BLM land in Colorado. Write a brief memo that outlines the GIS databases, hardware, and software related to precision techniques, and how they might benefit the management of BLM land in Colorado.

14.5 Distributed GIS. A timber company in the southeastern United States has just hired you as a field forester. You are eager to use the GIS skills you have learned in college to help yourself (and others) make informed forest management decisions. The timber company has a centralized GIS department and five remote field offices and they are in the midst of developing a system whereby personnel (foresters, biologists, hydrologists, etc.) in field offices (where you are located) can use desktop GIS software to make their own maps and perform their own analyses. What can you do to ensure that the distribution of responsibilities (map development and analysis) to your field office will be successful?

References

Beardslee, D.E. (2002). Do it! Or, call them before they call you. *Professional Surveyor, 22*(2), 20.

Bettinger, P. (1999). Distributing GIS capabilities to forestry field offices: Benefits and challenges. *Journal of Forestry, 97*(6), 22–6.

Bettinger, P., & Wing, M.G. (2004). *Geographic information systems: Applications in forestry and natural resources management.* New York: McGraw-Hill, Inc.

Cary, T. (2006). Geospatial digital rights management. *Geospatial Solutions, 16*(3), 18–21.

Clarke, K.C. (2001). *Getting started with geographic information systems* (3rd ed.). Englewood Cliffs, NJ: Prentice-Hall, Inc.

DiBiase, D., DeMers, M., Johnson, A., Kemp, K., Luck, A., Plewe, B., & Wentz, E. (Eds.). (2006). *The geographic information science and technology body of knowledge.* Washington, DC: University Consortium for Geographic Information Science, Association of American Geographers.

Environmental Systems Research Institute, Inc. (2006). *ArcPad - Mobile GIS software for field mapping applications.* Redlands, CA: Environmental Systems Research Institute, Inc. Retrieved May 8, 2007, from http://www.esri.com/software/arcgis/arcpad/index.html.

Faust, N. (1998). Raster based GIS. In T.W. Foresman (Ed.), *The history of geographic information systems: Perspectives from the pioneers* (pp. 59–72). Upper Saddle River, NJ: Prentice-Hall, Inc.

GeoEye. (2007). *GeoEye imagery products: IKONOS.* Dulles, VA: GeoEye. Retrieved May 9, 2007, from http://www.geoeye.com/products/imagery/ikonos/default.htm.

Google, Inc. (2007). *Google Earth.* Mountain View, CA: Google, Inc. Retrieved May 9, 2007, from http://earth.google.com.

ITT Corporation. (2007). *ENVI—The remote sensing exploitation platform.* Boulder, CO: ITT Corporation. Retrieved May 9, 2007, from http://www.ittvis.com/envi/.

Korte, G.B. (1997). *The GIS book* (4th ed.). Santa Fe, NM: OnWord Press.

Leica Geosystems, LLC. (2007). *Erdas Imagine.* Norcross, GA: Leica Geosystems LLC. Retrieved May 9, 2007, from http://gi.leica-geosystems.com/LGISub1x33x0.aspx.

Longley, P.A., Goodchild, M.F., Maguire, D.J., & Rhind, D.W. (2005). *Geographic information systems and science* (2nd ed.). Chichester, England: John Wiley & Sons, Inc.

National Aeronautics and Space Administration. (2007). *AVIRIS: Airborne visible / infrared imaging spectrometer.* Pasadena, CA: National Aeronautics and Space Administration, Jet Propulsion Laboratory, California Institute of Technology. Retrieved May 9, 2007, from http://aviris.jpl.nasa.gov.

Onsrud, H.J. (1999). Liability in the use of GIS and geographical datasets. In P.A. Longley, M. F. Goodchild, D.J. Maguire, and D.W. Rhind (Eds.), *Geographical Information Systems, Volume II* (2nd ed.) (pp. 643–52). New York: John Wiley & Sons, Inc.

Open Geospatial Consortium, Inc. (OGC). (2007). Wayland, MA: Open Geospatial Consortium, Inc. Retrieved May 21, 2007, from http://www.opengeospatial.org.

Wing, M.G., & Bettinger, P. (2003). GIS: An updated primer on a powerful management tool. *Journal of Forestry, 101*(4), 4–8.

Wing, M.G., & Sessions, J. (2007). Geospatial technology education. *Journal of Forestry, 105*(4), 173–8.

Chapter 16

Institutional Challenges and Opportunities Related to GIS

Objectives

This chapter offers some thoughts on the integration and use of GIS within natural resource management organizations. There are a number of issues related to the successful development of a GIS program within an organization, and programs in many organizations are continuously evolving as people, technology, and organizational structures change. At the conclusion of this chapter, readers should have an understanding of a number of issues that relate to the challenges facing the implementation and use of GIS in natural resource management, including:

1. an understanding of the potential challenges that are ahead for successful and efficient GIS applications within and among natural resource management organizations;
2. an understanding of the challenges that exist for organizations thinking of distributing GIS capabilities to field offices, a move becoming more prevalent as recent natural resource graduates likely will have GIS experience in coursework, and exposure or training in the field; and
3. an understanding of how to assess the benefit of using GIS, a measurement process that will likely be necessary to develop more efficient business operations, higher quality products, and more timely management decisions.

The development and use of GIS in natural resource management has evolved considerably in the past 20 years as field foresters, biologists, hydrologists, and other natural resource professionals have become empowered with this technology, and are able to perform many of their own spatial analyses and map production tasks (Wing & Bettinger, 2003). In addition, it is increasingly likely that natural resource professionals will have coursework involving GIS and related geospatial technologies during undergraduate or graduate studies (Wing & Sessions, 2007). While GIS technology continues to evolve, natural resource management organizations are faced with several lingering issues related to the GIS use, and several new challenges have arisen with the availability of GIS technology to field offices. Natoli et al. (2001) describe a set of challenges for GIS implementation and use within (mainly) municipal organizations, and Bettinger (1999) describes some of the challenges associated with implementing a distributed GIS system in forestry organizations. This chapter extends the discussion presented by these two sets of work, and adds to it some additional points that are relevant to forestry and natural resource management.

Sharing GIS Databases with Other Natural Resource Organizations

As mentioned in chapter 15, one aspect of the use of GIS in natural resource management is the notion that some GIS databases can be shared among other natural resource organizations. The potential to collaborate with other organizations, and thus the need to share GIS data, represents an interesting dynamic that is evolving in natural resource management. Public organizations, such as the USDA Forest Service and the USDI Bureau of Land Management, provide a wide variety of GIS databases to the public at no cost; as we discussed in chapter 3, many of these databases can be accessed over the Internet. State-level natural resource management organizations generally provide a more narrow range of GIS databases to the public, if they provide any at all. State-level GIS database clearinghouses, on the other hand, generally provide a wide variety of GIS databases (DEMs, digital orthophotographs, land cover, etc.) to the public at no (or low) cost. Private natural resource management organizations generally treat their GIS databases as proprietary—that is, they are not freely available to the general public (anyone outside of the organization). While the original source of the data for many of the private organization GIS databases may have been a public organization, once the private investment has been made in maintaining and updating the data, concern arises about how the cost can be recouped, and whether competitors may gain insight into the management practices being used (e.g., management costs, productive capacity), thus gaining a competitive edge in the marketplace. Although some may argue that organizations, public and private, should share data so that more informed decisions can be made for a landscape, the goals and objectives of each organization will guide the development of policies related to data sharing.

As a result, many private natural resource organizations do not share GIS databases with those outside of their organization, unless they can place a value on doing so. There are a number of cases, for example, of private organizations sharing GIS databases through cooperative planning and management efforts. For example, in Washington State, the Department of Natural Resources (WADNR) has taken an ownership role in the development of several state-level databases: roads, streams, and the public land survey to name a few. The WADNR has developed protocols whereby both public and private natural resource organizations can share their GIS data with the WADNR in an effort to improve the state-level GIS databases. While the process is open for contributions of information and knowledge from those more directly tied to the management of particular areas of land, it contains fairly rigid guidelines for contributors. The GIS databases shared with the WADNR, for example, must have been developed using standards and protocols developed by the WADNR, to maintain the quality of data in the WADNR-distributed GIS databases, and to minimize the costs associated with updating and managing the state-level GIS databases.

There are also a number of cases where private natural resource organizations actually have shared data with one another. For example, a number of watershed analyses were conducted in Washington State in the late 1990s. In performing a watershed analysis, one of the major landowners in the watershed generally coordinates the compilation of the GIS databases associated with the watershed. All other landowners interested in helping develop the watershed analysis can contribute GIS databases to this 'lead' organization. More than likely, the amount of GIS data shared amongst different private organizations will be limited to a level that is sufficient to complete the analysis. For example, the spatial extent of the GIS databases shared will likely be limited to the boundary of the watershed being analyzed, and the attribute data associated with landscape features may be reduced to a small subset of the total set of attributes prior to sharing the data. In addition, it is likely that the projection and coordinate systems for each shared GIS database will need to be changed to conform to the system used by the lead organization.

An example of multiple public organizations joining together to create a significant spatial database is the Puget Sound LiDAR consortium (PSLC, 2007). The PSLC formed in 1999 and now includes metropolitan, academic, county, state, and federal organizations. The group's original goal was to use LiDAR to develop public domain high-resolution topographic data initially for significant parts of Washington. More recently, public organizations within Oregon have also joined the PSLC, and large portions of land in Oregon are also included in data acquisition plans. Funding has come from the operating budgets of group members and also from grants that members have received to support landscape hazard mitigation efforts. This unique collaborative effort will result in LiDAR data being available for most of the population centers in Washington and Oregon, and surrounding lands. The joint efforts of these organizations have undoubtedly resulted in significant cost savings for LiDAR data, particularly in light of the fragmented ownership patterns represented by all groups. Aerial data acquisition is more efficient when continuous swaths of land can be imaged rather than disjointed sections, which require additional flight lines.

Sharing GIS Databases within a Natural Resource Organization

The issues of GIS database ownership, maintenance, distribution, and data quality within an organization are problematic, and partly a result of how a GIS database development and distribution system may have been designed. For example, a road engineer may be the most appropriate person to own and maintain a culvert GIS database, and a wildlife biologist may be most appropriate person to own and maintain a GIS database related to locations of threatened or endangered species. However, the readiness of each person to perform these tasks within GIS, and the time they have available to do so may be limited. Therefore assigning the ownership, maintenance, and other tasks to either a GIS technician, GIS manager, or a GIS contractor or consultant may be more logical. More than likely, GIS databases will be shared at some stage to facilitate GIS database management, as a verification step in the maintenance stages of GIS database management, or to facilitate natural resource management (after the GIS databases have been updated). If you were to assume that a natural resource management organization was structured in a traditional manner, with a central office and a set of field offices, the GIS database sharing process should be viewed as more than a one-way transaction between a centralized GIS department and everyone else (Figure 16.1), because the roles related to ownership, maintenance, distribution, and data quality may be shared. Defining the appropriate roles for each person in an organization can result in a difficult negotiation process, particularly when the roles are considered for change.

Establishing a process for sharing GIS databases at all levels within an organization is becoming more important as both field personnel and upper-level managers are becoming interested in using GIS technology. Data sharing can be as simple as routing a computer disk from person-to-person or office-to-office, or as sophisticated as placing all GIS databases on an organization's Internet (or Intranet) site or FTP server. In the latter cases, the ability to access the Internet sites and FTP servers may be limited to authorized personnel.

Technology and processes for sharing GIS data are advancing, however, leading to the potential for two-way transactions of data within a natural resource organization and nearly real-time updates of the information. For example, some organizations allow field managers to update GIS data directly in the corporate databases, a role traditionally held by a centralized office. These updates are sent electronically to the centralized office, and after data verification and consistency filters have been applied, redirected to other field offices. In an efficient system, the other field offices can acquire these updates in a matter of minutes. The importance of using up-to-date information is perhaps most important when dealing with time-

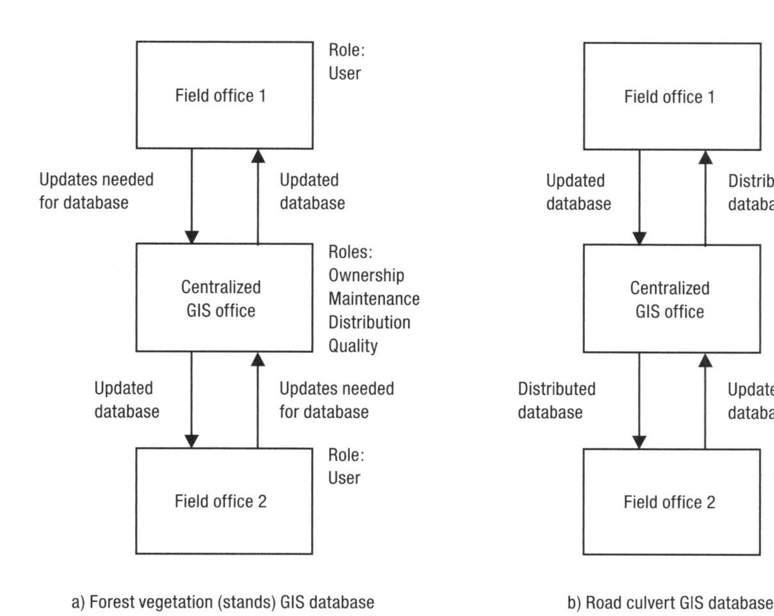

Figure 16.1 Example pathways of actions when sharing data within an organization, for (a) a typical forest vegetation (stands) GIS database and (b) a road culvert GIS database.

In Depth

The following story indicates not only the motivation of field office personnel to use GIS technology, but also a residual negative effect of the ability to share GIS databases within an organization.

One natural resource management organization the authors were familiar with had a central office where 30–40 'corporate' GIS databases were maintained: roads, streams, timber stands, and others. The organization had a distributed system, where GIS users in field offices could access a server and download (copy) the GIS databases to their field office computers. Field foresters, biologists, and hydrologists subsequently did just that—they copied the GIS databases to their personal computers and used them for a variety of purposes related to the management of the land, water, wildlife, and recreation resources. Some of these GIS databases were even updated with pertinent information that should have been incorporated into the corporate GIS databases. As the discussion grew related to sharing the modified corporate GIS databases with the corporate office, it was determined that there were over 1,000 variations of the corporate GIS databases stored across the organization's computer systems. Each GIS database required storage space on hard drive and it was likely that hundreds of megabytes of space were dedicated for this effort. In some cases, field personnel believed that their modified version of the corporate GIS database was of higher quality than other user's versions (even better than the corporate version of the GIS database). As expected, a number of discussions regarding the quality of these modified GIS databases arose, and the process of changing the GIS database maintenance policy for the corporate databases was begun.

sensitive decisions, such as those related to wildfire control and suppression. With GIS capabilities now available in hand-held computers, managers in the field can make more timely and informed decisions when necessary.

Distribution of GIS Capabilities to Field Offices

With the development of desktop GIS software the ability of every employee in an organization to use GIS is now possible. Bettinger (1999) has noted that the deployment of software and GIS databases to field offices is a significant challenge, and, as noted in chapter 15, that is a trend that will likely continue. Some of the challenges to natural resource management organizations interested in implementing a distributed GIS system, as noted by Bettinger (1999) include:

Where people are concerned

- Provide the appropriate training to reduce computer-related anxiety.
- Provide a statement of purpose behind implementing a distributed system.
- Implement methods to reduce the resistance to change.
- Provide motivation to use the system.

Where GIS databases are concerned

- Clarify GIS database ownership and maintenance issues.
- Outline data acquisition protocols.
- Develop data distribution systems.
- Develop quality control measures.

Where the technology is concerned

- Facilitate the acquisition of appropriate hardware and software.
- Control the purchase, update, and maintenance of software.

Where the leadership of the organization is concerned

- Demonstrate strong support for the program.
- Make a long-term budgetary commitment to the program.
- Adjust reward systems for added responsibilities of field managers.
- Document the direction and strategic goals for the system.

Most young professionals in natural resource management organizations will likely be called upon to provide assistance in the development of maps and GIS databases to facilitate the day-to-day management of a landscape.

This is a relatively new job expectation that may not ordinarily be placed on the more seasoned foresters and natural resource professionals. At the onset of a program such as this, you should acknowledge that it adds responsibility to the field manager for the development of his or her own management-related maps. Tasks traditionally performed by specialists in a centralized office will be transferred to field offices. This transfer of responsibility could lead to anxiety on the part of field managers, however the main advantage of the program is in reducing the amount of time required to produce the products (maps and analyses) needed to make management decisions. Once field personnel are sufficiently trained and become confident in their abilities, time savings across the organization should be realized.

Perhaps one of the most important contributions young professionals can make is to ask pertinent questions about the organization within which they work: What role do I play in each process that involves GIS databases? How are the GIS databases created or acquired? Who claims ownership over the maintenance and distribution of each GIS database? What technology (hardware) is available to display the results of an analysis? What technology (software) is available to perform an analysis? How and when does the organization value GIS analysis in supporting management decisions? By asking these questions, you indicate your willingness to understand how GIS has been implemented within an organization, and imply that you understand that GIS is a valuable tool in natural resource management.

Technical and Institutional Challenges

One of the most expensive and time-consuming aspects related to using GIS is the effort that is required to create a GIS database. Duplicating previous data collection efforts in the creation of a GIS database should always be avoided, thus a lack of awareness of existing GIS databases is a serious challenge to confront. To prevent duplicating GIS database development efforts, GIS users within an organization should be made aware of the types of GIS databases that the organization can easily access. This might include GIS database products that were developed internally within an organization, GIS databases developed by GIS contractors or land surveyors, or GIS databases that are available through agreements or relationships with other natural resource management organizations. Information regarding available GIS databases could be stored in a searchable database or it might be catalogued in a less formalized manner. Regardless of how this information is gathered and stored, personnel in an organization who use GIS should be able to easily identify and locate existing GIS databases that might facilitate the tasks that their jobs require.

Metadata, or information documenting the specifications and quality of landscape features in a GIS database, have become an important aspect of GIS databases in the past decade (Dobson & Durfee, 1999). In order to determine the fitness of a GIS database for a particular use, the metadata related to the GIS database should be considered. In particular, when a GIS database is acquired from another organization, the metadata should be relied upon to verify that the condition of the GIS database is what was expected when acquired. In many instances, however, little metadata exists to describe the qualities of GIS databases. The two hypothetical forests used extensively in this book are prime examples. In fact, more often than not you may find that GIS databases developed or maintained by non-federal organizations lack, or have insufficient, metadata. Thus, natural resource professionals must be careful when basing decisions on GIS databases where the level of quality is uncertain.

For organizations that are involved in producing and distributing GIS databases to other natural resource management organizations, guidelines or protocols should be in place that address all aspects related to the distribution of the GIS databases. Without guidelines, organizations are likely to be working with an inefficient distribution system, and may be prone to liability problems. Guidelines should include a pricing structure for all available GIS databases; this structure will need to reflect the organization's views on cost recovery. In the case of public organizations, there may be no need or desire to recover more than the delivery costs. Some public organizations, however, do utilize contractors to collect and develop GIS databases, and may need to recoup some of the costs of doing so. For most private natural resource management organizations, the pricing structure will likely need to reflect the actual costs of collecting the data. Organizations that distribute GIS databases will also need to develop a comprehensive liability policy to protect themselves from litigation. A liability policy will likely need to be tailored to each particular GIS database because the content, accuracy, and uses of databases will vary. A method of providing GIS databases to customers will also need to be identified. As discussed in chapter 15, organizations that provide GIS data to the public should make

GIS databases available in an expedient manner and use current Internet technologies such as creating new websites that allow users to browse database offerings and download both data and metadata. In addition, file transfer protocol (FTP) can be used to make data available, but this data-sharing system is less user-friendly.

For organizations that are not involved in producing and developing GIS databases for other natural resource management organizations, the high cost associated with creating GIS databases can result in a reluctance to share databases with other organizations. For example, certain GIS databases may contain sensitive information, and might reveal the location of landscape features (such as endangered species nesting locations or archeological sites) that might be disturbed or destroyed should the locations become public information. Two examples of these GIS databases include the location of endangered species nest sites, and the location of genetically-modified tree field trials. The GIS databases may also contain information about the status of landscape resources that would be of value to another organization with which it competes, providing the other organization an advantage in the market place. This is clearly important today as the number of land sales has skyrocketed, and potential (and perhaps hostile) investors desire complete information regarding a land asset.

A reluctance on the part of federal public organizations in the US to share or publicize data holdings is that all federal agencies are subject to the Freedom of Information Act (FOIA). FOIA was signed into law in 1966 and later amended in 2002. FOIA makes it possible for individuals to request access to federal agency records, except in cases where records are protected from disclosure. These requests can also be made for spatial databases that were produced by federal agencies. In some cases, spatial databases may contain information that the agency considers to be sensitive or potentially damaging to the resources it manages. One example of sensitive information might be spatial records of vandalism that occur in public recreation areas. Evidence of high vandalism occurrence might deter visitors from staying in affected areas, and potentially reduce revenue that is gathered from day-use permits. In addition, there may be hesitation to draw additional attention to 'hot-spots' of criminal activity in case the attention may encourage others to visit these locations out of curiosity or to cause additional vandalism. Other hesitations may involve the presence of unusual features, such as special habitat areas, or archeological sites, and the potential damage that too many visitors may bring to these areas. For various reasons, federal agencies may not openly advertise the types of spatial databases that they have produced. A primary hesitation to do so is likely because of the uncertainty of what will happen with the information within the databases should they become widely circulated.

Understanding that there are factors that may hinder an organization's willingness to share a GIS database is important prior to requesting the database. For data of a sensitive nature, it may be possible to enter into a confidentiality agreement to gain access to the GIS database. Ultimately, it may be necessary to pay a large sum of money ($5,000 to 10,000) for a GIS database, and for some organizations, it may be difficult to locate the necessary budget resources appropriate for this type of purchase.

In Depth

To facilitate a recent landscape analysis research project, a private natural resource management organization agreed to provide a highly detailed GIS database that described the management units within their ownership boundaries. The GIS database had been assembled at great cost and effort, and revealed considerable information about the natural resources that the private organization managed. The private organization required that a confidentiality agreement be signed prior to making the GIS database available to the research project. As noted in the confidentiality agreement, access to the GIS database was limited to the primary scientists of the research project, the sharing of the GIS database with others was prohibited, and protocols for distributing information drawn from the GIS database were outlined. Without the confidentiality agreement, which facilitated the sharing of the GIS database, knowledge of the status of resources located within the private natural resource management organization's ownership would likely have been less accurate, reducing the confidence placed on the landscape analysis results.

Benefits of Implementing a GIS Program

The decision to implement a GIS program (the entire suite of hardware, software, and personnel related to the use of GIS within an organization) can be intimidating for natural resource management organizations. There are many factors to consider, including investments in software, hardware, personnel, and GIS databases. Since natural resource management organizations typically rely on maps or mapped data to assist in making decisions, GIS can allow an efficient storage of maps, and can facilitate the generation of multiple versions of maps in a timely manner. In addition, GIS allows landscape features to be measured, analyzed, and integrated with other GIS databases in an expedient manner. New technology has provided the potential to convey information to field managers very quickly. These capabilities, if managed properly, can allow natural resource management organizations to make better management decisions, more accurately gauge the effort and cost of potential natural resource management projects, and increase the efficiency of tasks performed by their employees.

Successful GIS Implementation

Saving money, reducing the amount of time spent in the office analyzing options, and thus saving resources for other tasks and management activities are common goals of natural resource managers. Distributing GIS capabilities to field offices has been suggested as one way to address some of these issues. Successfully implementing and managing a GIS program can be difficult, as the costs of implementation and management vary from one organization to the next. Perhaps the strongest ingredient for success in implementing a GIS program is in establishing an organizational commitment within the upper levels of management of an organization. This commitment needs to view the GIS program as more than just a short-term experiment that can be discarded after early, disappointing results, since initial GIS products and experiences are likely to identify implementation problems. Unfortunately, upper-level managers tend to be less familiar with GIS technology than the actual GIS users. For this reason, GIS users should communicate their support of GIS in terms that are comprehensible to the upper-level managers and help them understand that when technical difficulties do arise, program implementation plans must be adjusted. Upper-level managers, in turn, should promote the GIS program as a way of making more efficient the tasks required of natural resource management.

Organizations must also be realistic about the time, effort, and budgetary resources that individual GIS projects or analyses will require. Proper planning requires that project objectives be clearly defined. Objectives provide a project mission and can help keep personnel focused, should setbacks occur. Project objectives are also important for establishing standards that allow you to gauge the success of a GIS project or analysis. Achievement benchmarks are also critical in justifying the continued and expanded use of GIS within an organization. Allowing users of GIS to become involved in the planning and implementation of GIS projects is also important, since they may be among the best qualified to assess whether GIS can accomplish the appropriate project tasks, which may lead to an improved level of efficiency in the management of natural resources.

Finally, GIS user training is an important consideration for the success of a GIS program within a natural resource management organization. Most recent college graduates from natural resource programs will likely have a rudimentary knowledge of how GIS can assist in natural resource management, but will likely lack the level of experience you gain from using GIS periodically for actual, on-the-ground, management purposes. To accelerate the development of personnel, organizations can provide GIS training internally, or can allow personnel to attend continuing education courses or pursue other training opportunities. Geospatial training courses and opportunities appear to be increasing within natural resource disciplines (Wing & Sessions, 2007). This investment in continuing education increases the knowledge level of personnel and demonstrates an organization's commitment to its personnel, which hopefully leads to increases in work efficiency and productivity.

Summary

What would the management of natural resources be like without a few challenges? With the introduction of new computer-related measurement technology and accompanying GIS databases, natural resource management organizations are facing numerous challenges related to GIS data management. Given that GIS-related technology continues to evolve to meet the increasing needs of society, an optimistic person might expect that the challenges described in this chapter can be overcome, even as new and varied issues arise. As we have noted, many of the challenges facing the use of GIS within natural resource management organizations are related to GIS databases, technology, people, and organizational issues.

Applications

16.1 Sharing GIS databases with people outside your organization. What issues should a natural resource management organization address when considering making GIS databases available to other organizations, including perhaps competitors?

16.2 Managing GIS technology. A seasoned professional (Mary Swarthmore) who manages another department (Accounting) in your natural resource organization is considering a career change. This change will result in Mary managing your organization's GIS department. Mary has never managed a GIS department before, nor been involved in the creation, acquisition, or use of GIS databases. She has approached you for some advice regarding the tough issues GIS managers face when concerned about successful implementation of a GIS program. What might you advise Mary?

16.3 Sharing GIS databases within an organization. You've heard that a colleague in another private natural resource management organization has developed a GIS database that may contain information that could assist you in some part of your job. Under what conditions should you expect your colleague to provide you access to the GIS database?

16.4 Distributed GIS program. You work for a large integrated forest products company that has a central office and five district offices. The company has been attempting to shift GIS capabilities to the field offices by purchasing the appropriate hardware and software resources, and insisting that field personnel (foresters, biologists, managers, and others) use it to make maps associated with their management activities. After two years, only one of the five offices has successfully implemented the program.

a) Why do you think the other four offices have been less than successful?
b) Why might the one office have been successful?

References

Bettinger, P. (1999). Distributing GIS capabilities to forestry field offices: Benefits and challenges. *Journal of Forestry, 97*(6), 22–6.

Dobson, J.E., & Durfee, R.C. (1998). A quarter century of GIS at Oak Ridge National Laboratory. In T.W. Foresman (Ed.), *The history of geographic information systems: Perspectives from the pioneers* (pp. 231–63). Upper Saddle River, NJ: Prentice-Hall.

Natoli, J.G., Pelgrin, W.H., Oswald, B., & Montie, K. (2001). Geographic Information Systems: The wave of the future for information analysis. *Public Works, May*, 22–9.

Puget Sound LiDAR Consortium (PSLC). (2007). *Puget Sound Lidar Consortium: Public-domain high-resolution topography for Western Washington*. Retrieved April 23, 2007, from http://pugetsoundlidar.ess.washington.edu/index.htm.

Wing, M.G., & Bettinger, P. (2003). GIS: An updated primer on a powerful management tool. *Journal of Forestry, 101*(4), 4–8.

Wing, M.G., & Sessions, J. (2007). Geospatial technology education. *Journal of Forestry 105*(4), 173–8.

Chapter 17

Certification and Licensing of GIS Users

Objectives

The progression of GIS use in natural resource management has been evolving from development and implementation of systems, to the distribution of analytical capabilities to field offices, to porting spatial technology into the resource setting. The evolution of GIS can be viewed from the perspective of a single organization or from the perspective of national or worldwide GIS communities. From a global perspective, GIS is facing one of its greatest challenges: that of implementing certification and licensing processes to define and recognize 'professional' GIS users. In recent years, concerns have been voiced from the land surveying and engineering community regarding the definition of surveying activities and whether GIS practitioners impede upon traditional surveying activities when collecting or mapping spatial data (Gibson, 1999). These concerns have fostered debate among surveyors, engineers, and GIS users regarding the types of measurement and analytical activities that might be required to competently perform specific activities in conjunction with the analysis. Thus, in an effort to gain credibility and recognition among other professions, the GIS community has pondered the issue of certification and licensing.

After reading this chapter and exploring the questions posed in the applications section, students should have an awareness of:

1. why certification and licensing of GIS users is being debated,
2. what organizations might be relevant in certification and licensing discussions, and
3. how certification and licensing issues might affect the typical GIS user in a natural resource organization.

During the last 10 years, one of the primary goals of the GIS community has been to educate other professionals and the general public about the power and usefulness of GIS beyond its map production capabilities. Many people in natural resource organizations (as well as academia) view GIS only as a map-making tool and may have limited understanding of its analytical power. The American Society for Photogrammetry and Remote Sensing (ASPRS) and the University Consortium for Geographic Information Science (UCGIS) are perhaps the most active groups in educating those who use GIS, as well as the public, about the capabilities of GIS. At the ASPRS 2007 National Convention in Tampa, one ASPRS member was heard to remark 'If you claim that GIS technology is vital to society you should also promote the need for certification and licensing among GIS users.' The UCGIS conducts national meetings twice a year to identify research and other activities that will identify and promote the use of GIS as a problem-solving tool for society. At a national meeting in June 2000, a member of the UCGIS asked the other delegates, 'Now that everyone seems to know about GIS, what are we going to do about it?' These seemingly innocent observations speak positively about society's growing awareness of GIS while also indicating a potential pitfall within natural resource management: GIS has been embraced by natural resource organizations in a way that facilitates open use by any professional who might be interested in the technology. Only recently has the GIS community begun to discuss in depth whether professional standards must put in place to ensure professional competency for data development, analysis, and other tasks. Given that other professions initiated these discussions, the leaders and champions of the use of GIS in natural resource management have found themselves at times in a defensive position when addressing issues related to the call for licensing and certification of GIS users.

Throughout the US, some professional land surveyors and engineers have perceived that GIS users were violating state surveying laws when using GPS to collect spatial data, and when reporting positional accuracies of collected measurements. Some of these GPS data collection activities have actually led to legal disputes, particularly when the collection and mapping of spatial data references land ownership locations. In California, state legislation was developed to clarify the activities involving spatial data collection (Korte, 1999). These activities fall into two categories: (1) those that constitute 'land surveying', and hence require professional licensing, and (2) all other activities, which do not require professional licensing. Most states have registration boards that license and regulate land surveyors and engineers. These boards often have the ability to interpret existing statutes and laws that govern the collection and representation of spatial data, and may also initiate, support, or approve legislation regarding spatial data collection activities. In addition, state-level professional land surveying societies are active in promoting or modifying laws regarding spatial data collection, and they may occasionally engage political lobbyists to assist in influencing the legislative process.

In contrast to the land surveying and engineering fields, the GIS community is not directly represented or controlled by a nationally- or state-recognized licensing board in most cases. There are GIS-related professional societies at the state or province level but these societies typically have not been in existence for very long, and are generally not very active in influencing legislation related to spatial data collection. The main objective of state-level and national-level GIS societies has been to communicate information related to the collection, maintenance, and analysis of GIS databases to interested users.

Current Certification Programs

In terms of nationally-recognized GIS certification programs in the US, there are primarily two current options. The more established and rigorous of the options is that made by the American Society for Photogrammetry & Remote Sensing (2006). The ASPRS has created a Mapping Scientist certification for GIS users and has also created certification programs for remote sensing and photogrammetry. The Mapping Scientist certification requires applicants to develop a statement of accomplishments, which are peer-reviewed, and to pass a written exam. Currently, only about 50 people are certified as Mapping Scientists through the ASPRS. The APSRS has also created technologist-certifications for GIS, remote sensing, and photogrammetry. These GIS technologist certifications require less work experience than the Mapping Scientist and other full certifications (three years as opposed to six). There are currently six certified GIS/LIS Technologists.

The Urban and Regional Information Systems Association (URISA) initiated a GIS professional certification program in 2004. The GIS Certification Institute (GISCI, 2007) manages the certification program, which results in qualified applicants becoming recognized as a certified geographic information systems specialist (GISP). Three categories of experience must be demonstrated in order to qualify. The primary experience necessary in gaining GISP certification is a documented work history involving GIS and other spatial tools. The second experience category is an education background that can be satisfied by attending conferences and workshops, as well as completing formal education programs or earning certificates. The third category is described as 'contributions' and includes GIS publications, conference planning or presentations, and volunteer activities related to GIS.

As of October 2007, there were 1,709 certified GISPs, giving the GISCI program visibility beyond the certification programs of the ASPRS. Although the creation of the GISP certification is noteworthy, the experienced-based portfolio approach to qualifying as a GISP lends itself to criticism (Longley et al., 2005). It remains to be seen whether a certification approach that does not include a written examination will receive respect and recognition from other professions and disciplines. In addition, there is no clear method for addressing unprofessional activities related to GIS, should they occur. Given the emphasis on self-reporting of experience, another issue of discussion is whether any applicants have been denied GISP status.

Many colleges and universities now offer certification degrees that are related to GIS, as well as other spatial data collection and analysis technologies, however no standards exist for what should be offered in those curricu-

In Depth

Besides the current course that you are taking (that hopefully uses this book), what other GIS courses does your university, college, or community college offer? Although there are many educational institutions that offer coursework or curriculum related to GIS, these programs can look quite different from one institution to the next. At most educational institutions, for example, GIS courses are located within the geography department. However, specialized GIS courses may be found within departments such as fisheries, wildlife, oceanography, forestry, soils, rangeland resources, and others. More than likely, GIS courses offered in departments other than geography will provide a different perspective on what is important to students pursuing natural resource degrees. If a university or college does not offer courses related to GIS, students can still learn about the capabilities of GIS through Internet courses, self-study of GIS texts, and volunteer work with local agencies or government offices.

lums. The National Center for Geographic Information & Analysis (2000) has produced, and suggested for use, a core curriculum to serve as a foundation for studies in GIS. Typically, GIS users can earn a GIS certificate after two years of part-time study. While this option does include organized coursework (and perhaps exams to evaluate competency), programs that offer certification degrees lack a recognized set of certification guidelines and are generally not accredited by a professional GIS or remote sensing society. As mentioned in chapter 15, the GIS&T Body of Knowledge (DiBiase et al., 2006) was recently published in order to define critical concepts and skills that relate to geographic information science and technology (GIST). This document was created through the joint efforts of many GIScience researchers and educators, and is an initial attempt to define the skills that you can use to describe geospatial competency. A second edition is intended that will provide detail for instructional activities that support important geospatial concepts and skills.

The NCEES Model Law

The National Council of Examiners for Engineering and Surveying (NCEES) has developed a set of guidelines described in a Model Law document to help states with licensing issues related to land surveying and engineering (National Council of Examiners for Engineering and Surveying, 2006). The Model Law contains reference to GIS activities associated with spatial data collection and use. The Model Law contains 29 sections that are designed to help state boards and other legislative bodies create or amend laws for land surveying and engineering.

The first section of the Model Law clarifies the necessity for guidelines by stating that the practices of land surveying and engineering are a matter of pubic interest. The decisions made (or recommended) by people employed in these professions can potentially affect the life, health, and property of the general public. Section 2 defines the tasks that are associated with surveying and engineering and no longer refers directly to GIS, as it did in an earlier version of the Model Law. Section 2 does state that mapping involves the configuration of the Earth's features, the subdivision of land, the location of survey control points, reference points, or property boundaries, and thus people performing 'mapping' are performing the 'Practice of Surveying'. The Model Law suggests that these people should be registered as professional surveyors before engaging in those activities.

The most direct pathway to becoming a professional surveyor is to first graduate from an accredited four-year college program in engineering or surveying. Then you must successfully pass an eight-hour written exam covering surveying fundamentals. Once the fundamentals exam has been successfully passed, four years of surveying experience under the supervision of a licensed surveyor must be accumulated before admittance is allowed to an eight-hour written comprehensive exam covering surveying principles and practice. Once this comprehensive exam has been successfully passed, and all other state-level requirements are satisfied, you are qualified to become a professional land surveyor. Those who graduate from four-year surveying curriculums that are not accredited must spend an additional two to four years working in the land survey profession before they can be admitted to the fundamentals exam. This process of attaining licensing is daunting, and requires a long-term commitment on the part of potential surveyors or others who wish to comply with the Model Law. For many current natural resource professionals who use GIS, active engagement in a career combined with family and community commitments offer few realistic opportunities to pursue an engineering degree or to work under the guidance of a land surveyor.

Earlier versions of the Model Law definition of surveying that contained direct mention of GIS were criticized as being too stringent and expansive in its description of GIS applications within surveying activities. Several prominent organizations related to surveying and GIS co-authored a report that suggested modifications to the Model Law (American Congress on Surveying and Mapping [ACSM] et al., 2001). The report urged the NCEES to drop explicit reference to GIS as a data manipulation and mapping tool, to refine the definition of surveying so that it was less broad, and to specifically include and exclude certain GIS-related activities in its definition of surveying. The NCEES appears to have acted at least in part to these recommendations. Previous to the Model Law changes, the broader definitions of surveying created difficulties for some states that tried to incorporate the Model Law's definitions of surveying into their own state statutes and regulations (Thurow & Frank, 2001).

The Need for GIS Certification and Licensing

Chief among the argument for GIS certification is an assessment of how GIS activities might impact society's welfare and safety (Gibson, 1999). The surveying and engineering professions have long been involved in determining how best to accurately and precisely collect and analyze Earth and structural measurements. Since land values in North America will likely continue to increase as the human population also increases, land areas that are incorrectly measured can result in large monetary losses or gains for land owners. Knowing the reliability of land measurements (expressed through uncertainty estimates) will allow land managers to make better decisions. While you may argue that a distinct part of natural resource data collection and analytical processes involves quantifying and expressing the uncertainty that is associated with those measurements, this quantification is rarely used to rate the reliability of measurements collected. Thus the surveying and engineering professions may be better positioned to provide information regarding the inherent uncertainty in land measurements.

Land surveyors are also charged with locating or establishing roads and other utilities such as power lines and fire hydrants. Surveyors have argued that it is inappropriate for unlicensed surveyors to operate GPS equipment for this purpose since locational errors can potentially affect public safety. In addition, since professional surveyors must absorb the on-going costs associated with maintaining licensing and liability insurance, they are likely to charge more for their services than unlicensed GPS operators. Thus, surveyors have stated that it is unfair for unlicensed GPS operators to compete with professional land surveyors in offering these types of data collection services.

Licensing varies throughout the world and may also vary by state or province. In addition, it appears that the number of licensed professions is in transition. In Canada, professions that require licenses are referred to as regulated occupations. There are approximately 50 different regulated professions in Canada (Government in Canada, 2007). Within the US, the number of professions that require a license for participation has been gradually increasing. Doyle (2007) found that about 20 per cent of all professions in the US require licensing, up from about 5 per cent during the early 1960s. About 50 professions have a registration process that is recognized in all 50 states. Some people criticize the licensing process and claim that it results in higher prices for services without an equivalent gain in the quality of the service or good provided. In addition, some people see entry into the profession as being prohibitively limited once licensing is in place. While Kleiner (2000) found that incomes from licensed occupations were higher for those occupations that required more education and training, faster employment growth was evidenced in many licensed professions, such as engineering and law, when compared to non-licensed professions. Kleiner (2000) also reports that empirical evidence addressing whether licensing results in greater societal goods, such as increased safety, is currently lacking.

Issues of control and enforcement are additional aspects of licensing and certification. Most states have clear definitions of what constitutes acceptable surveying practices. These rules typically cover what clients should expect from a professional land surveyor's services, in terms of products, as well as ethical considerations. Ethical considerations not only address the surveyor–

client relationship but also offer advice on the professional relationships that should exist between surveyors. Most states also have developed a process to facilitate the submittal of complaints against land surveyors. All state licensing boards have the power to revoke a surveying license if disciplinary infractions occur as a result of complaints. This power encourages most land surveyors to become familiar with board rules for professional conduct and to adhere to the rules as they engage in survey activities. Such rules do not exist in general GIS use in natural resource management.

Criticism has also been weighed against unlicensed GIS and GPS users due to the lack of an accredited educational curriculum. Most professions, including forestry and wildlife, have identified an educational and professional background that is necessary for accreditation or licensing *within their fields*. The accreditation process usually suggests the coursework, minimum competency standards, professional standards, and integration with other disciplines that should be provided to students pursuing degrees in those fields (Huxhold, 2002). An accreditation process for GIS could be developed (and perhaps is currently under development). However, the dilemma is that if you view GIS as a field of study, the students in those curricula will graduate with a geography degree, and be considered professional geographers. Most natural resource organizations hire biologists, foresters, soils scientists, or other associated professionals, as well as geographers, to assist in the management of natural resources. These organizations increasingly expect all of their personnel to utilize GIS, not just those who have obtained a degree from an accredited geography program.

GIS Community Response to Certification and Licensing

As you might expect, some members of the GIS community have voiced opposition to the suggestion that certain GIS activities be included in the list of functions only to be performed by land surveyors. Criticism has been directed toward the sometimes-broad definitions of land surveying, and whether GIS databases developed and distributed by public agencies should require management by a licensed surveyor (Joffe, 2001). In addition, some proposals for certification and licensing have also been viewed as exclusionary, and could prevent natural resource professionals from performing the GIS activities that they historically performed. Thus the opposition to certification and licensing from the natural resource GIS community, largely composed of natural resource managers, biologists, foresters, and others, is to be expected.

Defining those areas of spatial data collection and mapping that potentially affect public welfare and safety is easy in some cases, and challenging in others. Clearly, GIS databases used for activities such as navigation, locating underground facilities when excavating, or ensuring that property boundaries have been accurately located and used in calculating land areas, should fall under the purview of a professional land surveyor or engineer. Activities that involve displaying data for illustrative purposes, recreational guides, or in activities that may only affect the person(s) or agency responsible for the associated decision, could perhaps fall outside the purview of a professional land surveyor or engineer. Although these examples illustrate distinctions between activities that clearly affect public welfare and safety and those that do not, there are many other examples that are less clear and will require further discussion before agreements between the surveying and GIS communities are reached. In cases where mapping products might affect public welfare and safety, several scenarios for making map consumers aware of potential limitations have been suggested. Suggestions include requiring that maps explicitly identify the source documents, associated metadata, appropriate use of map content, and any positional adjustments (Joffe, 2001). Examples of these caveats and disclaimers were introduced in chapter 4. Whether these caveats will continue to fall short of requiring a certified or licensed GIS professional to develop the mapping products will likely be an area of discussion.

Finally, and interestingly, some people have argued that the surveying profession has traditionally not required training in GIS in order to obtain professional surveying certification (American Congress on Surveying and Mapping, 1998). For this reason, they argue, it may not be appropriate for land surveyors to manage the development, maintenance, and use of GIS databases. However, the current exam syllabi for the national land surveying exams (both the fundamental and professional principles and practice) do include GIS and land information systems as potential exam topics.

MAPPS Lawsuit

In June 2006, the Management Association for Private Photogrammetric Surveyors (MAPPS) and three other professional engineering associations filed a lawsuit against the US government. The lawsuit was filed on behalf of the Federal Acquisition Regulation (FAR) Council and is referred to as the MAPPS lawsuit. The actual title for the lawsuit is *MAPPS et al. v. United States of America*. The lawsuit requests changes in interpretation of the 1972 Brooks Act (US Public Law 92-582) which is intended to direct federal government policy in selecting providers to perform architectural, engineering, and related services. More specifically, the lawsuit seeks to modify how the selection process for government contractors is evaluated as it relates to mapping activities. The results of the lawsuit may impact how some US federal agencies award government contracts. The MAPPS lawsuit has been a considerable concern for many GIS-oriented organizations and has once again fueled the debate over the definition of activities that can be considered surveying and engineering, which activities influence public welfare and safety, and the appropriate geospatial certification and licensing requirements. The participation of prominent organizations on both sides of the lawsuit is evidence that the MAPPS lawsuit is not a trivial legal exercise.

Plaintiffs of the MAPPS lawsuit included the American Society of Civil Engineers (ASCE), National Society of Professional Engineers (NSPE), and Council on Federal Procurement of Architectural and Engineering Services (COFPAES). The Brooks Act established that price alone should not be used in the awarding of government contracts to individuals or firms. Instead, Qualifications-Based Selection (QBS) is to be used, which involves evaluating professional qualifications and experience, in addition to seeking a 'fair and reasonable' cost to the government, in awarding contracts. Negotiations for an acceptable price should begin with the most qualified firm. If negotiations fail, then negotiations should proceed with the second most qualified firm, and so on. The FAR Council applies the rules and laws related to the Brooks Act, and is supposed to ensure that the intentions of the act are upheld. Although the Brooks Act includes 'surveying and mapping' among the list of architectural and engineering services that it is intended to cover, federal agencies have not been consistent in their interpretation and adherence to stated protocols. The MAPPS lawsuit seeks to compel the FAR council to more rigorously interpret and apply the Brooks Act in the selection of contractors for surveying and mapping services. In states where licensed surveyors or engineers are required to perform surveying and mapping activities, the lawsuit contends that the FAR council should direct that licensed professionals be selected for government contracts involving surveying and mapping. The implications of the MAPPS lawsuit are then potentially significant given that many states do not draw a clear distinction between surveying and mapping in their laws and rules that govern surveying.

The American Association of Geographers (AAG), GISCI, Geospatial Information & Technology Association (GITA), UCGIS, and URISA jointly submitted a briefing to the US District Court in Virginia (Alexandria Division) that opposed the MAPPS lawsuit (Association of American Geographers, 2007). The brief stated that a successful lawsuit could cause serious concern not only for the GIS community but also for other related activities and professions. Other related activities included GPS data collection, Internet mapping, geospatial analysis, remote sensing, academic research that involved mapping, and the broad activities encompassed within cartography. The briefing claimed that the lawsuit's impact would greatly affect many activities and industries that involve or rely on mapping activities and information.

The US District Court ruled against MAPPS in June 2007 and issued a summary judgment in favor of the US Government. The judge in the case stated that the MAPPS plaintiffs failed to 'establish that an injury in fact was suffered by the individual surveyors or their firms'. In order for a case to be tried, a plaintiff (those filing the lawsuit) must establish standing. Standing is indicated by suffering a loss of some sort, be it monetary or otherwise. The judge found that MAPPS and the other plaintiffs had not established sufficient standing to support the lawsuit. Although there are opportunities to appeal the judge's decision, the judge's ruling appears to be strong enough that an appeal was considered unlikely.

There was great uncertainty in determining how a successful MAPPS lawsuit would have impacted the GIS community and many community members expressed relief that the lawsuit came to an end. Doubtlessly, there will be similar disputes and uncertainty in the future regarding licensing that involves the court system. Although this represents a US example of the conflict that has arisen with the widespread use of GIS, you could reasonably envision this happening in other countries with an established land records system and regulations related to engineering and land surveying practices.

Summary

In some circumstances, certification and licensing may be necessary for those involved in the development and management of GIS databases, to ensure that minimum standards of competency exist and that standards are being met in the development, maintenance, and application of GIS databases. Although many GIS activities related to natural resource management may have no bearing on public welfare and safety, some GIS activities result in maps or mapped data being sold or made available to the public, and thus may have direct or indirect implications on public welfare and safety. GIS has evolved into its own discipline, and is being integrated with other fields of study. Guidelines should be developed for determining the experience, educational background, professional conduct, and continuing education that defines the qualifications appropriate for those disciplines. Guidelines should also be developed to define the extent to which persons who are qualified within disciplines can appropriately develop, manage, and use certain GIS databases. Until a national certification or licensing program for GIS gains credibility and respect within society, GIS users in natural resource fields will find themselves struggling with local or national regulatory groups, the legal system, and other professions for control of GIS activities.

Applications

17.1. Licensing. Assume that the state or province in which you work is considering the development of a licensing board to oversee the licensing of GIS users, and to regulate the use and management of GIS databases related to your field of natural resource management.

a) What benefits might a licensing board provide for professionals engaged in GIS activities?
b) What disadvantages for GIS users might result from the development of a licensing board?
c) Describe three key dimensions of the licensing process that the board should develop.

17.2. GIS certification programs. The owners of the Brown Tract have recently become aware that GIS is used extensively in the management of the forest. They are concerned about the credibility of their land management team, and have suggested that all employees obtain GIS certification.

a) What are the potential benefits of GIS certification for GIS users?
b) What are the potential drawbacks of GIS certification for GIS users?
c) What elements would be necessary in order for a GIS certification to be more widely recognized and respected within society?

17.3. NCEES Model Law. Does the NCEES Model Law process seem like a reasonable or rational approach to clarifying the issue of GIS licensing? Why or why not?

17.4. The need for licensing or certification. Identify three GIS or GIS-related activities that could affect public welfare, and that might suggest that those developing, maintaining, or using the supporting data should be licensed or certified.

17.5. GIS community response to certification and licensing issues. Many professional (foresters, wildlife biologists, hydrologists, engineers, etc.) working in natural resource management currently use GIS to assist in making management decisions. Why might they be concerned about the issue of GIS certification and licensing?

References

Association of American Geographers (AAG). (2007). *Amicus Curiae Brief of the Association of American Geographers, GIS Certification Institute, Geospatial Information & Technology Association, University Consortium for Geographic Information Science, and Urban and Regional Information Systems Association in opposition to plaintiff's notion for summary judgment.* Retrieved on June 6, 2007, from http://www.aag.org/Donate/links.html.

American Congress on Surveying and Mapping (ACSM). (1998). Should surveyors supervise GIS? *ACSM Bulletin, November/December*, 26–31.

American Congress on Surveying and Mapping (ACSM), American Society of Civil Engineers-Geomatics Division, American Society for Photogrammetry and Remote Sensing, Management Association for Private Photogrammetric Surveyors, National Society of Professional Surveyors, National States Geographic Information Council, & Urban and Regional Information System Association. (2001). GIS/LIS addendum to the report of the task force on the NCEES Model Law for surveying. *Surveying and Land Information Systems, 61*(1), 24–34.

American Society for Photogrammetry & Remote Sensing (ASPRS). 2006. *Certification and recertification guidelines for the ASPRS certification program.* Bethesda, MD: American Society for Photogrammetry & Remote Sensing. Retrieved June 7, 2007, from http://www.asprs.org/membership/certification/certification_guidelines.html#Certified_Mapping_Scientist_GIS-LIS.

DiBiase, D., DeMers, M., Johnson, A., Kemp, K., Luck, A., Plewe, B., & Wentz, E. (Eds.). (2006). *The geographic information science and technology body of knowledge.* Washington, DC: University Consortium for Geographic Information Science, Association of American Geographers.

Doyle, R. (2007). By the numbers: license to work. *Scientific American, February 9*, 28.

Gibson, D.W. (1999). Conversion is out, measurement is in—are we beginning the surveying and mapping era of GIS? *Professional Surveyor, 19*(7), 14–18.

GIS Certification Institute. (2007). *Are you a GIS practitioner or a GIS professional?* Retrieved October 5, 2007, from http://www.gisci.org/.

Government in Canada. (2007). *Types of work.* Retrieved October 5, 2007, from http://www.goingtocanada.gc.ca/Types_of_work_in_Canada-en.htm.

Huxhold, W.E. (2002). GIS professionals—get a profession! *Geospatial Solutions, 12*(2), 58.

Joffe, B.A. (2001). Surveyors and GIS professional reach accord. *Surveying and Land Information Systems, 61*(1), 35–6.

Kleiner, M. (2000). Occupational Licensing. *Journal of Economic Perspectives, 14*(4), 189–202.

Korte, G.B. (1999). The current controversy: GIS and land surveying legislation, Part 1. *Point of Beginning, 24*(6), 70–3.

Longley, P.A., Goodchild, M.F., Maguire, D.J., & Rhind, D.W. (2005). *Geographic information systems and science* (2nd ed.). Chichester, England: John Wiley & Sons.

National Council of Examiners for Engineering and Surveying (NCEES). (2006). *Model law.* Clemson, SC: NCEES.

National Center for Geographic Information & Analysis (NCGIA). (2000). *The NCGIA core curriculum in GIScience.* Santa Barbara, CA: National Center for Geographic Information & Analysis. Retrieved June 5, 2007, from http://www.ncgia.ucsb.edu/education/curricula/giscc/.

Thurow, G., & Frank, S. (2001). Coming to terms with the Model Law: The search for a definition of surveying in New Mexico. *Surveying and Land Information Systems, 61*(1), 39–43.

Appendix A

GIS Related Terminology

The following represents a brief treatment of the terminology common to all types of GIS software programs and processes. An attempt has been made to avoid the variations of definitions that tend to be more descriptive of the processes related to one (or more) particular GIS software programs. It is important that natural resource professionals gain an understanding of a common, generic language. The lines of communication between those highly versed in GIS and those with a cursory knowledge of GIS need to be clear when making natural resource management decisions.

Accuracy: The ability of a measurement to describe a landscape feature's true location, size, or condition.

Adjacency: A spatial relationship indicating which landscape features share boundaries, or are within certain distance of other landscape features. For example, adjacency relationships for timber stands may indicate which stands touch other stands, allowing one to model green-up requirements.

Annotation: Text or strings of characters and numbers used to describe landscape features on a map.

Arc: A single string of X,Y coordinates (vertices), that when connected, form a line. A single line may contain many arcs, but a single arc may only represent one line, or part of a line. Arcs have starting and ending nodes to allow analysis of directional travel, such as in stream systems, where water enters one end of the arc (the start-node, or 'From-node') and leaves the other end of the arc (the end node, or 'To-node').

ASCII: American Standard Code for Information Interchange. Many types of text files are commonly referred to as ASCII files; some contain comma-delimited data (e.g., 1, 12, 3.45, 65.2, 0.45), others contain space-delimited data (e.g., e.g., 1 12 3.45 65.2 0.45), and still others use other delimiters to separate individual pieces of data. See *Comma-delimited text file* and *Space-delimited text file*.

Attribute: A characteristic of a landscape feature. Attributes can be represented by characters or numbers (or a combination of both), and they describe various characteristics of landscape features. For example, the attributes of a timber stand could include the following: stand number (or code), basal area, trees per acre, volume per acre, dominant tree species, and so on. The attributes of a research plot (a point) could include: study type, installation date, date last remeasured, date of next remeasure, etc. See *Field*.

Azimuth: A degree of a circle, with North being 0° (or 360°), East being 90°, South being 180°, and West being 270°.

Azimuthal projection: A projection system where the directions from a central point of origin are all preserved.

Bearing: An angle of 90° or less originating from either the North or South (and directed towards the East or West). Thus an azimuth of 353° represents a bearing of N7°W.

Boolean expression: A type of expression used in a query or computer code that requires a yes or no response. AND, OR, and NOT are the three most common Boolean expressions. For example, a query using a

Boolean expression could take this form:

stand_age ≥ 50 AND land_allocation
= 'Even-Aged'

BASIC computer code using a Boolean expression would look like this:

```
If (stand_age ≥ 50) AND (land_allocation
= 'Even_aged') Then
. . .
End if
```

Buffering (or buffer): A type of spatial analysis of proximity, where zones of a given distance are generated around selected landscape features. The result of a buffering operation is one or more polygons that represent the area within a specific distance (fixed, or variable, as defined through an attribute field) around landscape features.

Buffer zone: A set of one or more polygons that represent the area within a specific distance around landscape features.

Cartesian coordinate system: A system that allows one to locate any point on a planar surface divided by a set of grid lines.

Cell: see *Grid cell.*

Character: A single attribute describing a landscape feature, such as a letter, number, or special symbol, or a type of data that indicates the attribute should be considered a piece of text (even though numbers and special symbols may be valid attributes).

Clipping process: The process of extracting from one GIS database only those landscape features within the bounds of another GIS database (which could contain a single polygon or a set of polygons). This is an action that essentially acts like a cookie-cutter.

Column: A set of cells in a spreadsheet or database that are vertically aligned, usually representing a single attribute of every feature (record) in the database. Columns could contain column headers, or terms that describe the data in the column, or in spreadsheets, simply be represented by characters such as A, B, C, etc. Often called 'Fields'; see *Field.* Related to 'Records'; see *Record.*

Combine process: A process of eliminating the shared edges or intersections between two landscape features.

Comma-delimited text file: A text file created in a word processing system, a text editor, spreadsheet, or database, and saved in a format where commas separate items. The following, for example, could indicate habitat suitability index values for specific timber stand polygons, with the first item of each line identifying the polygon, and the second item listing the habitat suitability index:

```
1,0.657
2,0.433
3,0.298
```

Completeness: A description of the types and extent of landscape features included in a GIS database, and conversely, those that are omitted.

Conformal projection: A projection system where angles on the Earth's surface are represented by approximately the same angles on a map, thus the angles related to map features are preserved. Therefore, the scale around any single point on a map using this projection system is the same in every direction.

Conic projection: A projection system where a cone is positioned so that it cuts through the Earth's surface at one latitude, and comes out at another (usually the equator), and mapped features are projected onto the cone's surface, based on lines radiating outward from the center of the Earth.

Contour interval: The vertical distance that distinguishes neighboring contour lines.

Contour lines: Lines that are connected, and represent locations of equal elevation.

Cylindrical projection: A projection system where mapped features are projected onto a cylinder, then the cylinder is unrolled and the map becomes a planar surface.

Database: A collection of information that is managed and stored as a single entity. A spatial database includes information regarding the spatial coordinates of all of the landscape features in the database, as well as information regarding the attributes of each landscape feature, usually in a tabular (spreadsheet) form.

Database conversion: The translation of a database from one format to another. For example, to convert ArcView shapefiles to MapInfo tables would require a database conversion.

Datum: A mathematical representation of the Earth's surface forming a control surface upon which an ellipsoid and other location data are referenced.

Destination table: One of two tables used in a Join operation, the one where the information resides after the join operation. In the case of linking two database tables together, it is the table that is active just before they are linked.

Digital elevation model (DEM): A GIS database formed of features (typically regularly-spaced as in a grid) that contain X,Y as well as Z (elevation) coordinates. Usually it represents a topographic database.

Digital orthophotograph: An image, usually a scanned photograph, that has had removed some of the displacement normally caused during the aerial photographic process (tilt, terrain relief).

Digital raster graphics (DRGs): Digitally scanned representations of USGS 7.5 Minute Series Quadrangle maps.

Digitize (digitizing): To record the X,Y coordinates of landscape features in a computer file or system as digital data. Digitizing can occur using a digitizing tablet, where maps are taped down, reference points are noted, and features (points, lines, or polygons) are then traced, and their locations in space are known with some level of precision and accuracy. A looser form of digitizing can occur when using the 'heads-up' technique, where you use your computer's mouse to delineate landscape features on a computer screen, perhaps with a digital orthophotograph as a backdrop. The heads-up technique is quicker than regular digitizing, but usually comes with a cost reflected in lower precision or accuracy, or both.

Dissolve: To remove the boundaries (arc, lines) between adjacent polygons, keying on the fact that some of the polygons have the same value for some attribute, thus they should be logically combined.

Dynamic segmentation: A vector data analysis process that centers on the use of line segments, and attempts to link a network of lines, based on a common attribute, so that the lines are grouped or joined into categories of interest.

Easting: A measure of distance east of a coordinate system's origin.

Editing: The process of modifying either the spatial shape or location of landscape features, or the tabular data that describes the attributes of each landscape feature.

Elevation contours: Lines that indicate vertical elevation distances, or changes in elevation, across a landscape.

Ellipsoid: A spheroidal figure used to describe the shape of the Earth.

Equal area projection: A projection system where the Earth's features are represented on a map using a constant scaling factor, thus the area of land features is preserved.

Erasing process: The process of extracting from one GIS database only those landscape features outside the bounds of another GIS database (which could contain a single polygon or a set of polygons). This is an action that essentially acts like a cookie-cutter with objects outside the cookie-cutter being retained.

Extent: The limits of the locations of all landscape features in a GIS database. Coordinates are used to define the lower-left and upper-right corners of a rectangular area that would include all of the landscape features.

Field: As in attribute field, a column in a tabular database that represents some characteristic of the landscape features in that database. For example, in a polygon database representing timber stands, stand volume, trees per acre, basal area, and dominant species type could all be fields in a tabular database. In an owl location database, a field could be developed to represent the first sighting, last sighting, and number of owls at certain points on the landscape.

File Transfer Protocol (FTP): A widely used method of transferring data over the Internet, allowing one to transfer computer files to other remote computers, or to download computer files from a remote computer.

Fixed-width buffers: Buffers that do not vary in size and are applied uniformly to all landscape features.

Flattening ratio: A ratio used to describe the difference between the shape of the Earth and a perfect sphere, described by the relationship $(a-b)/a$, where a is the equatorial (or semi-major) radius and b is the polar (or semi-minor) radius of the Earth.

Font: The type of text style being used, such as Times New Roman, Arial, or `Courier`. It may also represent types of symbols commonly used in word processing or graphics programs.

From-node: One of the two end nodes of a line or arc, the first one of the two that was digitized. The other is the to-node.

Geodesy: An area of mathematics that involves determining the precise shape and size of Earth features, as well as positions of features on the Earth's surface.

Geographic Information Science (GIScience): The identification and study of issues that are related to GIS use, affect its implementation, and that arise from its application (Goodchild, 1992).

Geographic Information System (GIS): A system designed to capture, store, edit, manipulate, analyze, display, and export data related to geographic features, and includes not only the hardware and software necessary to accomplish these tasks, but also the databases acquired or developed, and the people performing the tasks.

Geoid: An irregular shape that approximates Earth's mean sea level perpendicular to the forces of gravity.

Global Positioning System (GPS): A system allowing one to locate their position on the Earth's surface by receiving signals from satellites. The signals are used to calculate a position based on trilateration, and perhaps differential correction, processes.

Graticule: The collection of meridians and parallels superimposed on the Earth's surface.

Grid: A geographic database made up of raster features, commonly called grid cells. Can also refer to a graticule.

Grid cell: The smallest unit in a raster database, or pixel. Usually these are represented by squares; however, any regular form that fully covers an area (square, rectangle, triangle, hexagon, polygon, etc.) could be considered a grid cell.

Gross error: Sometimes referred to as human error, it is a blunder or other mistake made somewhere in the data collection, map creation, or editing processes.

Heads-up digitizing: A method of developing vector GIS databases (of points, lines, or polygons) where a user creates landscape features using the computer's mouse, generally with a raster image as a backdrop so as to draw (or point out) those landscape features that are important. For example, displaying a digital orthophotograph as a backdrop, one might use the computer's mouse to draw roads, or timber stands. This method of data development is quicker than traditional digitizing, but usually less accurate, and does not require a digitizing table.

Identity process: The acquisition of information within a GIS database concerning the area represented by another GIS database. Here, like with an intersect process, one GIS database is physically laid onto another, yet the resulting third (new) GIS database is defined by the area of coverage of one of the input GIS databases.

Intersect process: The acquisition of information concerning the overlapping areas of two GIS databases. In utilizing an intersect process, a third, new GIS database will be created that consists of only those areas where the two original GIS databases overlap, no more, no less.

Intervisibility: A description (from each unit of land's perspective) of the number of viewpoints that can be seen from a land unit.

Intranet: A network of computers similar to the Internet, except access is limited to a set of authorized people (usually internal to an organization).

Join process: A process designed to either (1) associate an external table (with no landscape features) to a table of a spatial database using a join item, or (2) to associate features from one spatial database to another spatial database, using point-in-polygon or nearest neighbor techniques. When joining tables, the type of join can be a one-to-one join, a one-to-many join, a many-to-one join, or a many-to-few join.

Join item: A field (column) in a tabular database that contains similar information as a field (column) in a second tabular database. For example, a field in one database may be called 'stand_number' and contain timber stand numbers, while a field in a second database may be called 'stand', and again contain timber stand numbers. While the field names do not have to be the same, the information within each field must be similar, such as in this example, where both contain timber stand numbers. The join would then link the two databases using timber stand number as the common element, so data for stand 1234 from one database will be matched with data for stand 1234 in the second database.

Labels: Like annotation, which is text or strings of characters and numbers, these are used to describe landscape features on a map. They usually arise from some attribute field, or column, in the spatial database's tabular database, where for each landscape feature, a description of that attribute (such as stand number, volume per acre, trail type, road type, etc.) exists.

Legend: That part of a physical map that contains the reference information necessary to interpret the colors of polygons, and the styles and colors of lines and points.

Line: A string of X,Y coordinates (vertices) that make up a continuous linear feature. A single line feature (a road) can contain many small arcs that are topologically linked at their nodes (end-points), since each arc starts and ends with a node.

Link: A line that connects points, as defined by nodes and vertices (sometimes called an arc).

Link process: A process designed to either associate landscape features from one database to another database. This process allows the selection and view of landscape features from database, and the inspection of associated (linked) data in the other database. Sometimes called a relate.

Logical consistency: A description of how well the relationships of different types of data fit together within a system.

Logical operators: The operators 'and', 'or', and 'not' that allowing one to develop a complex query without having to perform several single criterion queries in sequence.

Map scale: The ratio of map distance to ground distance represented on the map. For example, 1:250,000 map scale indicates that 1 unit (inch, perhaps) on a map represents 250,000 units (inches) on the ground represented by the map.

Merge process: A process that creates a single GIS database from a set (or subset) of one or more previously-developed GIS databases. Point, line, and polygon databases can be merged together, however, database types are generally mixed. The resulting merged GIS database may contain landscape features that overlap.

Metadata: Data that summarize the characteristics of databases (or 'data about data').

Network: A collection of lines connected via their nodes, and representing possible paths from one location to another. For example, a stream network would include all of the streams, where the smaller headwater streams connect to wider streams with more water flow, and so on to rivers, and perhaps the oceans. A road system is another network that is more complex in nature, because there may not be a logical flow of traffic from one road to the next. Yet to determine optimum paths or alternative routes, you would need to know which roads connect to which other roads, what types of roads they are, and what restrictions may be placed on them.

Node: One (of two) end-points of a line.

Northing: A measure of distance north of a coordinate system's origin.

Overlay analysis: The process of analyzing or combining multiple layers of information at one time.

Pan: To slide the viewable image to one side (left, right, up, down, or some combination of these), allowing a view of some portion of the landscape not viewable with the previous arrangement. Actively called 'panning,' or 'grabbing' when users are performing this action.

Photogrammetry: The act of collecting measurements of landscape features from a image or photograph.

Pixel: The smallest unit in a raster database, or a grid cell. Formed by combining the two terms 'picture' and 'element'.

Planar coordinates: X,Y coordinates that relate to, or are positioned on, a planar or horizontal surface.

Plane coordinate system: Rectangular coordinates that reference landscape features as displayed on a flat map of the Earth's surface.

Point: A single X,Y coordinate that represents a feature on the landscape (such as a research plot, culvert, owl nest, or spring). These features are usually deemed too small (by an organization) to represent as a line or polygon.

Polygon: A multi-sided, closed spatial object that has an area. Polygons are formed by connecting lines (arcs) until a closed area is formed. They may define the boundaries of timber stands, soils types, riparian buffers, wildlife habitat, and so on, using a set of logic consistent across the landscape. Squares, triangles, rectangles, and hexagons can be considered polygons, yet the term usually applies to irregularly shaped objects.

Precision: The degree of specificity to which a measurement is described. Can also refer to consistency.

Public Land Survey System (PLSS): Established in 1785 by the US Congress as a national system for the measurement and subdividing of public lands. At 24-mile intervals north and south of each baseline, standard parallels are established that extended east and west of the principal meridian. Guide meridians were also established at 24-mile intervals east and west of the principle meridian. The grid of meridians and parallels creates blocks, each nominally 24 miles square. Townships are created within each block by forming range lines (running north and south) and township lines (running east and west) both at six-mile intervals. Each township is six miles square. Each township is divided into sections, with each section measuring one square mile; there are 36 sections within a township. Sections are numbered 1–36 starting at the upper right hand corner of a township.

Query: To select a subset of landscape features from a larger set using some selection criteria. These statements can include a single piece of logic (stand_age ≤ 40), or multiple pieces of logic connected by Boolean operators. For example, suppose you have 10,000 timber stands, and you are interested in those that could be commercially thinned. You could query the GIS database for those stands of a certain age, with a query such as (stand_age ≥ 30 AND stand_age ≤ 40). Queries can be rather simple, or quite complex.

Random error: A natural by-product of how one measures and describes landscape features. No matter how well maps are developed or data are collected, error will usually exist in the representation of landscape features.

Raster: A data structure based on cells organized in rows and columns. This is a grid-based structure where an entire area is represented by a cell, and a single landscape feature (timber stand, road, research plot) can be represented by one or more cell.

Rasterization: The process of converting vector data to raster data, usually by scanning.

Record: Each feature (point, line, or polygon) in a GIS database is represented by a record in a tabular database. A record is a row in the tabular database, thus each row represents a feature. Associated with each record are one or more fields (columns), which contain the characteristics of each feature.

Region: A data structure that consists of vector features (lines or polygons), yet allowing overlapping areas.

Remotely sensed data: Raster data acquired by a sensor (camera, satellite) that is some distance from the landscape features being sensed.

Root mean square error (RMSE): A measure of the error between a mapped point and its associated true ground position. Commonly used when assessing spatial accuracy or digitizing a map, RMSE measures the positional error inherent in the registration points on the hardcopy map.

Satellite imagery: Data captured by a remote sensing device housed in a satellite that is positioned above the Earth's surface. Generally, the data consists of values representing the relative degree of reflectance of electromagnetic energy in certain wavelength categories (bands). The imagery is stored as raster data with a spatial resolution that can range from 1 meter to 1 kilometer.

Scale: The relationship between a map displayed on a computer screen or printed on a type of media (paper, mylar, etc.), and the actual physical dimensions of the same area. For example, a township (36 square miles, or 23,040 acres) drawn on a map where 1 inch on the paper represents 1 mile on the ground, is displayed at a scale where 1 inch represents 63,360 inches. This scale can then be expressed as a fraction (1:63360) or a equivalence (1″ = 1 mile).

Scanning: The process of extracting features from a map or photograph and storing reflected values generally as a raster database. Line following and text recognition processes are promising methods for converting analog maps to vector GIS databases.

Secant projection: A projection system where the Earth's surface intersects a map surface in more than one location.

Selection: A set of one or more landscape features (points, lines, or polygons) from a single GIS database, chosen based on a Query or by manual methods (pointing and clicking with a computer mouse). The landscape features selected are usually done so for some reason, perhaps to perform some spatial operation on them (such as buffer some selected streams), or to simply visually inspect those landscape features with a characteristic of interest (such as being curious about where the culverts over 20 years old are located in a forest).

Shaded relief map: A map intended to simulate the sunlit and shaded areas of a landscape when assuming that the sun is positioned at some location in the sky.

Slivers: Very small polygons that result during overlay operations (union, identity, intersection) or during clipping or erasing processes. Sometimes these occur when common borders are represented differently, from separate digitizing processes, and sometimes these occur simply as a result of the spatial process that was attempted.

Slope class: The gradient of a portion of a landscape, as described by a distinct class (e.g., 0–10 per cent, 11–20 per cent, etc.).

Space-delimited text file: A text file created in a word processing system, a text editor, spreadsheet, or database, and saved in a format where items are separated by blank spaces. The following, for example, could indicate habitat suitability index values for specific timber stand polygons, with the first item of each line identifying the polygon, and the second item listing the habitat suitability index:

1 0.657
2 0.433
3 0.298

Spatial database (or data): A database containing some information about an area or landscape, the relationships among the features in the landscape, and perhaps some tabular or attribute (non-spatial) data about each feature. These databases are usually stored in some known coordinate system, thus each landscape feature has one or more spatial coordinates that define where it exists.

Splitting process: A process of creating multiple landscape features from a single landscape feature.

Spurious polygons: Small fractions of polygons created as a result of a GIS process.

State plane coordinate system: A coordinate system developed in the 1930s by the US Coast and Geodetic Survey to create a unique set of planar coordinates for each of the 50 United States.

Systematic error: Sometimes referred to as instrumental error, it is propagated by problems in the processes and tools used to measure spatial locations or other attribute data.

Tabular data: The data in a GIS database that describe the attributes of each landscape feature. Usually displayed as rows (records) and columns (fields), where each row represents a landscape feature, and each column represents an attribute of that feature. When displayed, a tabular database often looks as if it is in a spreadsheet.

Tangent projection: A projection system where the Earth's surface touches a map surface at one location (a tangent).

To-node: One of the two end nodes of a line or arc, the last one of the two that was digitized. The other is the from-node.

Topology: An expression of the spatial relationships among landscape features in a GIS database.

Triangular irregular network (TIN): A vector data model that describes the landscape using triangles. Each corner of each triangle is described by a set of values, such as elevation, aspect, and coordinates.

Union process: The acquisition of information within two GIS databases concerning the area represented by both GIS databases. Here, like with an intersect process, one GIS database is physically laid onto another, yet the resulting third (new) GIS database is defined by the area represented by both of the input GIS databases.

Universal Transverse Mercator (UTM): The most common coordinate system used in the US, which divides the Earth into 60 vertical zones, each zone covering 6° of longitude. The zones are numbered 1–60 starting at 180° longitude (the international date line) and proceeding eastward.

Update interval: The period of time between the performance of subsequent update processes on a GIS database.

Update process: The methods used to maintain the current status and description of landscape features contained in GIS databases.

Variable-width buffers: Buffers that vary in size based on some attribute of the landscape features being buffered.

Vector: A data structure commonly used to represent points, lines, or polygons. This is a coordinate-based structure (not a regular grid), which may not entirely fill an area, and each landscape feature is represented by X,Y coordinate pairs. Attributes can be associated with each feature (point, line, or polygon).

Vectorization: The process of converting raster data to vector data.

Verification process: The processes that one would use to find landscape features or attributes requiring editing. With a verification process, the goal is to ensure that a particular set of data is appropriate (or reasonable, or within some standard)

Vertex (***Vertices** pl.*): One of a set of X,Y coordinates that delineate an arc or line.

Viewshed analysis: The process of understanding the portions of a landscape visible from specific landscape features of interest.

X,Y coordinates: A set of values that represent a point in space, relative to the coordinate system being employed. A single X,Y coordinate could represent a point feature. Lines (arcs) are characterized by a series of X,Y coordinates. Polygons are formed by a collection of lines (arcs), thus have many X,Y coordinates to define their shape and position on a landscape.

Z coordinate: A third value associated with an X,Y coordinate usually indicating the elevation of the point in space above some reference (such as mean sea level). X,Y coordinates do not necessarily need to have a Z coordinate to be useful, whereas Z coordinates need their X,Y associates to be of use.

Zoom: To focus more closely on a smaller or larger area of a spatial database, or to enlarge or make smaller an area of a spatial database, showing more or less detail. *Zoom-in* refers to focusing more closely on a portion of a spatial database. *Zoom-out* refers to focusing less closely on a portion of a spatial database.

References

Goodchild, M.F. 1992. Geographical information science. *International Journal of Geographical information Systems.* 6(1):31–45.

Appendix B

GIS Related Professional Organizations and Journals

Compiled by Rongxia (Tiffany) Li

The following is a list of GIS-related professional organization and peer-reviewed journals. We apologize in advance for any omissions, and will gladly add any other organizations or journals to the list as they are brought to our attention.

Organizations

American Association for Geodetic Surveying. 6 Montgomery Village Avenue, Suite #403 Gaithersburg, MD 20879 USA. (http://www.aagsmo.org/)

American Association of Geographers. 1710 Sixteenth Street NW, Washington, DC 20009-3198 USA. (http://www.aag.org)

American Congress on Surveying and Mapping. 6 Montgomery Village Avenue, Suite #403 Gaithersburg, MD 20879 USA. (http://www.acsm. net/)

American Geophysical Union. 2000 Florida Avenue NW, Washington, DC 20009-1277 USA. (http://www.agu.org/)

American Planning Association. 122 S. Michigan Ave., Suite 1600 Chicago, IL 60603 / 1776 Massachusetts Ave., NW, Washington, DC 20036-1904 USA. (http://www.planning.org/)

American Society of Landscape Architects. 636 Eye Street, NW, Washington, DC 20001-3736 USA. (http://www.asla.org/)

American Society for Photogrammetry & Remote Sensing. 5410 Grosvenor Lane, Suite 210, Bethesda, MD 20814-2160 USA. (http://www.asprs.org/)

Association of American Geographers. 1710 16th Street, NW, Washington, DC 20009-3198 USA (http://www.aag.org/)

British Cartographic Society. BCS Administration, 12 Elworthy Drive, Wellington, Somerset, TA21 9AT, England, UK. (http://www.cartography.org.uk/)

Canadian Association of Geographers. McGill University, Burnside Hall 805 Sherbrooke St. West, Room 425, Montreal, Quebec, Canada H3A 2K6 (http://www.cag-acg.ca/en/)

Cartography and Geographic Information Society. 6 Montgomery Village Avenue, Suite #403 Gaithersburg, MD 20879 USA. (http://www.cartogis.org/)

Geographic and Land Information Society. 6 Montgomery Village Avenue, Suite #403 Gaithersburg, MD 20879 USA. (http://www.glismo.org/)

Geographical Society of New South Wales. PO Box 162 Ryde NSW 1680, Australia. (http://www.gsnsw.org. au/)

Geospatial Information and Technology Association. 14456 East Evans Avenue, Aurora, CO 80014 USA. (http://www.gita.org/)

Management Association for Private Photogrammetric Surveyors. 1760 Reston Parkway, Suite 515, Reston, VA 20190 USA. (http://www.mapps.org/)

National Council of Examiners for Engineering and Surveying. 280 Seneca Road, Clemson, SC 29633-1686 USA. (http://www.ncees.org)

National Society of Professional Surveyors. 6 Montgomery Village Avenue, Suite #403 Gaithersburg, MD 20879 USA. (http://www.nspsmo.org/)

National States Geographic Information Council. 2105 Laurel Bush Road, Suite 200 Bel Air, MD 21015 USA. (http://www.nsgic.org/)

New Zealand Geographical Society. Department of Geography, The University of Waikato, Private Bag 3105, Hamilton, New Zealand. (http://www.nzgs.co.nz/)

Remote Sensing and Photogrammetry Society. c/o Department of Geography, The University of Nottingham, University Park, Nottingham NG7 2RD, United Kingdom. (http://static.rspsoc.org/)

Society of Cartographers. Mr Brian Rogers, Membership Secretary, Cartographic Resources Unit, Dept of Geographical Sciences, University of Plymouth, Drake Circus, Plymouth PL4 8AA, United Kingdom (http://www.soc.org.uk/)

University Consortium for Geographic Information Science. PO Box 15079, Alexandria, VA 22309 USA (http://www.ucgis.org)

Urban and Regional Information Systems Association (URISA). 1460 Renaissance Dr., Suite 305 Park Ridge, IL 60068 USA. (http://www.urisa.org/)

Journals

Annals of the Association of American Geographers (http://www.blackwellpublishing.com/journal.asp?ref=0004-5608)

Asian Geographer (http://geog.hku.hk/ag/default.htm)

Asian Journal of Geoinformatics (http://www.a-a-r-s.org/ajg/index.htm)

Australian Geographer (http://www.tandf.co.uk/journals/carfax/00049182.html)

Cartographica (http://www.utpjournals.com/carto/carto.html)

Cartography and Geography Information Science (http://www.cartogis.org/publications)

Computational Geosciences (http://www.springerlink.com/content/1573-1499)

Computers and Electronics in Agriculture (http://www.elsevier.com/locate/compag)

Computers & Geosciences (http://www.elsevier.com/locate/cageo)

Cybergeo, European Journal of Geography (http://www.cybergeo.presse.fr/)

Environmental Modeling and Software (http://www.elsevier.com/locate/envsoft)

Geocarto International (http://www.geocarto.com/geocarto.html)

Geographical Analysis (http://www.blackwellpublishing.com/journal.asp?ref=0016-7363)

Geographical and Environmental Modelling (http://www.tandf.co.uk/journals/carfax/13615939.html)

Geographical Research (http://www.blackwellpublishing.com/journal.asp?ref=1745-5863&site=1)

Geographical Systems (http://link.springer-ny.com/link/service/journals/10109/tocs.htm)

Geography Compass (http://www.blackwellpublishing.com/journal.asp?ref=1749-8198&site=1)

GeoInformatica (http://www.springer.com/west/home?SGWID=4-40109-70-35704166-0)

GIScience & Remote Sensing (http://www.bellpub.com/ msrs)

IEEE Transactions on Geoscience and Remote Sensing (http://ieeexplore.ieee.org/xpl/RecentIssue.jsp?punumber=36)

International Journal of Geographical Information Science (http://www.tandf.co.uk/journals/tf/13658816.html)

International Journal of Remote Sensing (http://www.tandf.co.uk/journals/tf/01431161.html)

ISPRS Journal of Photogrammetry & Remote Sensing (http://www.itc.nl/isprsjournal/)

Journal of Geographic Information and Decision Science (http://www.geodec.org/)

Journal of Geographical Systems (http://link.springer.de/link/service/journals/10109/index.htm)

New Zealand Geographer (http://www.nzgs.co.nz/Journals Online.aspx)

Norwegian Journal of Geography (http://www.tandf.no/ngeog)

Photogrammetric Engineering & Remote Sensing (http://www.asprs.org/publications/pers/index.html)

Remote Sensing of Environment (http://www.elsevier.com/locate/rse)

Spatial Cognition and Computation (http://www.wkap.nl/journalhome.htm/1387-5868)

Surveying and Land Information Systems (http://www.acsm.net/salisjr.html)

The Bulletin of the Society of Cartographers (http://www.soc.org.uk/bulletin/bulletin.html)

The Canadian Geographer (http://www.blackwellpublishing.com/CG)

The Cartographic Journal (http://www.maney.co.uk/journals/carto)

The Professional Geographer (http://www.blackwellpublishing.com/PG)

Transactions in GIS (http://www.blackwellpublishing.com/journals/tgis)

Transactions of the Institute of British Geographers (http://www.blackwellpublishing.com/journal.asp?ref=0020-2754&site=1)

URISA Journal (http://www.urisa.org/urisajournal)

Appendix C

GIS Software Developers

Compiled by Rongxia (Tiffany) Li

The following is a list of organizations—governmental, university, and private—that develop and distribute GIS-related software programs. Included are many of the common GIS software programs as well as contact information (e.g., website addresses) for each, however, the list is not exhaustive. In most cases, GIS software programs must be purchased either from the developers, or from software distributors, who are not listed below. Sales representatives associated with the developers may be able to direct you to a local software distributor. We apologize in advance for any omissions, and will gladly add any other products to the list as they are brought to our attention.

GIS Software Program, Distributor, and Website

ArcGIS(Environmental Systems Research Institute, Inc. [ESRI], 380 New York Street, Redlands, CA 92373-8100 USA) http://www.esri.com

ArcInfo (Environmental Systems Research Institute, Inc. [ESRI], 380 New York Street, Redlands, CA 92373-8100 USA) http://www.esri.com

ArcView (Environmental Systems Research Institute, Inc. [ESRI], 380 New York Street, Redlands, CA 92373-8100 USA) http://www.esri.com

ATLAS (Environmental Systems Research Institute, Inc. [ESRI], 380 New York Street, Redlands, CA 92373-8100 USA) http://www.esri.com

Auto CAD (Autodesk Media & Entertainment, Mumbai, 400 052, India) http://www.autodesk.com

ERDAS Imagine (Leica Geosystems Geospatial Imaging, 5051 Peachtree Corners Circle Norcross, GA 30092-2500 USA) http://gi.leica-geosystems.com/LGISub1x33x0.aspx

Geomatica (PCI Geomatics, 50 West Wilmot Street, Richmond Hill, Ontario, Canada, L4B 1M5) http://www.pcigeomatics.com

GeoMedia (Intergraph Corporation, Huntsville, AL 35894 USA) http://www.intergraph.com

GRASS (Geographic Resources Analysis Support System) http://grass.itc.it

IDRISI (Clark Labs, Clark University, 950 Main Street, Worcester, MA 01610-1477 USA) http://www.clarklabs.org

ILWIS (International Institute for Geo-Information Science and Earth Observation, [ITC], 7500 AA Enschede, The Netherlands) http://www.itc.nl/ilwis/

Manifold System GIS (Manifold Net Ltd., 1945 North Carson Street, Suite 700, Carson City, NV 89701 USA) http://www.manifold.net

MapInfo (MapInfo Corporation, One Global View, Troy, NY 12180-8399 USA) http://www.mapinfo.com

MGE Products (Intergraph Corporation, 170 Graphics Drive, Madison, AL 35758 USA) http://www.intergraph.com

SuperMap GIS (SuperMap GIS Technologies, Inc., 7th Floor, Tower B, Technology Fortune Center, Xueqing Road, Haidian District, Beijing, China, 100085) http://www.supermap.com

Index